Mastering
Technical
Mathematics

Mastering Technical Mathematics

Norman H. Crowhurst

TAB Books

Division of McGraw-Hill, Inc.

New York   San Francisco   Washington, D.C.   Auckland   Bogotá
Caracas   Lisbon   London   Madrid   Mexico City   Milan
Montreal   New Delhi   San Juan   Singapore
Sydney   Tokyo   Toronto

This book is a revision of *Basic Mathematics* © 1961 by Norman H. Crowhurst, which was published by John F. Rider Publisher Inc., a division of Hayden Publishing Company, Inc., New York, New York.

pbk 4 5 6 7 8 9 10 11 12 13 FGR/FGR 9 9 8 7 6 5
hc   1 2 3 4 5 6 7 8 9 10 FGR/FGR 9 9 8 7 6 5 4 3 2 1

Library of Congress Cataloging-in-Publication Data

Crowhurst, Norman H.
    Mastering technical mathematics / by Norman H. Crowhurst.
        p. cm.
    Includes index.
    ISBN 0-8306-6438-6   ISBN 0-8306-3438-X (pbk.)
    1. Mathematics.   I. Title.
QA37.2.C76   1991
510—dc20   91-18193
                                                    CIP

Acquisitions Editor: Roland S. Phelps
Book Editor: Andrew Yoder
Director of Production: Katherine G. Brown
Book Design: Jaclyn J. Boone   WT1
Cover Design and Illustration: Greg Schooley, Mars, Pa.   3438

Contents

Part 2. Introducing algebra, geometry, and trigonometry as ways of thinking in mathematics

8. First notions leading into algebra 123

9. Developing "school" algebra 140

10. Quadratics   154

11. Finding short cuts   173

20. Combining calculus with other tools 347

Introduction

I wrote my first math book series over 30 years ago. It used what for the time was a new approach, though really it was not new. It taught math the way it should be taught: so the student always understands. Most text books and teachers try to inculcate math as if it's just something you do—you never understand it! That's more true now than when I wrote that book.

Over 60 years ago, I began teaching math and science. Even then there was a "new math." Here is the difference:

Old way	New way
3829	3829
× 324	× 324
15316	11487
7658	7658
11487	15316
1240596	1240596

Here is what happens when such "new" math is introduced. Mathematicians who make the change do so for a good reason. However most teachers use the difference to see if kids do their own homework, or if their parents "help" them!

When I began teaching a college freshman class a student asked, "How do you want us to do long multiplication?" Not knowing his problem, I quipped, "Any way that gets a right answer!" Immediately, I saw I had not answered his question. My class explained that in high school, even if they had the right answer, but used the "wrong" method, the teacher would mark it wrong!

I took time to show them what was definitely not college freshman math: a dozen ways are possible to get a right answer in multiplication! Those stupid questions (to me, not the students) should be out of the way first, rather than make them more confused. Next year, I decided to try that. My colleagues thought my students wouldn't complete that year's curriculum and that I'd be fired for incompetence.

Instead, by Christmas recess, my students reached where the other classes took till Easter to get to. My students were learning at least three times as fast now. They didn't forget anything during holidays, either. They learned more instead! They loved math. My whole class achieved honors grades in their finals. With 30 years experience, I wrote the books so that others could do it that way.

Thirty years ago, calculators were young. For $1200, you could buy a pocket model that wouldn't do half what one for under $5 will now do. I saw calculators as a way to help learn math, but I could only say, "A calculator would do it this way—if you had one!" When TAB wanted this book redone, we agreed that calculators (everyone has a pocket one nowadays) could make learning math easier.

The rewrite went well up to division and fractions. I found an unexpected problem. Divide 30 by 7 with a calculator; what do you get? We would have had quotient 4, remainder 2 (2/7). That was hard enough, the way they taught it then. Decimals came next. For most, they were "impossible!" Now, use a calculator to divide 30 by 7: it reads 4.285714, plus however many digits it "supports." Most students haven't the faintest idea what that string of numbers mean.

The sensible thing to do would be to reverse the old way. Don't teach decimals as a special kind of fraction. First explain why the calculator gets all those figures, then get to fractions by the "back way."

That's not all. Algebra always was another bugaboo. Now a more important reason exists to clarify it. A computer programmer needs a whole different kind of algebra from what he probably never understood in school. I found a lot more problems as I went through the book.

The resulting revision lasted longer than I planned. Now that it's done, it will let people understand work with computers and how they function! Only people have brains. From my past experience, I know you'll have fun and not find it a struggle.

Just a word about using this book: traditional courses require lessons to be learned in strict sequence. I write my books knowing how many people use them: they jump in wherever they want to. Everyone has a different way in which he or she learns best, so each should do it his or her own way.

Sometimes, if you follow a strict sequence, you might find parts "foggy" at first. Either jump ahead to where it is explained or don't worry about it until it comes clear.

Part 1

Arithmetic as an outgrowth of learning to count

1
CHAPTER

From counting to addition

We've all seen people count. You put a number of things in one group, move them over to another group one at a time, and count as you go, "One, two, three . . ."

We learn to save time counting by spotting patterns. Here are several ways in which you can arrange seven things.

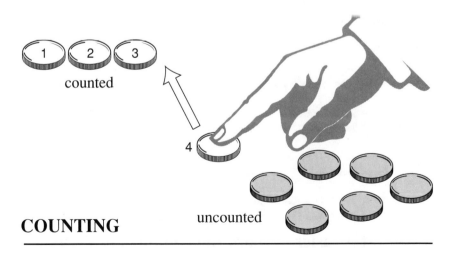

counted

4

COUNTING

uncounted

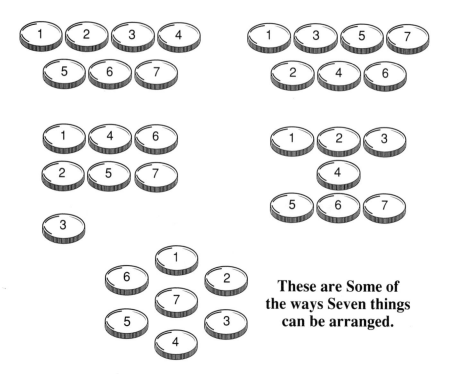

These are Some of
the ways Seven things
can be arranged.

Counting in tens and dozens

When you have a large number of things to count, putting them into separate groups of convenient size makes the job easier. People in most of the world use the number 10 as a basis or "base" for such counting. It is called the decimal system, from the Latin *decem*, which means ten. Thus, 2 groups of 10 are 20; 3 are 30; 4 are 40; and so on.

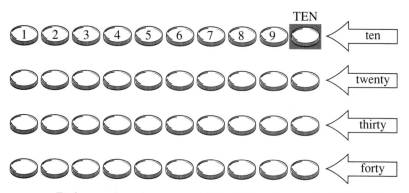

It is easier to count Big Numbers in TENS

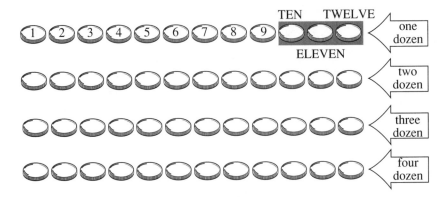

or in DOZENS
not much used nowadays

Tens aren't the only size of group (base) that people have used. At one time, many things were counted in dozens (twelves). Eggs and other things are still bought in dozens. This system is called the duodecimal.

Writing numbers greater than 10

When we have more than ten, we state the number of complete ten groups with the extras left over. Thus, 35 means 3 tens and 5 ones left over. The numbers are written side by side. The left-hand number is tens and the right hand number is ones: 35.

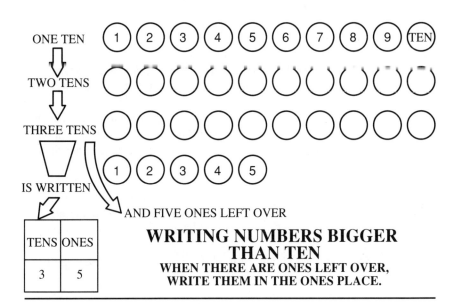

WRITING NUMBERS BIGGER THAN TEN
WHEN THERE ARE ONES LEFT OVER, WRITE THEM IN THE ONES PLACE.

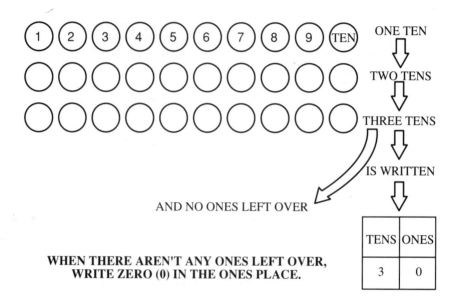

ONE TEN

TWO TENS

THREE TENS

IS WRITTEN

AND NO ONES LEFT OVER

**WHEN THERE AREN'T ANY ONES LEFT OVER,
WRITE ZERO (0) IN THE ONES PLACE.**

TENS	ONES
3	0

Why zero is used in counting

If we have an exact count of tens and no ones are left over, we need to show that the number is in tens, not ones. To do this, we write a zero (0) as the right-hand number in the ones place, which shows an exact number of tens, because there is nothing left over for the ones place. Zero means "none."

Man's earliest computer: the abacus

Various kinds of abacuses have been around for thousands of years. The one shown has a number of rows of beads, separated so that one bead is in one space and 4 beads are in another, all in the same row. First, we show how to count with it.

Start with the bottom row. All of the beads are pushed to the left. You count 1 and move one bead to the right. Notice the little diagrams underneath that show what it looks like to count to 9. After you've moved all 4 beads to the right on the "4" count, push them all back to the left and bring the one bead to the right for the "5" count. So, the one bead represents 5. To count 6, start moving the 4 beads over again.

If you want to count 9, what do you do? The successive rows of beads represent "registers." Move a bead to the right in the next row up, and return all of the first row to the left. That bead represents 10. The second row contains the tens beads.

Other kinds of abacuses might be used differently, but the idea is the same.

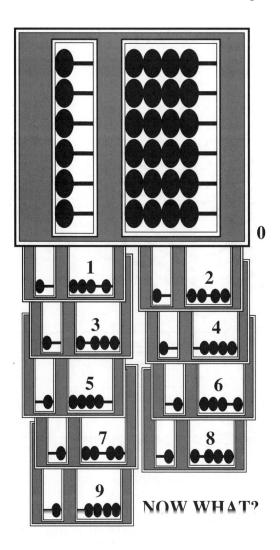

By tens and hundreds to thousands

The abacus represents numbers well. A better way to visualize numbers is to think of packing many things into boxes. This box holds 10 things (apples, for example) each direction. So, each layer (this box would be rather large) contains 100 apples.

If you have 10 of these layers, with 100 in each layer, the box contains 1,000 apples.

Just imagining packing in this manner helps us to understand numbers. Thus, the 2 full boxes each contains 1000. The one part-filled box contains 5 full layers (500), 6 full rows on the next layer (60), and 3 ones in an incomplete row. The whole number adds up to two thousand, five hundred and sixty three (2,563).

COUNTING IN THOUSANDS

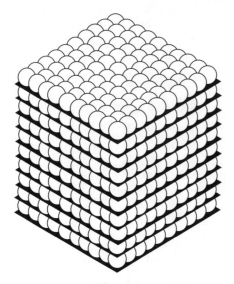

TEN ROWS OF TEN
IN EACH LAYER IS
1 HUNDRED

TEN LAYERS OF
ONE HUNDRED IS
TEN HUNDRED OR
1 THOUSAND

thousands	hundreds	tens	ones
2	**5**	**6**	**3**

Two thousand five hundred sixty three

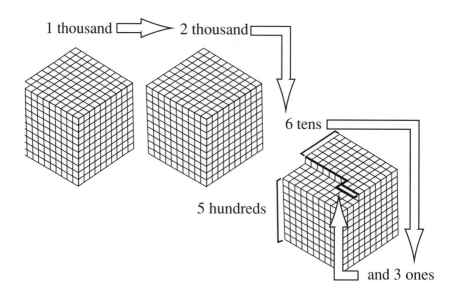

1 thousand ➡ 2 thousand

6 tens

5 hundreds

and 3 ones

Don't forget the zeros

When a count has leftover layers, rows and parts of rows with this systematic arrangement idea, you will have numbers in each column. However, if you have no complete hundred layers (as at A) the hundreds place will be a zero. That is three thousand and sixty five (3,065). You might have no ones left over (as at B) or no tens (as at C), or even no tens or hundreds (as at D).

In each case, it's important to write a zero to keep the other numbers in their proper places. For this reason, zero is called a "placeholder." I repeat, don't forget to use zeros!

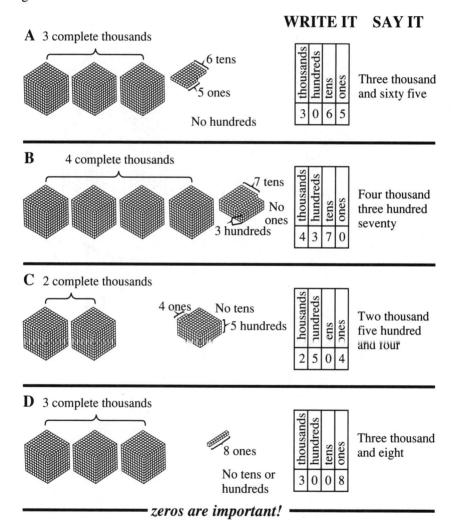

WRITE IT   SAY IT

A 3 complete thousands

6 tens
5 ones
No hundreds

thousands	hundreds	tens	ones
3	0	6	5

Three thousand and sixty five

B 4 complete thousands

7 tens
No ones
3 hundreds

thousands	hundreds	tens	ones
4	3	7	0

Four thousand three hundred seventy

C 2 complete thousands

4 ones   No tens
5 hundreds

thousands	hundreds	tens	ones
2	5	0	4

Two thousand five hundred and four

D 3 complete thousands

8 ones
No tens or hundreds

thousands	hundreds	tens	ones
3	0	0	8

Three thousand and eight

zeros are important!

Beyond thousands: millions and more

Maybe you can imagine stacking thousands of boxes so that the boxes represent a whole new set of counting. Here one complete box that contains one thousand apples is magnified in a stack of many similar boxes. In this picture, the million stack is nearly complete.

In the million stack, each layer contains one hundred thousand (100,000), each row contains ten thousand (10,000), and each box contains one thousand (1,000). So think of those commas as marking off according to the size of "box" you count in for the time being.

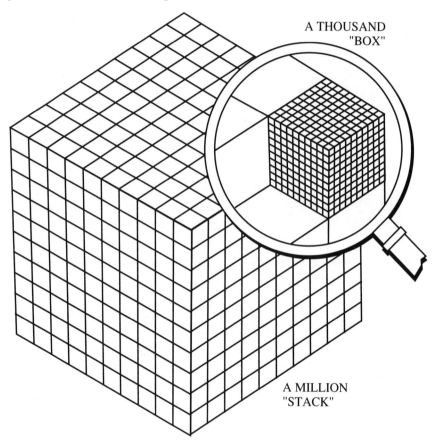

A THOUSAND "BOX"

A MILLION "STACK"

Different ways of viewing big numbers

Take another look at the abacus to see how useful it is. Each row represents a successively higher counting group, or register, by 10 times. Thus, with only 6 rows you can count to one million (actually, up to 999,999, which is 1 short of one million). If you had 9 rows, you could count up to one billion. Each three digits are marked with a comma to "keep track" of the number.

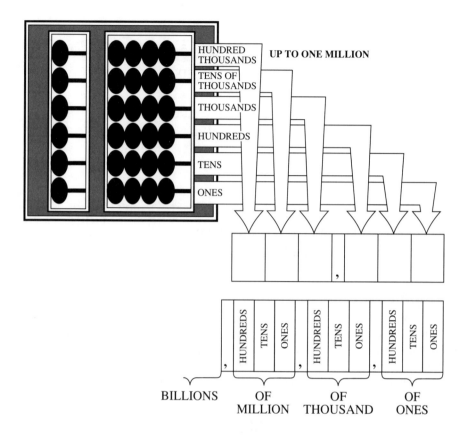

Addition is counting on

Now that a method of counting is established, a method of calculating can be developed. The first step is addition. Suppose you've already counted 5 in one group and 3 in another group. You put them together or add them and what do you have? The easiest way to picture this situation is to count on. People count on their fingers all the time if they don't have their "addition facts" memorized.

If you memorize your addition facts, that's fine. But nothing is wrong with counting on: it just takes longer. Some make an addition table, like a multiplication table (such as in chapter 3) and use that till they remember all the addition facts. Do what's best for you.

To Add

FIVE   THREE

continued

Count On

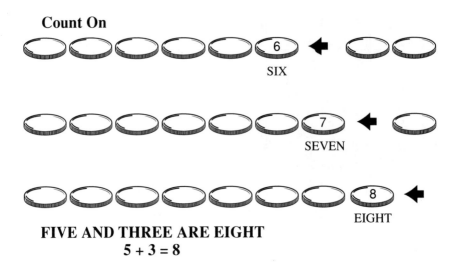

FIVE AND THREE ARE EIGHT
5 + 3 = 8

Adding three or more numbers

Here is a principle that those who invented the "new mathematics" gave a fancy name. Put simply, it says that you can add three or more numbers in any order. Suppose you have to add 3 and 5 and 7. Whatever order you add these three numbers, the answer is 15. This principle extends to however many numbers you might have to add. It becomes more important when we start adding together numbers with more than just the one digit.

add together

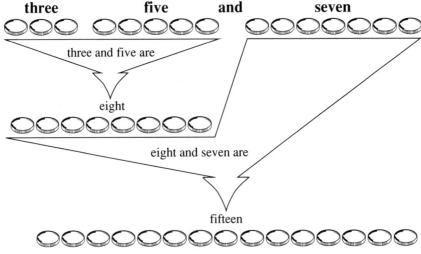

three, five, and seven are fifteen
no matter which two you add first!

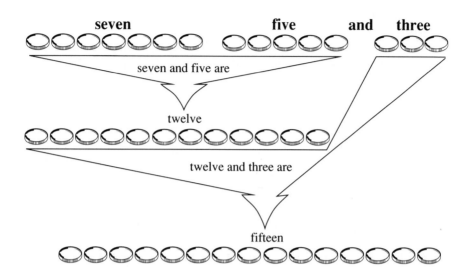

seven — five — and — three

seven and five are

twelve

twelve and three are

fifteen

BIG NUMBERS ARE ADDED IN THE SAME WAY

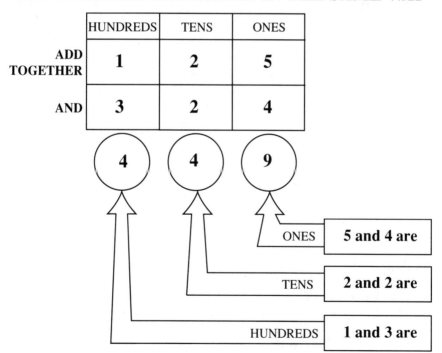

	HUNDREDS	TENS	ONES
ADD TOGETHER	1	2	5
AND	3	2	4

4	4	9

ONES	5 and 4 are
TENS	2 and 2 are
HUNDREDS	1 and 3 are

THE TOTAL IS 449
Four Hundred and Forty Nine

Adding larger numbers

So far, you have added numbers with only a single figure or digit—ones. Bigger numbers can be added in just the same way, but be careful to add only ones to ones, tens to tens, hundreds to hundreds, and so on.

Just as 1 and 1 are 2, so 10 and 10 are 20, 100 and 100 are 200, and so on. We can use the counting-on method or the addition table for any group of numbers, so long as all the numbers in the group belong. That is, they are all in the same place: one, tens, hundreds, or whatever.

So, let's add 125 and 324. Take the ones first: 5 and 4 are 9. Next the tens: 2 and 2 are 4. Last the hundreds: 1 and 3 are 4. Our result is 4 hundreds, 4 tens, and 9 ones: 449.

Notice that we are taking short cuts. We no longer count tens and hundreds one at a time, but in their own group, tens or hundreds. If you added all those as ones, you would have 449 chances of skipping one, or of counting one twice. So, the short cut not only makes it quicker, it also reduces the chances of making a mistake.

Carrying

In that example, we deliberately chose numbers in each place that did not add up to over 10, to make it easy. If any number group or place adds to over 10, you must "carry" it to the next higher group or place.

Suppose you had to add 27 and 35. Take the ones first: 7 and 5 are 12. That is, 1 ten and 2 ones. The 1 belongs in the tens' place. Now, instead of just 2 and 3 to add in the tens' place, you have the extra 1 that resulted as ten "carried" from adding 7 and 5. The 1 is said to be carried from the ones' place.

This carrying goes on any time the total at a certain place goes over ten. For example, add 7,358 and 2,763. Starting with the ones: 8 and 3 are 11: we write 1 in the ones' place and carry 1 to the tens' place. Now, the tens: 5 and 6 are 11, and the 1 carried from the ones makes 12. Write 2 in the tens' place and carry 1 to the hundreds' place. Now, the hundreds: 7 and 3 are 10, and 1 carried from the tens' makes 11 hundreds. Again, write 1 in the hundred's place and carry 1 to the thousands' place. Now, the thousands: 7 and 2 are 9, and 1 carried from the hundreds make 10 thousands. Since neither of the original numbers had any ten thousands, write 10 thousands and finish, because nothing is left to add to the 1 carried this time. The answer is 10,121.

Another example: suppose you now have to add 7,196 and 15,273. Start with the ones: 6 and 3 are 9. Write nine in the ones' place and nothing is left to carry to the tens'. Next, 9 and 7 are 16. Write the 6 and carry the one to the hundreds. Now, the hundreds: 1 and 2 are 3, and the 1 carried makes 4. Again, none to carry to the thousands. So, in the thousands: 7 and 5 are 12. Now, carry 1 to ten thousands, where only one number already has 1. 1 and 1 are 2 for the ten thousands' place. The answer is 22,469.

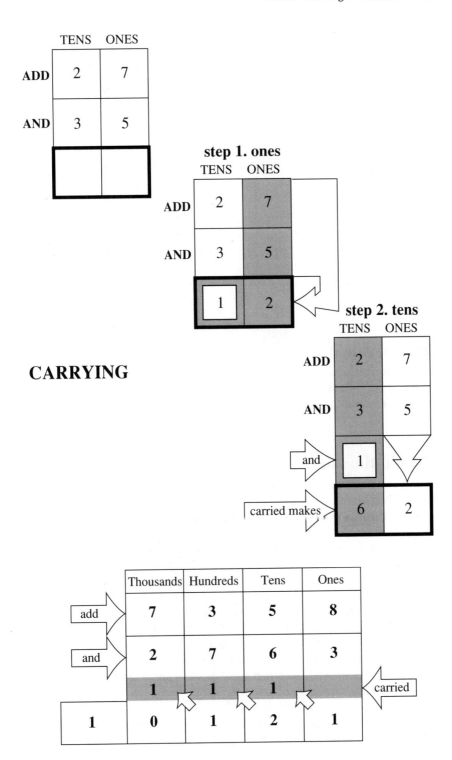

CARRYING

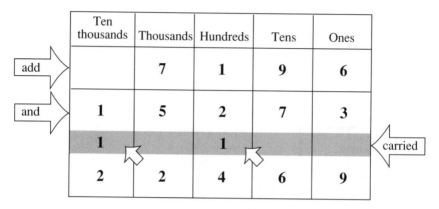

	Ten thousands	Thousands	Hundreds	Tens	Ones
add		7	1	9	6
and	1	5	2	7	3
	1		1		carried
	2	2	4	6	9

**CARRY LEFT-OVER NUMBERS TO
THE "PLACE" AT THE LEFT**

Successive addition

A calculator performs addition by adding each number by counting on, like this book showed you how to do. But it does it ever so much more quickly than people can. Trace it through to see how it does it, then do it on your own calculator.

The old-time way

Older people learned to add by columns. Each way gets the right answer if you don't make a mistake. At the top of page 17, we take the same five numbers that were added on the previous section and add them the old way.

First, the ones. 6 and 3 are 9; 9 and 3 are 12; 12 and 2 are 14; 14 and 6 are 20; write 0 in the ones' place and carry 2 to the tens' place.

Maybe it's best to count the carried number first so that you don't forget it. Some people add it last, just be sure. Now, the tens: 2 carried and 7 are 9; 9 and 2 are 11; 11 and 4 are 15; 15 and 4 make 19; 19 and 7 make 26; write 6 in the tens' place and carry the 2 to the hundreds' place.

Do the hundreds' place the same way: 2 carried and 4 are 6; 6 and 5 are 11; skip the number that has no hundreds' place; 11 and 5 are 16; 16 and 3 are 19; write 9 in hundreds' and carry 1 to the thousands'.

Now, the thousands': 1 carried and 3 are 4; 4 and 5 are 9; skip the next; 9 and 6 are 15; and no thousands are in the last number; write 5 in the thousands' place and carry 1 to the ten thousands'.

Finally the ten thousands: 1 carried and 1 are 2; 2 and 2 are 4; and that's all there is. So, the answer is 45,960, the same as we got the other way.

ADD TOGETHER

Ten thousands	Thousands	Hundreds	Tens	Ones
1	**3**	**4**	**7**	**6**
2	**5**	**5**	**2**	**3**
			4	**3**
	6	**5**	**4**	**2**
		3	**7**	**6**

	Ten thousands	Thousands	Hundreds	Tens	Ones	
	1	3	4	7	6	
Step 1 and are and	2 3	5 8	5 9	2 9 4	3 9 3	
		1	1	1		◄ carried
Step 2 are and	3	9	0 5	4 4	2 2	
	1					◄ carried
Step 3 are and	4	5	5 3	8 7	4 6	
			1	1		◄ carried
Step 4 are	4	5	9	6	0	

This is how an Adding Machine adds.

Successive Addition (contd.)

ADD TOGETHER

Ten thousands	Thousands	Hundreds	Tens	Ones	
1	3	4	7	6	
2	5	5	2	3	
			4	3	
	6	5	4	2	
		3	7	6	
1	1	2	2		◄ carried
4	**5**	**9**	**6**	**0**	

Or you can add the same numbers this way.

Checking answers

Already, the same five numbers have been added in two different ways. Before calculators made it so easy, bookkeepers would use two methods, usually those shown here as the 2nd and 3rd, to check themselves. First, they would add the numbers by columns, starting at the top and working down, as you did in the last section. Then, they would add the same numbers starting at the bottom and working up.

In the units' column that would go 6 and 2 make 8; 8 and 3 are 11; 11 and 3 are 14; 14 and 6 are 20. Each column should have the same answer, whether you add from the top or bottom. Now, calculators are used so much more, but that is no guarantee that you won't make a mistake. A good plan is to add up with the calculator as well as to add down.

376 and 6,542 are 6,918; 6,918 and 43 are 6,961; 6,961 and 25,523 are 32,484; and 32,484 and 13,476 are 45,960.

You see the advantage of using more than one method. The partial sums that you move through on the way are different. You only reach the same answer at the end. The likelihood that you would enter the same wrong number twice under these conditions is much reduced. If you get different answers, work each one again until you find where you made your mistake.

Three Ways to Add – Two Ways to Check
your Answer

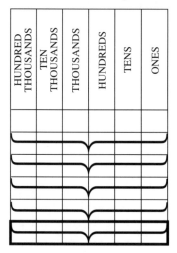

1st
ADD WHOLE
NUMBERS

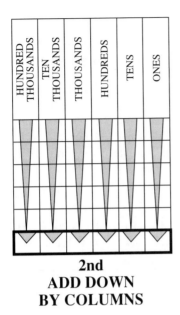

2nd
ADD DOWN
BY COLUMNS

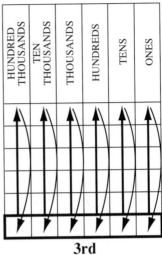

3rd
ADD UP
BY COLUMNS

Weights

Modern scales read weight digitally. You might have seen another kind of scale in a doctor's office. The old-fashioned grocer's scales are now antique items, but knowing how they worked helps understand math.

Those scales had two pans supported from a beam pivoted across a point (*fulcrum*) at an equal distance on either side of the fulcrum. When the weight in both pans was equal, the scales balanced: the pans were level with one another. When the weights were unequal, the pan with the heavier weight dropped and the other rose. To use such scales, the grocer needed a set of standard weights, shown here.

Standard (*avoirdupois*) weight, still used in English speaking countries, does not follow the metric or "10 times" scale. Instead, it has 16 *drams* to an *ounce*, 16 ounces to a *pound*, 28 pounds to a *quarter*, and 4 quarters (112 pounds) to a *hundredweight*, 20 hundredweights (or 2240 pounds) to the ton, often called the *long ton* because a ton of 2000 pounds is used more today.

A set of weights consisted of those shown at the top of this diagram—just 12 of them, unless the grocer wanted to measure more than 15 pounds. With these weights, if the scale was sensitive enough, he could weigh anything to the nearest dram.

Suppose you have to weigh a parcel. First, put the parcel in the pan on the left. Then, put standard weights on the other pan until the scale tips the other way.

If a 1-pound weight doesn't tip it, a 2-pound weight is tried. It still doesn't tip. But the 2 and 1 together, making 3 pounds, does tip it. So, the parcel weighs more than 2 pounds, but less than 3. He leaves the 2-pound weight in the pan and starts using the ounce weights.

8 ounces doesn't tip the scale. If 4 ounces are added, to make 12 ounces, it doesn't tip. But the 2-ounce weight, which brings the weight up to 2 pounds 14 ounces, tips it. If the 1-ounce weight is used instead of the 2-ounce weight, the scale still doesn't tip. So, the parcel is more than 2 pounds 13 ounces and less than 2 pounds 14 ounces. If you want to be more accurate, follow this method until it balances with 2 pounds, 13 ounces and 3 drams.

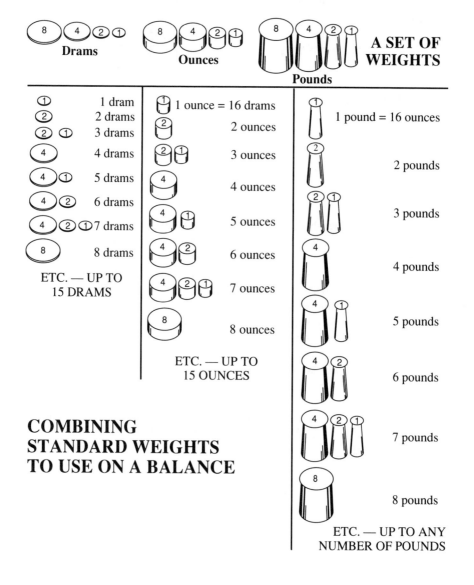

Drams

Ounces

Pounds

A SET OF WEIGHTS

①	1 dram
②	2 drams
② ①	3 drams
④	4 drams
④ ①	5 drams
④ ②	6 drams
④ ② ①	7 drams
⑧	8 drams

ETC. — UP TO
15 DRAMS

①	1 ounce = 16 drams
②	2 ounces
② ①	3 ounces
④	4 ounces
④ ①	5 ounces
④ ②	6 ounces
④ ② ①	7 ounces
⑧	8 ounces

ETC. — UP TO
15 OUNCES

①	1 pound = 16 ounces
②	2 pounds
② ①	3 pounds
④	4 pounds
④ ①	5 pounds
④ ②	6 pounds
④ ② ①	7 pounds
⑧	8 pounds

ETC. — UP TO ANY
NUMBER OF POUNDS

COMBINING
STANDARD WEIGHTS
TO USE ON A BALANCE

PARCEL WEIGHS . . .

more than 2 pounds

less than 3 pounds

more than 2 pounds
12 ounces

less than 2 pounds
14 ounces

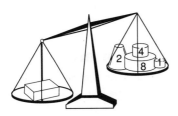

more than 2 pounds
13 ounces

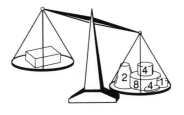

less than 2 pounds
13 ounces and 4 drams

more than 2 pounds
13 ounces and 2 drams

BALANCES AT 2 POUNDS
13 OUNCES AND 3 DRAMS

WEIGHING A PARCEL

Liquid and dry measures

The common measures of quantity (bulk), both liquid and dry, are pints, quarts, and gallons. Although they are not the same, unless you are measuring water (which isn't dry), each measure has 2 pints to a quart, and 4 quarts to a gallon. The metric measures of all these units will be covered later.

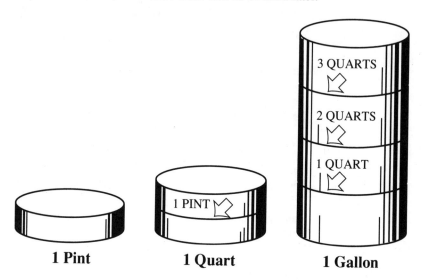

1 Pint   **1 Quart**   **1 Gallon**

Questions and problems

1. Does it make any difference in the final answer whether you count them (a) one by one, (b) in groups of ten, and (c) in groups of twelve?

2. Why do we count larger numbers in hundreds, tens, and ones, instead of one at a time?

3. The figure zero (0) means none. Why then should we bother to write down this number?

4. What are (a) 10 tens and (b) 12 twelves?

5. What are (a) 10 hundreds, (b) 10 thousands, and (c) 1,000 thousands?

6. By counting on, add the following groups of numbers, then check your results by adding the same numbers together in reverse order. Finally, use your pocket calculator:

    (a) 3 + 6 + 9   (b) 4 + 5 + 7   (c) 2 + 7 + 3
    (d) 6 + 4 + 8   (e) 1 + 3 + 2   (f) 4 + 2 + 2
    (g) 5 + 8 + 8   (h) 9 + 8 + 7   (i) 7 + 1 + 8

7. Add together the following groups of numbers: in each use a manual method (without using a calculator) first, then verify your answer with a calculator. Practice using different methods:

    (a)   35,759 + 23,574 + 29,123 + 14,285 + 28,171
    (b)   235 + 5,742 +   4 + 85,714 + 71,428
    (c)   10,590 +   423 + 6,129 +   1 +   2
    (d)   12,567 + 35,742 +   150 + 90,909 + 18,181
    (e)   1,000 +   74 +   350 + 9,091 + 81,818

8. How does adding money differ from adding numbers?

9. Add together the following weights: 1 pound, 6 ounces, and 14 drams; 2 pounds, 13 ounces, and 11 drams; 5 pounds, 11 ounces, and 7 drams. Check your result by adding them in at least three ways. If you get different answers, find your mistakes.

10. What weights would you use to weigh out each of the quantities in question 9, using the system of weights. Check your answers by adding up the weights you name for each object weighed.

11. In weighing a parcel, the 4-pound weight tips the pan down, but the 2- and 1-pound weights do not raise the parcel pan. What would you do next to find the weight of the parcel (a) if you want it to the nearest dram; (b) if you have to pay postage on the number of ounces or fractions of an ounce?

12. How many inches are in 2 yards? (First, add the number of inches in 3 feet to make 1 yard, then add the inches in another yard to make 2 yards.)

13. A fleet of cars needs oil changes. Three of the cars require 5 quarts each, two require 6 quarts each, four require one gallon each. How many gallons of oil does the owner need? Write down the number of quarts for each car (4 for the ones that use a gallon), add them up, and count off 4 quarts for each gallon.

14. If the owner can buy quarts at 90 cents and gallons at $3.50, to be economical, how will he buy the oil?

15. A woman buys three dresses at $12.98 each, spends $3.57 on train fare to get to town, and $5.00 on a meal while she is there. How much did she spend altogether?

2
CHAPTER

Subtraction

Subtraction is counting away

Just as addition is counting on, subtraction is counting away, or counting back. Start with the total number, count away the number to be subtracted, then see how many of the original number remain.

Many people have more trouble with subtraction than addition. If you do, make a subtraction table, like the one shown at the top of page 25. From what you know so far, you can subtract only a smaller number from a larger one. That will not always be true. More information on tablemaking is included in the multiplication chapter.

SUBTRACTION IS COUNTING AWAY

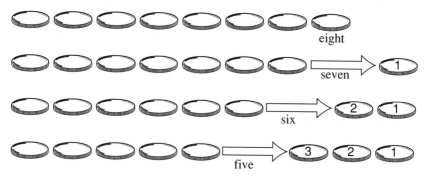

THREE FROM EIGHT IS FIVE

$8 - 3 = 5$

	1 is	2 is	3 is	4 is	5 is	6 is	7 is	8 is	9 is
1 from	0	1	2	3	4	5	6	7	8
2 from		0	1	2	3	4	5	6	7
3 from			0	1	2	3	4	5	6
4 from				0	1	2	3	4	5
5 from					0	1	2	3	4
6 from						0	1	2	3
7 from							0	1	2
8 from								0	1
9 from									0

SUBTRACTION TABLE

CHECK SUBTRACTION BY ADDITION

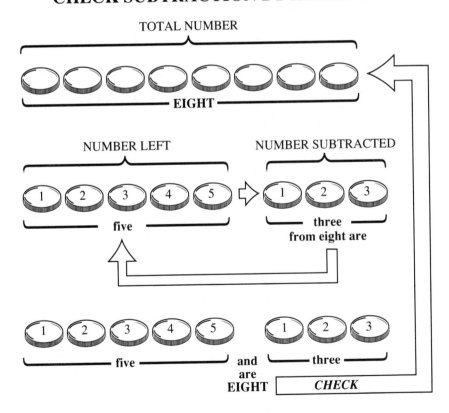

TOTAL NUMBER

EIGHT

NUMBER LEFT

NUMBER SUBTRACTED

five

three
from eight are

five

and
are
EIGHT

three

CHECK

Checking subtraction by addition

It is most important, all through mathematics, to be sure that you have the right answer. That is why we use at least two ways of adding, one to check the other. In subtraction, an easy way to check the answer is to reverse the process by addition, to see if we get back the number we began with.

Borrowing

In addition, if the one's figures added to ten or over, it carried into the tens' figure, and so on. In subtraction, this process is reversed. Suppose you must subtract 17 from 43.

First, subtract the ones. But 7 is larger than 3. You could subtract 7 from 13. So, "borrow" a ten to make that 3 into 13. 7 from 13 leaves 6. Taking away the 1 that was borrowed from the ten's column leaves only 3 from which we can take the 1. So, 17 from 43 is 26.

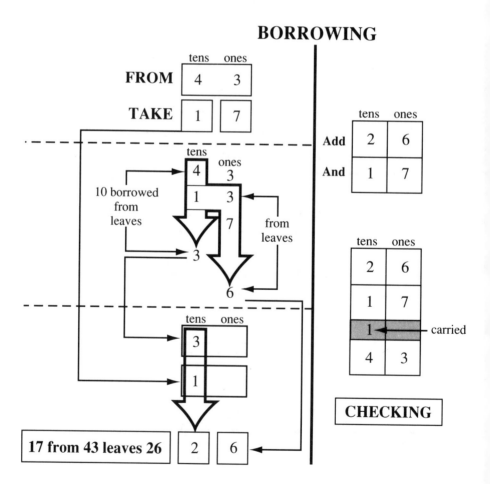

BORROWING

CHECKING

Now, check the result by adding 26 and 17. In the ones, 6 and 7 are 13. Write the 3 and carry the 1 to the ten's column. In the ten's column, 1 carried and 2 are 3, and 1 is 4. So, 26 and 17 are 43, which checks back with the number you began with. Your answer should be right because you would have to make two very special mistakes to return to the correct original figure. Usually, if you made two mistakes, the final result would be even further from the correct answer than if you only made one.

Subtracting with larger numbers

Now that you have the idea, try some really big numbers. Here 17,583 is subtracted from 29,427. Work through it yourself. You will have to borrow from the hundreds for the tens, and again from the thousands for the hundreds. Go over this carefully. When borrowing, some people like to cross out the original figure and reduce it by 1. Thus, at the hundred's figure, you are subtracting 5 from 3, which, with 1 borrowed from the thousand's, is 13. Then, in the thousands, subtract 7 from 8, not 9.

Finally, (as shown on the next page) turn it around and add 17,583 and 11,844. Carry in the same places that you borrowed and you should return to the original number in the top line of the subtraction.

	TEN THOUSANDS	THOUSANDS	HUNDREDS	TENS	ONES
FROM	2	9	4	2	7
TAKE	1	7	5	8	3

	TEN THOUSANDS	THOUSANDS	HUNDREDS	TENS	ONES
	2	9	4	2	7
		1	1			◀ **borrowed**
	1	7	5	8	3
	1	1	8	4	4

TEN THOUSANDS	THOUSANDS	HUNDREDS	TENS	ONES
1	7	5	8	3
1	1	8	4	4
	1	1		
2	9	4	2	7

CHECKING
⇩
ADD

← carried

THE SAME IS TRUE OF LARGER NUMBERS

Subtracting cash

Cash is no more difficult to subtract than other numbers are. The only difference is in the dollars and cents. Start at the cents and work back. If you take a larger number of cents from a smaller number, you have to borrow a dollar.

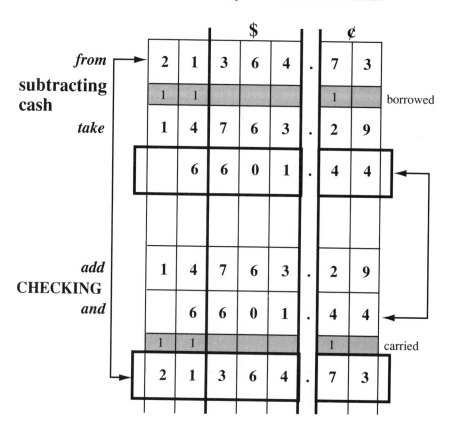

Making change

This idea of counting on, or using addition to check subtraction, is often used by sales clerks when making change. Suppose you bought something for $3.27 and use a $5 bill for payment. Subtraction will show that you should get $1.73 in change. The clerk figures the bill (or maybe the cash register does it—most new ones do), then "proves" it by giving you change, as shown.

Making Change

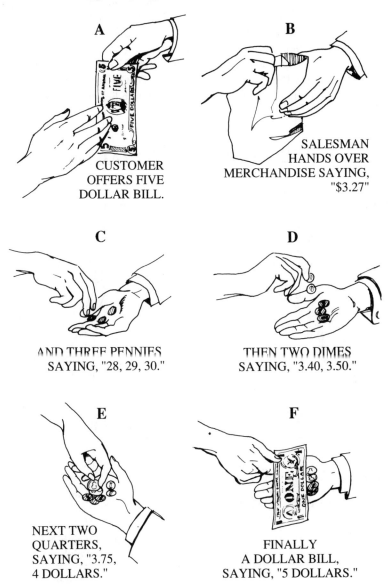

A
CUSTOMER
OFFERS FIVE
DOLLAR BILL.

B
SALESMAN
HANDS OVER
MERCHANDISE SAYING,
"$3.27"

C
AND THREE PENNIES
SAYING, "28, 29, 30."

D
THEN TWO DIMES
SAYING, "3.40, 3.50."

E
NEXT TWO
QUARTERS,
SAYING, "3.75,
4 DOLLARS."

F
FINALLY
A DOLLAR BILL,
SAYING, "5 DOLLARS."

Putting 3 pennies in your hand, he says, "$3.27, 28, 29, 30." Then, he puts 2 dimes in your hand, saying "$3.40, 50." Next 2 quarters, saying "$3.75, $4.00." Finally, he gives you a dollar bill, saying "$5," which was the amount you tended. What he did was check the change by adding it to the cost of what you bought, to come to the amount you tended in payment.

Subtracting weights

Suppose a mother wants to weigh her baby, who is too big for baby scales and too wriggly to get a reading on ordinary scales. The mother weighs herself holding the baby in her arms, then weighs herself without the baby. The difference, obtained by subtraction, is the baby's weight.

If she weighs 156 pounds holding the baby and 121 pounds without the baby, then the baby weighs 156 − 121, or 35 pounds. That minus (−) sign is the sign used to indicate subtraction, just as a plus (+) sign indicates addition.

WEIGHING BY SUBTRACTION

156
pounds

121
pounds

Mother and baby weigh	156 pounds
Mother only weighs	121 pounds
So baby weighs	35 pounds

check

Mother weighs	121 pounds
Baby weighs	35 pounds
So mother and baby weigh	156 pounds

Using a balance

You might not see the balance used often today. If you can get some old-fashioned scales, it's a good exercise. If not, you must imagine it.

Suppose you are weighing something that eventually you find weighs 3 pounds 14 ounces. In the traditional method, you would put the parcel in one pan and a selection of weights in the other pan. You'll have 1- and 2-pound weights, and 8-, 4-, and 2-ounce weights.

The other method uses subtraction (backwards addition, however you want to view it). The parcel weighs just under 4 pounds. So, put small weights in the pan with the parcel and find that it balances with the 2-ounce weight in the parcel pan, and the 4-pound weight in the weight pan. So, the parcel weighs 2 ounces less than 4 pounds or the weight of the parcel plus 2 ounces equals 4 pounds.

There are 2 ways to use a balance

METHOD 1

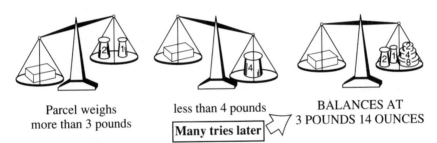

Parcel weighs
more than 3 pounds

less than 4 pounds

Many tries later

BALANCES AT
3 POUNDS 14 OUNCES

METHOD 2

Parcel weighs
more than 3 pounds

less than 4 pounds

PARCEL AND 2 OUNCES
BALANCE 4 POUNDS
EXACTLY

So the parcel must weigh 3 POUNDS 14 OUNCES

Subtracting liquid and dry measures

Another application for subtraction is when figuring the mileage of a car. You cannot measure the gas used on a trip, because you no longer have it. For example, you began the trip with 20 gallons in the tank. If you could measure what is left in the tank, you would find 11 gallons. But the gas gauge might not be that accurate.

At the beginning of the journey, you filled the tank. You took the trip. Then, you filled the tank again. It needed 9 gallons, so you used 9 gallons. You automatically check the subtraction when you refill the tank.

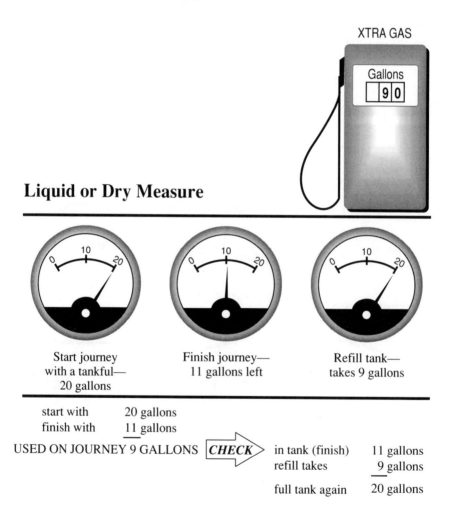

XTRA GAS

Gallons
9 0

Liquid or Dry Measure

| Start journey with a tankful—20 gallons | Finish journey—11 gallons left | Refill tank—takes 9 gallons |

start with   20 gallons
finish with   11 gallons

USED ON JOURNEY 9 GALLONS   *CHECK*

in tank (finish)   11 gallons
refill takes   9 gallons

full tank again   20 gallons

Questions and problems

1. Make the following subtractions and check your results by addition, both by hand and using your calculator.

(a)	69	(b)	123	(c)	543	(d)	762	(e)	509
	− 46		− 81		− 37		− 371		−410

(f)	263	(g)	4,321	(h)	6,532	(i)	11,507
	− 74		−1,234		−2,356		− 8,618

2. A refrigerator's list price is $659.95. A local discount store offers a $160 discount on this item. How much would you pay at this store? Check your result by addition.

3. A lady purchased items priced at $2.95, $4.95, $3.98, $10.98, and $12.98. After adding up the bill, the store clerk offers to knock off the extra cents, so she would pay only the dollar amount. The lady has a better idea. Why not knock the extra cents off each item? How much more would she save if the store clerk accepted this suggestion?

4. A child wants to weigh her pet cat. The cat won't stay on the scales long enough to get a reading. So she weighs herself with the cat, then without it. With the cat, she weighs 93 pounds. Without it, 85 pounds. How much does the cat weigh?

5. A recipe calls for 1 pound 12 ounces of rice. You have an old fashioned scale to measure it with. All the pound weights are there, but the only ounce weights that have not been lost are the 1- and 4-ounce weights. How can you weigh the 1 pound 12 ounces? Prove it by showing that the scale balances.

6. You are studying a road map on which town D is between towns A and C. The map shows only selected distances. Between A and B, it shows 147 miles. Between A and C, it shows 293 miles. If you want to go from B to C, how far is it?

7. A freight company charges partly on weight and partly on mileage. The distance charge is based on direct distance—even if the company's handling might necessitate taking it further. A package addressed to town B had to travel from A to C, which is 1,200 miles, and back to B, on the direct route between A and C, which is 250 miles. What distance is the charge based on?

8. A man has a parcel of land along a 1-mile frontage of highway. He has sold pieces with frontages of 300 yards, 450 yards, 210 yards, and 500 yards. How much frontage does he have left to sell?

3
CHAPTER

Multiplication

A short cut for repeated addition

Suppose you go into a store and buy 7 articles for $1 each. The cost is $7. That's easy, but suppose the articles cost $3 each. Now, to find the total, you must count $3 seven times.

It Takes Less Time to MULTIPLY Than to Add

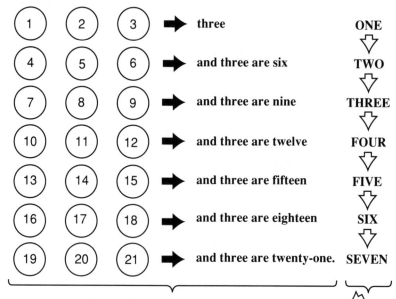

(1) (2) (3)	➡	three	ONE	
(4) (5) (6)	➡	and three are six	TWO	
(7) (8) (9)	➡	and three are nine	THREE	
(10) (11) (12)	➡	and three are twelve	FOUR	
(13) (14) (15)	➡	and three are fifteen	FIVE	
(16) (17) (18)	➡	and three are eighteen	SIX	
(19) (20) (21)	➡	and three are twenty-one.	SEVEN	

Seven threes added together are twenty-one
or
Seven times three are twenty-one
$7 \times 3 = 21$

This problem leads to the next step in calculating. This short cut remembers what so many counts of any particular number add up to. At one time, children learning arithmetic would spend hours memorizing printed multiplication tables, often without knowing why they'd need such things. If you were lucky enough to remember that 7 times 3 equal 21, you got the answer more quickly. However, many people who "knew" that answer could not tell you why 7 times three are 21.

Use of tables

The multiplication table was one of man's earliest computers: a ready way of getting answers without having to do all that adding. So to understand it, and see more about how numbers "work," make your own multiplication table, such as the one shown here. Start with the numbers along the top and down the left side.

Now, count in twos, and write the results in the next column, under "2 times." Each next figure down the column is 2 more than the one above it. Now, do the same thing with the "3 times" column, adding 3 for each next figure down the column. Continue until you have done the "9 times" column.

MULTIPLICATION TABLE

	2 times	3 times	4 times	5 times	6 times	7 times	8 times	9 times
2 are	4	6	8	10	12	14	16	18
3 are	6	9	12	15	18	21	24	27
4 are	8	12	16	20	24	28	32	36
5 are	10	15	20	25	30	35	40	45
6 are	12	18	24	30	36	42	48	54
7 are	14	21	28	35	42	49	56	63
8 are	16	24	32	40	48	56	64	72
9 are	18	27	36	45	54	63	72	81

Patterns in numbers

If you counted carefully enough, your table should be right. The idea of making your own is so you know that the table is true. That way, using it will not be "cheating," which is what some teachers say about students who use printed tables. You will be using what you have already done and verified. So, how do you verify it?

The simplest check is odds and evens. An even number is one whose one's figure is 2, 4, 6, 8, or 0. An odd number has a one's figure of 1, 3, 5, 7, or 9. Notice that the only places where you have odd numbers are where both the number at the top of the column and the beginning of the line are odd. If either of them are even, the number in that space is even.

Now notice the column and line opposite 5. These are shown by themselves in Part A. All the numbers have a one's figure of either 0 or 5.

Another thing you might notice is that any multiplication can be found in two places, except where the number is multiplied by itself. Thus, 3 times 7 is the same as 7 times 3. This rule is true for every combination of two different numbers.

Another interesting set is the column or line against 9. Notice that each successive number has one more in the tens and one less in the ones, and also that adding the two digits together always makes 9.

	2	3	4	5	6	7	8	9
				times				
2				10				
3				15				
4				20				
5	10	15	20	25	30	35	40	45
6				30				
7				35				
8				40				
9				45				

(A) CHECKING THE FIVES

	2	3	4	5	6	7	8	9
				times				
2								
3						21		
4								
5								
6								
7		21						
8								
9								

(B) CHECKING BY SYMMETRY

	2	3	4	5	6	7	8	9
				times				
2								18
3								27
4								36
5								45
6								54
7								63
8								72
9	18	27	36	45	54	63	72	81

(C) CHECKING THE NINES

	2	3	4	5	6	7	8	9
				times				
2	4		8					
3		9		15				
4	8		16		24			
5		15		25		35		
6			24		36		48	
7				35		49		63
8					48		64	
9						63		81

(D) CHECKING BY DIAGONALS

Some PATTERNS in numbers make useful CHECKS

Now take the diagonal where numbers are multiplied by themselves. These numbers are called "squares." We'll see why later. Complete the box of figures by writing a "1" at top left. Now, notice the difference between successive places down the "square" diagonal. 1 and 3 are 4. 4 and 5 are 9. 9 and 7 are 16. 16 and 9 are 25. 25 and 11 are 36. 36 and 13 are 49. 49 and 15 are 64. 64 and 17 are 81. Each step adds the next odd number from the one you added last.

Now, take diagonals the other way. 1 either way from 9 finds 8. 1 either way from 16 finds 15, and so on. Moving away from the "square" diagonal along the other diagonal always drops by 1. If you pursue this direction, you would find that the next answer drops by 3, and so on. This information is very useful later on.

How calculators multiply

The table we just examined only goes up to 9 times 9. Years ago, children had learned up to 12 times 12, or even 24 times 24. What a chore! In the tens (or "decimal") system, all we really need is up to 9 times 9. Multiplying by 10 merely adds a zero to the end. 5 times 6 equal 30. 5 times 60 are 300. 50 times 60 are 3000. Whichever number you multiply by 10, the product gets multiplied by 10 (using an extra 0), too.

Calculators use this fact. Actually a calculator does not multiply. It keeps on adding. But because it does so much faster (one addition in about a millionth of a second) it adds more quickly than you can multiply.

If you enter 293 times 135, here is what the calculator does inside (see page 38). Starting from zero in both "registers," it takes the first digit of the multiplier, here 1 (representing 100), moves the product up by the multiplicand (that's the fancy name for 293 in this case) in the product register, at the hundreds place and by 100 in the multiplier register. Since no more hundreds are left, it now moves to the tens, which it moves up by 3 additions. Finally, it moves up 5 times for the units place in the multiplier to find the product 39,555. The calculator has merely done addition to complete what people do by multiplication.

Putting together how people did it

Simple multiplication, just using the facts in the table you constructed, is easy. But people have to multiply bigger numbers, like the one at the top of page 39. To do it, you must multiply every part of one number, the *multiplicand* (as it used to be called) by every part of the other number, the *multiplier*. You can multiply in any order, so long as you do it systematically, so as not to miss any. You could multiply as another exercise and let your calculator assemble it.

Let's take it in order, from left to right in each number, the multiplicand first in each case. Few people ever learned to do it that way. Maybe they would have understood it better if they had, before learning the usual ways of shortening what you write down.

.

How calculators multiply

| 2 | 9 | 3 |

×

| 1 | 3 | 5 |

PRODUCT						MULTIPLIER		
0	0	0	0	0		0	0	0
2	9	3	0	0	+	1	0	0
2	9	3	0	0	=	1	0	0
	2	9	3	0	+		1	0
3	2	2	3	0	=	1	1	0
	2	9	3	0	+		1	0
3	5	1	6	0	=	1	2	0
	2	9	3	0	+		1	0
3	8	0	9	0	=	1	3	0
		2	9	3	+			1
3	8	3	8	3	=	1	3	1
		2	9	3	+			1
3	8	6	7	6	=	1	3	2
		2	9	3	+			1
3	8	9	6	9	=	1	3	3
		2	9	3	+			1
3	9	2	6	2	=	1	3	4
		2	9	3	+			1
3	9	5	5	5	=	1	3	5

Putting together how people did it

Multiplicand		Multiplier		Subproduct	Cumulative Total
200	×	100	=	20 000	20 000
90	×	100	=	9 000	29 000
3	×	100	=	300	29 300
200	×	30	=	6 000	35 300
90	×	30	=	2 700	38 000
3	×	30	=	90	38 090
200	×	5	=	1 000	39 090
90	×	5	=	450	39 540
3	×	5	=	15	39 555

Carrying in multiplication

In addition, carrying is used as a way of saving what had to be written down. Multiplication can do the same thing. It also helps us to be systematic.

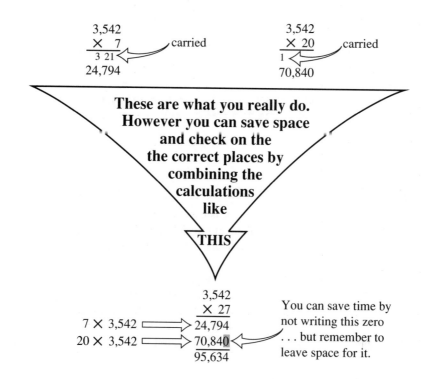

Here's another example that will show this rule better: 3,542 multiplied by 27. First the ones: 7 times 2 are 14; write 4, carry 1. The tens: 7 times 4 are 28, with 1 carried makes 29; write 9 carry 2. The hundreds: 7 times 5 are 35, with 2 carried is 37; write 7, carry 3. The thousands: 7 times 3 are 21, with 3 carried is 24. That finishes 3,542 × 7: 24,794.

Now do the tens part, which equals 70,840, in the same way. Sometimes you don't write the zero, just space the last digit (4) over, so it's in the tens column. The whole thing is written down in one "piece" or *algorithm*, as the professional mathematicians call it. This is essentially how people performed multiplication before we had computers.

A matter of order

As the Introduction showed, at one time "new math" consisted in multiplying from the other direction. Here, the same multiplication has been performed in the reverse order: first the 20, then the 7 where, on the previous page, the 7 was multiplied first, then the 20. The answer is the same either way, provided no mistakes are made.

The important thing in long multiplication by hand is to do it systematically. If the multiplier has three or more digits, work consistently, either from left to right or from right to left.

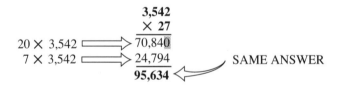

$$
\begin{array}{r}
3,542 \\
\times\ 27 \\
\end{array}
$$

20 × 3,542 ⟹ 70,840
7 × 3,542 ⟹ 24,794   SAME ANSWER
                95,634

IT DOESN'T MATTER WHICH YOU DO FIRST
YOU'LL GET THE SAME ANSWER

Using your pocket calculator to verify this process

When you have a pocket calculator, it is easy to punch in one number, then the multiply sign, then the other number, and finally the equals sign. Bingo, you have the answer, all complete. This method doesn't help you see how it's done. Here's how you can do that.

Assume you have a calculator with a single memory, which is the simplest type. Multiplying 7 by 2, it gives you 14, which you enter in memory with the MS button. Now, multiply 7 by 40, which gives you 280. Add this number to the 14 with the M+ button. You can read what you already have by pressing the MR button. Finish multiplying by 7 and the MR button will list your answer "longhand:" 24794.

7	7	×	7	2	2	=	14	MS	14		
7	7	×	7	40	40	=	280	M+	280	MR	294
7	7	×	7	500	500	=	3500	M+	3500	MR	3794
7	7	×	7	3000	3000	=	21000	M+	21000	MR	24794
20	20	×	20	2	2	=	40	M+	40	MR	24834
20	20	×	20	40	40	=	800	M+	800	MR	25634
20	20	×	20	500	500	=	10000	M+	10000	MR	35634
20	20	×	20	3000	3000	=	60000	M+	60000	MR	95634

Now, you can go on multiplying by 20. With a single memory, you don't see the "times 20" part separately, as you did in longhand. The final answer is the same. If you have a calculator with more than one memory, you can store each part in a separate memory and then add the contents of the two memories.

Skipping zeros

When you multiply longhand, don't forget the zeros, if your multiplier has a zero in it. When you pass zero, there is no point in writing a line of zeros, because zero times anything is still zero. But don't forget that, in this case, after multiplying by 20 (the tens figure), the next one is the thousands figure, which moves to the left two places instead of one, because the multiplier has no hundreds figure.

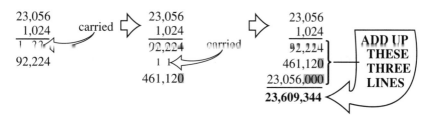

**In multiplication too,
zeros are used to keep figures in their "places."**

Either number can be the multiplier

We tend to use the "shorter" number as the multiplier, because it looks like less work. It isn't: you still have to multiply every digit in one number by every digit in the other, but it looks simpler. Here are two examples where the same two numbers are multiplied together, swapping places for multiplicand and multiplier.

Of course, you can verify this very easily with your pocket calculator. Whichever way you punch in the numbers, e.g., 3542 × 27 or 27 × 3542, you will get the same answer, 95634. This principle is given a fancy name in one version of the so-called "new math." Don't bother with it, just remember that it's true.

CHECKING

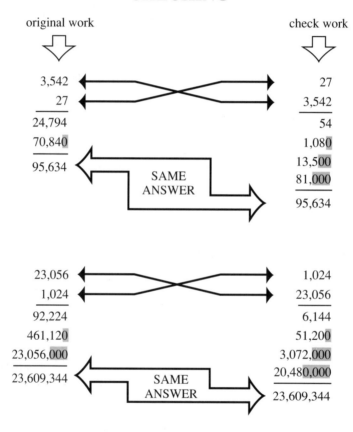

Using subtraction in multiplication

Sometimes it's convenient to make complicated calculations in your head, instead of relying on a calculator. Although a calculator isn't "interested" in short cuts, they can be a big help when you do things in your head. It is also a useful way to verify your result—even in your head!

Here are two examples of using subtraction, instead of addition, in multiplication. When I show some people this "trick," they say "I didn't know you could do that!"

The multiplier 29 is 1 less than 30. So, multiply by 30, which is much easier than multiplying by 29. Then, subtract once times the original number, which is itself. Check it in the usual way; you will find the same answer.

In the second example, 98 is 2 less than 100. This example is even easier. Multiplying by 100 just puts two zeros to the right of the original number. Subtract twice the original number, to represent -2 in the multiplier, and you have the answer. Multiplying by 2 is much less work than multiplying by 8 and then 9 and adding them together, which is what has been done to verify the result.

If you don't like all that work longhand, do it on your pocket calculator.

By subtraction		By addition	
	47,392		47,392
	29		29
30 × 47,392 ➡	1,421,760	9 × 47,392 ➡	426,528
minus		plus	
1 × 47,392 ➡	47,392	20 × 47,392 ➡	947,840
1 × 47,392 ➡	1,374,368	29 × 47,392 ➡	1,374,368
	SAME ANSWER		
	63,257		63,257
	98		98
100 × 63,257 ➡	6,325,700	8 × 63,257 ➡	506,056
minus		plus	
2 × 63,257 ➡	126,514	90 × 63,257 ➡	5,693,130
98 × 63,257 ➡	6,199,186	98 × 63,257 ➡	6,199,186
	SAME ANSWER		

Multiplying by factors

Here's another way that works very well with certain numbers. Suppose the multiplier is 35. That happens to be 5 × 7. Instead of multiplying the whole original number by 5 and by 30 and adding the results, you can multiply first by 5, then multiply that result by 7. As is shown here, both ways give the same answer. You can verify this answer on your pocket calculator, too.

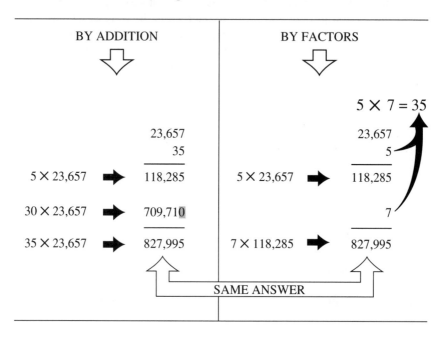

BY ADDITION		BY FACTORS
		5 × 7 = 35
	23,657	23,657
	35	5
5 × 23,657 ➡	118,285	5 × 23,657 ➡ 118,285
30 × 23,657 ➡	709,710	7
35 × 23,657 ➡	827,995	7 × 118,285 ➡ 827,995

SAME ANSWER

Multiplying with weights

When you use systems that are nondecimal (not based on the 10 system) multiplication is complicated a little. Some pocket calculators are equipped to make such conversions, but you need to know what to do if yours isn't when you encounter such a problem.

Suppose you have to find what 25 times 1 pound 3 ounces is. You can multiply the pounds by 25, and you can multiply the ounces by 25. But 25 pounds, 75 ounces isn't the way you'd normally express that weight! Anything over 16 ounces must be converted into pounds.

Convert 16 ounces to an extra pound and subtract the 16 ounces. Doing this procedure 4 times converts the 75 ounces to 4 pounds and 11 ounces. Now, add the 4 pounds to the 25 pounds, so the total is 29 pounds 11 ounces. You could convert the 75 ounces to pounds by dividing by 16. However, the remainder would still have to be converted back to ounces.

Your calculator will convert 75 ounces to 4.6875 pounds. If you subtract the 4, that leaves 0.6875 as the part over 4 pounds. Multiply that by 16 to convert back to ounces, and you have 11.

PROBLEM IN WEIGHT
25 × 1 pound 3 ounces

25 times	3 ounces is	75 ounces
		−16
	or 1 pound	59 ounces
		−16
	or 2 pounds	43 ounces
		−16
	or 3 pounds	27 ounces
		−16
	or 4 pounds	11 ounces
25 times 1 pound is		25 pounds

25 times 1 pound 3 ounces is 29 Pounds 11 Ounces

Multiplying lengths

Multiplying lengths is similar, except that the English system uses 12 inches to the foot, 3 feet to the yard, etc., instead of the decimal or metric system. Where necessary, we have to make the same kind of conversions between inches, feet, and yards. See the example on page 46.

Multiplying measures

In the same way, if you multiply a measure by a fairly large number, it is usually convenient to change the unit of measure in which we express it. See example at the bottom of page 46 and top of page 47.

Suppose you multiply 3 pints by 250. The answer is quite easily found to be 750 pints. But quantities this large are usually given in gallons, not in pints. Remembering that 8 pints are in a gallon, you can proceed to count off in eights. This procedure is quite long (unless your calculator does it). From 750, 8 can be subtracted 93 times and 6 pints are left over. If you divide 750 by 8, you have 93 3/4 pints. The calculator would read 93.75.

MULTIPLYING LENGTHS

How much lumber is needed
to cut off 5 pieces 10 inches long?

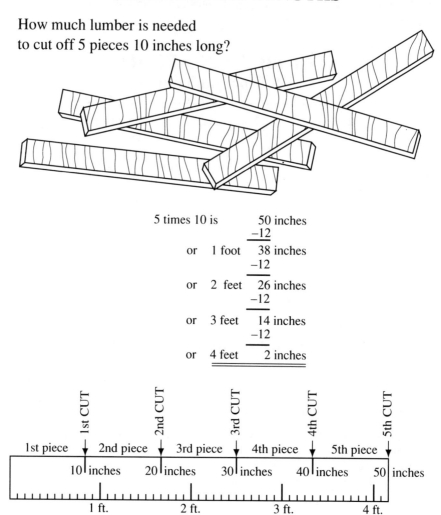

5 times 10 is   50 inches
                       −12

or   1 foot   38 inches
                       −12

or   2 feet   26 inches
                       −12

or   3 feet   14 inches
                       −12

or   4 feet   2 inches

MULTIPLYING MEASURES

 One motor crankcase takes 3 pints of oil

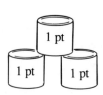 

crankcase

 How much is needed for 250 motors?

250 times 3 pints is		750 pints
	− 8	
or 1 gallon		742 pints
	− 8	
or 2 gallons		734 pints
	− 8	
or 3 gallons		726 pints
	− 8	
or 4 gallons		718 pints

AFTER SUBTRACTING 8
FROM PINTS AND ADDING 1
TO GALLONS 93 TIMES (IF YOU
DIDN'T MAKE A MISTAKE) . . .

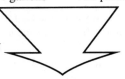

or 93 gallons	6 pints
or	3 quarts
or	$\frac{3}{4}$ gallon

$$93\frac{3}{4} \text{ gallons}$$

 There MUST be an easier way!

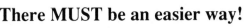

Questions and problems

1. Multiply the following pairs of numbers together as shown: check your results by using the upper number as the multiplier in each case.

(a) 357	(b) 243	(c) 24	(d) 37	(e) 193	(f) 187
× 246	× 891	× 36	× 74	× 764	× 263

2. Multiply the following pairs by using subtraction to make the working simpler. Verify your results by the more usual method.

(a) 2,573	(b) 7,693	(c) 4,927	(d) 5,396	(e) 7,109
× 19	× 28	× 18	× 59	× 89

3. Multiply the following pairs by using factors of the second number. Check your results by long multiplication.

(a) 1,762	(b) 7,456	(c) 8,384	(d) 9,123	(e) 1,024
× 45	× 32	× 21	× 63	× 28

4. An airline runs 4 flights per day between two cities, every day except Sunday, when it runs 2. How many flights a week is this?

5. The same flight (question 4) is also made only twice on the 12 public holidays of the year. How many flights are made per year (based on 52 weeks)?

6. A mass-produced item costs 25 cents each to make, and $2 to package. The package cost is the same for a packet of one or of many thousand. What is the cost for packets of:

(a) 1 (b) 10 (c) 25 (d) 100 (e) 250 (f) 1,000 (g) 5,000 (h) 10,000?

7. If you ignore the manufacturing cost, how much is saved by packaging the items of question 6 in packets of 250 instead of 100? (Assume a total quantity of 500.)

8. On a commuter train, single tickets sell for $1.75. You can buy a 10-trip ticket for $15.75. How much does this save over the single-ticket rate?

9. On the same train, a monthly ticket costs $55. If a commuter makes an average of 22 round trips a month, how much will he save by buying a monthly ticket?

10. An employer offered an employment contract beginning at $500 a month, with a raise of $50 a month every year for 5 years. The contract expired after the 6th year. Employees bargained for a starting figure of $550 a month, with a raise of $20 a month every 6 months. Which rate of pay was higher at the beginning of the sixth year? Which rate resulted in the greatest total earnings per employee for 6 years? By how much?

11. A manufacturer prices parts according to how many are bought at once. The price is quoted per hundred pieces in each case, but the customer must take the quantity stated to get a particular rate. The rates are $7.50 apiece for 100, $6.75 apiece for 500, $6.25 apiece for 1,000, $5.75 apiece for 5,000 and $5.50 apiece for quantities over 10,000. The rate for any in between quantity is based on the next lower number. What is the difference in total cost for quantities of 4,500 and 5,000?

12. Small parts are counted by weighing. Suppose 100 of a particular part weigh 2 1/2 ounces and you need 3,000 of the parts. What is the weight?

13. A bucket that is used to fill a tank with water holds 4 gallons and 350 bucket-fulls are required to fill the tank. What is the tank's capacity?

14. A freight train has 182 cars each loaded to the maximum which, including the weight of the car, is 38 tons. What is the total weight that the locomotive has to haul?

15. A car runs 260 miles on one tankful of gas. It has an alarm that lets the driver know when it needs refilling. If the journey requires 27 fillings and at the end, the tank is ready for another, how long was the journey?

16. Two railroads connect the same two cities. One charges 10 cents per mile, the other 15 cents per mile. The distance between the cities is 450 miles by the first railroad, but 320 miles by the second. Which company offers the cheaper fare? By how much?

17. One airline offers rates based on 15 cents per mile for first-class passengers. Its distance between two cities is 2,400 miles. Another airline offers 10 cents a mile for coach, but uses a different route, marking the distance at 3,200 miles. Which fare is cheaper? By how much?

18. The first airline (question 17) offers a family plan. Each member of a family, after the first, pays a rate that is based on 9 cents per mile. Which way will be cheaper for a family (a) of 2? (b) of 3? and by how much?

19. A health specialist recommends chewing every mouthful of food 50 times. One ounce of one food can be eaten in 7 mouthfuls. A helping of this food consists of 3 ounces. How many times will a person have to chew this helping to fulfill the recommendation?

20. An intricate pattern on an earthenware plate repeats 9 times around the edge of the plate. Each pattern has 7 flowers in the repetition. How many flowers are around the edge of the plate?

4
CHAPTER

Division

Division began as counting out

Before the days of electronic calculators and computers, division was a short cut for counting out. To divide 28 items among 4 people, you would keep giving one to each of the four until the items were all gone and then see how many each person had. This form is shortened by using the division sign (÷), and this equation is written: $28 \div 4 = 7$.

Divide 28 items among 4 people

① ② ③ ④ ⑤ ⑥ ⑦ ⑧ ⑨ ⑩

⑪ ⑫ ⑬ ⑭ ⑮ ⑯ ⑰ ⑱ ⑲ ⑳

㉑ ㉒ ㉓ ㉔ ㉕ ㉖ ㉗ ㉘

28 ÷ 4

First Method

 ① ⑤ ⑨ ⑬ ... ② ⑥ ⑩ ⑭ ... ③ ⑦ ⑪ ⑮ ... ④ ⑧ ⑫ ...

etc. until we finish

① ⑤	② ⑥	③ ⑦	④ ⑧
⑨ ⑬ ⑰	⑩ ⑭ ⑱	⑪ ⑮ ⑲	⑫ ⑯ ⑳
㉑ ㉕	㉒ ㉖	㉓ ㉗	㉔ ㉘
SEVEN	SEVEN	SEVEN	SEVEN

28 divides into 4 groups, with 7 in each group

Second Method

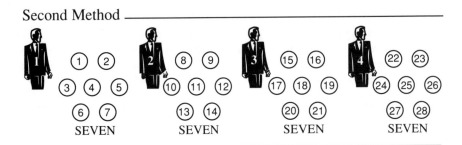

① ②	⑧ ⑨	⑮ ⑯	㉒ ㉓
③ ④ ⑤	⑩ ⑪ ⑫	⑰ ⑱ ⑲	㉔ ㉕ ㉖
⑥ ⑦	⑬ ⑭	⑳ ㉑	㉗ ㉘
SEVEN	SEVEN	SEVEN	SEVEN

How a calculator does it

A calculator uses a repeated process of subtraction to divide 45355 by 193. 100 times 193 is 19300. For the moment, leave out the comma to mark thousands. Subtracting 19300 from 45355 leaves 26055. *That's 100 times.* 26055 is still greater than 19300, so it subtracts the same number again, leaving 6755. *That's 200 times.* Now, it drops by 10 to 1930. That subtracts from 6755 three times, leaving, in succession, 4825, 2895, and 965, while adding 10 to the quotient "*column*" (called a *register* in calculator parlance) each time. *That's 230 times.* To finish, it subtracts the plain number, 193, five times and *the quotient column is 235.*

DIVIDEND

4	5	3	5	5
1	9	3	0	0
2	6	0	5	5
1	9	3	0	0
	6	7	5	5
	1	9	3	0
	4	8	2	5
	1	9	3	0
	2	8	9	5
	1	9	3	0
		9	6	5
		1	9	3
		7	7	2
		1	9	3
		5	7	9
		1	9	3
		3	8	6
		1	9	3
		1	9	3
		1	9	3

DIVISOR

1	9	3
1	0	0
1	0	0
1	0	0
2	0	0
	1	0
2	1	0
	1	0
2	2	0
	1	0
2	3	0
		1
2	3	1
		1
2	3	2
		1
2	3	3
		1
2	3	4
		1

QUOTIENT 2 3 5

Division is multiplication in reverse

Just as a calculator performs multiplication by repeated addition, it also performs division by repeated subtraction. In fact, this pattern in mathematics is useful to follow through. Each process that we learn has a reverse and each reverse process provides a way to check the one it reverses.

Addition	Subtraction
$4 + 7 = 11$	$11 - 4 = 7$
four plus seven equals eleven	eleven minus four equals seven
four and seven are eleven	four from eleven leaves seven

Subtraction is the reverse of Addition

Multiplication	Division
$4 \times 7 = 28$	$28 \div 4 = 7$
four times seven equals twenty-eight	twenty-eight divided by four equals seven
four sevens are twenty-eight	four goes into twenty-eight seven times
	$\dfrac{28}{4} = 7$ or $4\,\overline{)28}\,^{7}$

Division is the reverse of Multiplication

Dividing into longer numbers

Though few people do this, it's useful to have a multiplication table for the divisor, in this case 7, at the side. This table enables you to subtract the number in the quotient all in one "bite," rather than one piece at a time (like the calculator does). The remainder each time is less than the divisor, so "bring down" the next digit or figure and continue for the next place in the quotient.

Here is what you really do, then how it is usually written.

IN DIVISION WE START WITH THE LARGEST FIGURE AND WORK TOWARD THE SMALLEST

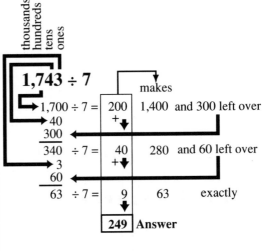

MULTIPLICATION TABLE		
↓ ➡		7 times
2 are		14
3 are		21
4 are		28
5 are		35
6 are		42
7 are		49
8 are		56
9 are		63

thousands hundreds tens ones

1,743 ÷ 7   makes

1,700 ÷ 7 = | 200 | 1,400 and 300 left over
→ 40
300 ◄
340 ÷ 7 = | 40 | 280 and 60 left over
→ 3
60 ◄
63 ÷ 7 = | 9 | 63   exactly

249 | Answer

Usual Way of Writing

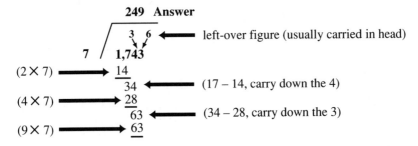

249 Answer

7 / 1,743   left-over figure (usually carried in head)

(2 × 7) ➡ 14
34 ◄ (17 – 14, carry down the 4)
(4 × 7) ➡ 28
63 ◄ (34 – 28, carry down the 3)
(9 × 7) ➡ 63

Multiplication checks division

You can always check division by multiplication, whether you do it the old-fashioned way or on a calculator. Using a calculator, you would punch in *1743*, then ÷, then 7, then =, and the calculator would read 249. Now, with 249 still reading, punch ÷ then 7, then = again, and it will read 1743.

"Why do this?" you might ask. This procedure confirms that you hit the correct keys. If the last figure isn't the one you began with, since the calculator doesn't make mistakes, you probably hit a wrong button somewhere.

CHECK YOUR DIVISION BY MULTIPLICATION

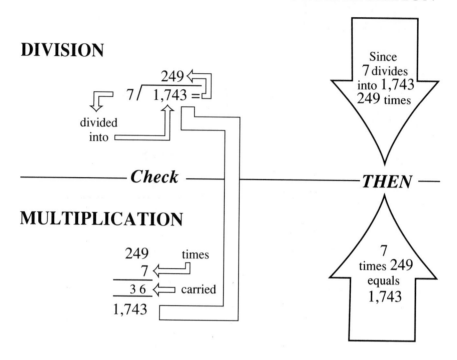

DIVISION

249

7 / 1,743 =

divided into

Since 7 divides into 1,743 249 times

—————— *Check* —————— *THEN* ——

MULTIPLICATION

    249   times
      7
     3 6   carried
  1,743

7 times 249 equals 1,743

More about how a calculator does it

Look back to *How a calculator does it*. The calculator has to "know" that its first subtraction will be 100 times. To do this procedure, it begins with 1 times, then tries 10 times and 100 times. If it finds that the dividend is larger, it will increase to 1000 times. If it finds that divisor is larger, it will drop back to 100 times.

After subtracting 100 times twice, it tries the third time, finds that the divisor is bigger, so it drops back to 10 times. The same thing happens when it has subtracted 10 times three times. The fourth time it finds that the divisor is bigger, so it drops back to the plain number as shown on opposite page.

Dividing by larger numbers (the people way)

Here is an example of why we suggested having the multiplication table at the left. 23 is not in your regular multiplication table. Of course, you might try at each place, without the table, and save a little time. But it is more "methodical" to prepare the table first, so you can see at a glance of the table what the next digit is. For example, 23 into 149 seven times is 161. That number is too large, so it tries six times, which is 138. Then 138 is larger than 119, so it must be 5 times. Finally, the last digit (conveniently) is exactly twice. What happens when it isn't exact? The answer follows shortly as shown on page 56 at the top.

DIVIDEND

	4	5	3	5	5
			1	9	3
		1	9	3	0
	1	9	3	0	0
1	9	3	0	0	0
	1	9	3	0	0
	2	6	0	5	5
	1	9	3	0	0
		6	7	5	5
	1	9	3	0	0
		1	9	3	0
		4	8	2	5
		1	9	3	0
		2	8	9	5
		1	9	3	0
			9	6	5
		1	9	3	0
			1	9	3
			7	7	2
			1	9	3

DIVISOR

	1	9	3
			1
		1	0
	1	0	0
1	0	0	0
	1	0	0
	1	0	0
		1	0
		1	0
		1	0
			1
			1

QUOTIENT ACCUMULATOR

1	0	0
2	0	0
2	1	0
2	2	0
2	3	0
2	3	1

23 × 2	46		652
23 × 3	69	23 / 14,996	
23 × 4	92		138
23 × 5	115		119
23 × 6	138		115
23 × 7	161		46
23 × 8	184		46
23 × 9	207		

Multiplication as a "check again"

With larger numbers as divisors, it's more important to check your techniques because you have more chances of making a mistake. These mistakes occur both when you do it by hand and when you use a calculator.

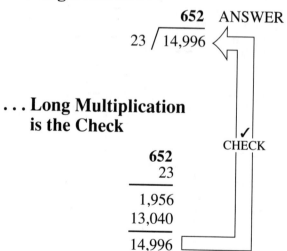

In Long Division . . .

652   ANSWER

23 / 14,996

. . . Long Multiplication is the Check

✓ CHECK

652
  23

1,956
13,040

14,996

Division by factors

In multiplication, you used multiplication by factors. It also works in division, if the divisor has factors. For example, 28 is 4 times 7. So you can divide the dividend by 4, then by 7. You can also multiply by factors to check the answer.

DIVIDING BY FACTORS

Divide 37,996 by 28
4 × 7 = 28 so . . .

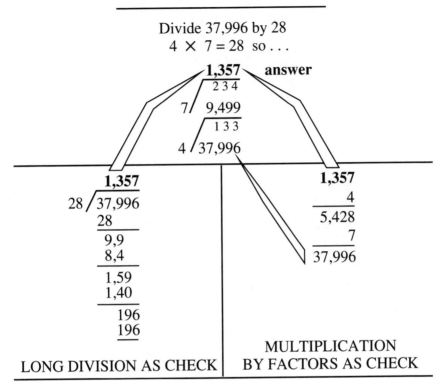

LONG DIVISION AS CHECK | MULTIPLICATION BY FACTORS AS CHECK

Which method is best

The previous page showed that, whether you used long division or factors, the same answer was produced. Any number in the multiplication table (and many more) can use factors. So, which method is best?

It's a matter of preference. With a calculator, maybe there's not much point in using factors. With longhand, you have another way to check your work (see example on page 58).

When a remainder is left

In the section, *Dividing by bigger numbers (the people way)*, I asked what happens when division doesn't "come out" exactly? Try it on a calculator: you'll see. Doing it longhand, for example, 37 divides into 10,050 271 times, with a remainder 23. What does that mean? How does it relate to the string of figures that a calculator reads out? See example at top of page 59.

With these numbers, you have a choice: Factors or Long Division

MULTIPLICATION TABLE

TIMES ARE →	2	3	4	5	6	7	8	9
2	4	6	8	10	12	14	16	18
3	6	9	12	15	18	21	24	27
4	8	12	16	20	24	28	32	36
5	10	15	20	25	30	35	40	45
6	12	18	24	30	36	42	48	54
7	14	21	28	35	42	49	56	63
8	16	24	32	40	48	56	64	72
9	18	27	36	45	54	63	72	81

It's so much quicker

I always feel safer this way

AIRLINES

TRAINS

TO NEW YORK

TO NEW YORK

It depends on which you like the best!

Division doesn't always come out exactly even

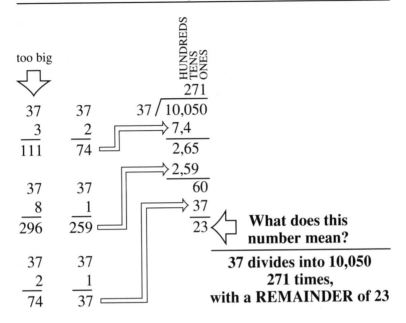

too big

37
  3
―――
111

37
  2
―――
74

$37\overline{)10,050}$ with quotient 271 (HUNDREDS TENS ONES)

7,4
2,65
2,59
60
37
23

37
  8
―――
296

37
  1
―――
259

37
  2
―――
74

37
  1
―――
37

What does this number mean?

37 divides into 10,050 271 times, with a REMAINDER of 23

What does the remainder mean?

Take a simpler example. Pursue the dividing-out picture that the chapter started with. Dividing 25 into 6 shares gives each share 4, with one over the remainder number 25. What do you do with the remainder?

Suppose they were pies. You'd take the 25th pie and divide it into 6 equal parts, then give each of the 6 people 1 part, in addition to 4 whole pies.

This is the origin of fractions. The part, one sixth, is written as 1/6. The bottom number is called the *denominator*, the number of possible pieces. The top number is called the *numerator*, the number of pieces that you actually have.

Divide 25 by 6

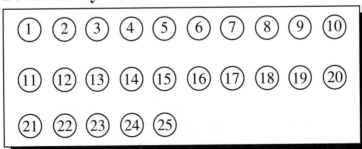

continued

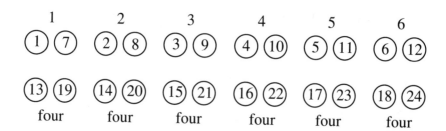

Using 24 makes 4 each in 6 groups.

WHAT ABOUT THE ODD ONE ㉕ ?

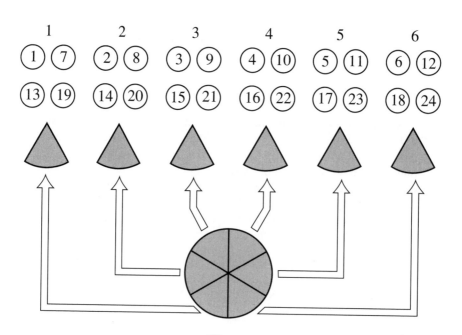

Divide Number ㉕ into six equal parts.

Each part is one-sixth of the whole.
The fraction one-sixth is written $\frac{1}{6}$

$$25 \div 6 = 4\frac{1}{6}$$

How a calculator handles fractions

Now, let's start on what a calculator does with fractions. If you divide 25 by 6, it will read out 4.16666. . . . Here's what the little thing does. Down till the quotient accumulator reads 4, it follows what you already know. However, it just doesn't stop. Had you punched in 2500 divided by 6, it would read 416.66. . . . Following the description in *What does the remainder mean?*, your answer would be 416 1/6.

The calculator adds a decimal point when it goes beyond zero in the ones place. Look at it this way: 25 is the same as 25.0000. Call that "25 point zero, zero, zero, zero." The zeros only make a difference where they "hold a place," with "real numbers" before and after them, or where they keep the real numbers away from the ones place.

A division such as 6 into 2500 would go on forever. Each 4 can be divided by another 6.

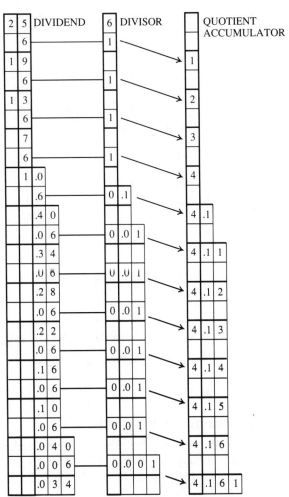

Fractions that have multiple parts

In *What does the remainder mean?*, the 1 left over divides into 1/6 part each. Here, we do the same thing again, with 30 divided by 7. You get to 28, with 4 each, and the remainder is 2—that is, 2 parts are left over. So, the fraction is 2/7: Numerator 2, denominator 7. 7, the bottom number, indicates that each whole is divided into 7 parts. The 2 on top indicates that each gets 2 of those one-seventh parts. That's all that fractions really mean.

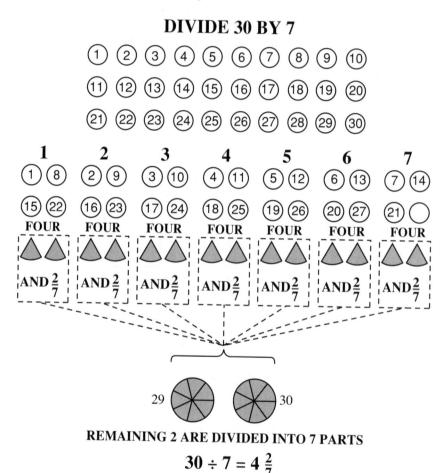

DIVIDE 30 BY 7

REMAINING 2 ARE DIVIDED INTO 7 PARTS

$$30 \div 7 = 4\frac{2}{7}$$

Decimal equivalents of fractions

Previous pages have shown how a calculator turns a fraction into a decimal. That's what it does, naturally—if anything is natural about a calculator! Now, look at what a calculator does with a few simple fractions. Here are a few that convert to relatively simple decimals:

$$\frac{1}{2} \quad \vdots \quad 2\overline{\smash{\big)}\,1.0}^{\,0.5} \qquad so \qquad \frac{1}{2} = 0.5$$

$$\frac{1}{4} \quad \vdots \quad 4\overline{\smash{\big)}\,1.00}^{\,0.25} \qquad so \qquad \frac{1}{4} = 0.25$$

$$\frac{1}{5} \quad \vdots \quad 5\overline{\smash{\big)}\,1.0}^{\,0.2} \qquad so \qquad \frac{1}{5} = 0.2$$

$$\frac{1}{8} \quad \vdots \quad 8\overline{\smash{\big)}\,1.000}^{\,0.125} \qquad so \qquad \frac{1}{8} = 0.125$$

$$\frac{1}{10} \quad \vdots \quad 10\overline{\smash{\big)}\,1.0}^{\,0.1} \qquad so \qquad \frac{1}{10} = 0.1$$

More difficult fractions

Those fractions turned into relatively simple decimals. The harder ones are those that always run off the end of the calculator's space. 4 1/6 was one. Here's another that looks simple: 1/3. But when you divide 3 into 1, after bringing down zero after zero, you get an unending string of 3s. Earlier generations had a useful trick to avoid having to keep writing 3s; they put a dot over the 3 to indicate that it keeps repeating.

Here we show some more fractions that end up with one number that keeps repeating.

$$\frac{1}{3} \quad \vdots \quad 3\overline{\smash{\big)}\,1.0000000}^{\,0.3333333\ldots} \qquad \frac{1}{3} = 0.\dot{3} \quad \begin{array}{l}\text{point three}\\ \text{recurring}\end{array}$$

$$\frac{1}{6} \quad \vdots \quad 6\overline{\smash{\big)}\,1.0000}^{\,0.1666\ldots} \qquad \frac{1}{6} = 0.1\dot{6}$$

$$\frac{1}{9} \quad \vdots \quad 9\overline{\smash{\big)}\,1.00}^{\,0.111\ldots} \qquad \frac{1}{9} = 0.\dot{1}$$

$$\frac{1}{12} \quad \vdots \quad 2\overline{\smash{\big)}\,0.1\dot{6}666666}^{\,0.08\dot{3}333\ldots} \qquad \frac{1}{12} = 0.08\dot{3}$$

Some decimals would go on forever with one number . . .

$$\frac{5}{6} \quad \vdots \quad \textbf{1st Method} \quad 6\overline{\smash{\big)}\,5.0000}^{\,0.8333}$$

$$\frac{5}{6} = 0.8\dot{3}$$

2nd Method

$$5 \times \frac{1}{6} \qquad \begin{array}{r}0.1\dot{6}6666\ldots\\ 5\\ \hline {}^{\diagdown}3\text{ carried}\\ \hline 0.8\dot{3}333\ldots\end{array}$$

3rd Method $\quad \dfrac{5}{6} = \dfrac{2}{6} + \dfrac{3}{6} = \dfrac{1}{3} + \dfrac{1}{2}$

$$\frac{1}{3} = 0.\dot{3}33$$

$$\frac{1}{2} = 0.5$$

$$\frac{5}{6} = 0.8\dot{3}33$$

Where more figures repeat

In all the fractions in the last section, the decimal equivalent, which is what your calculator always gives you, ended up with one figure that kept repeating. Another kind of fraction first shows up in the decimal equivalent for 1/7. Here, 6 digits repeat. Interestingly, for numerators 1 through 6, the decimal equivalent uses the same sequence of repeating digits, starting at a different place. The old way of indicating the repeating digits is to put a dot over the digits that repeat.

Notice that when you multiply the repeating decimal for 1/7 by 7, you get a row of repeating nines. This number virtually is the same as 1, because it falls short of 1 by an infinitely small number. This problem leads to an idea for converting repeating decimals to equivalent fractions.

. . . and some keep it going with groups of numbers

$$\frac{1}{7} : \quad 7\overline{\smash)\underset{1.000000000000}{\overset{0.142857142857}{}}} \qquad \frac{1}{7} = 0.\dot{1}4285\dot{7}$$

$$\frac{2}{7} : \quad 7\overline{\smash)\underset{2.000000000000}{\overset{0.285714285714}{}}} \qquad \frac{2}{7} = 0.\dot{2}8571\dot{4}$$

$$\text{or } 2 \times \frac{1}{7} = 2 \times 0.\dot{1}4285\dot{7} = 0.\dot{2}8571\dot{4}$$

$$\frac{7}{7} = 7 \times 0.\dot{1}4285\dot{7} = 0.\dot{9}9999\dot{9} = 0.\dot{9}$$

$$0.\dot{9} + 0.\dot{0} = 1 \quad \text{and} \quad 0.\dot{0} = 0$$
$$\text{so} \quad 0.\dot{9} = 1$$

Decimal for one eleventh and others

When your denominator is 11 and the numerator is 1, .09 repeats. If the numerator is 2, .18 repeats, and so on, until 10/11 = .90 repeating. As before 11/11, being the same as 1, comes to .99 repeating. Here is a clue. 1/11 is the same as 9/99, because 9 × 11 is 99. Do you begin to see how to handle this problem?

The decimal for 1/7 has 6 recurring digits. Here are the recurring decimals for 1/17 and 1/19. Notice that each has one less number of digits than the denominator. The one for 1/17 has 16 digits. The one for 1/19 has 18 digits. Do you wonder why?

$$\frac{1}{11} : \quad 11\overline{\smash)\underset{1.0000}{\overset{0.0909}{}}} \qquad \frac{1}{11} = 0.\dot{0}\dot{9}$$

$$\frac{2}{11} : \quad 11\overline{\smash)\underset{2.0000}{\overset{0.1818}{}}} \qquad \frac{2}{11} = 0.\dot{1}\dot{8}$$

Decimals recur in Pairs for Elevenths

$$\text{OR } 2 \times \frac{1}{11} = 2 \times 0.\dot{0}\dot{9} = 0.\dot{1}\dot{8}$$

$$\frac{3}{11} = 0.\dot{2}\dot{7} \qquad \frac{4}{11} = 0.\dot{3}\dot{6} \qquad \frac{5}{11} = 0.\dot{4}\dot{5} \qquad \frac{6}{11} = 0.\dot{5}\dot{4}$$

$$\frac{7}{11} = 0.\dot{6}\dot{3} \qquad \frac{8}{11} = 0.\dot{7}\dot{2} \qquad \frac{9}{11} = 0.\dot{8}\dot{1} \qquad \frac{10}{11} = 0.\dot{9}\dot{0}$$

$$\frac{11}{11} = 0.\dot{9}\dot{9} = 0.\dot{9} = 1$$

$$\frac{1}{13} = 0.\dot{0}7692\dot{3} \qquad\qquad \frac{2}{13} = 0.\dot{1}5384\dot{6} \qquad\qquad \frac{3}{13} = 0.\dot{2}3076\dot{9}$$

These numbers used for	These numbers used for	(six numbers for thirteenths)
$\frac{1}{13}, \frac{3}{13}, \frac{4}{13}, \frac{9}{13}, \frac{10}{13}, \frac{12}{13}$	$\frac{2}{13}, \frac{5}{13}, \frac{6}{13}, \frac{7}{13}, \frac{8}{13}, \frac{11}{13}$	

$$\frac{1}{17} = 0.\dot{0}58823529411764\dot{7} \qquad \begin{array}{c}\text{(sixteen numbers} \\ \text{for seventeenths)} \\ \text{and}\end{array}$$

$$\frac{1}{19} = 0.\dot{0}5263157894736842\dot{1} \qquad \begin{array}{c}\text{(eighteen numbers} \\ \text{for nineteenths)}\end{array}$$

Converting recurring decimals to fractions

Although recurring decimals are easier to handle when you know what they mean, using old-fashioned fractions is often easier. How can we convert a recurring decimal to a fraction? Shown here is how to do it when only one digit recurs.

Nonrecurring digits represent that many tenths, hundredths, thousandths, or whatever place the digit is in. Then, a recurring digit represents that many ninths—zeros after the 9 in the denominator to put it in the right place value.

Always check your result by dividing top by bottom, calculator style, to see if you get the decimal that you began with

Using 9's to convert recurring decimals to fractions

$$0.\dot{1} = \frac{1}{9} \qquad\qquad \frac{1111111\,...}{9999999\,...}$$

$$0.\dot{3} = \frac{3}{9} = \frac{1}{3}$$

$$0.1\dot{6} = \frac{1}{10} + \frac{6}{90} = \frac{9}{90} + \frac{6}{90} = \frac{15}{90} = \frac{1}{6}$$

continued

$$0.08\dot{3} = \frac{8}{100} + \frac{3}{900} = \frac{24}{300} + \frac{1}{300} = \frac{25}{300} = \frac{1}{12}$$

$$0.8\dot{3} = \frac{8}{10} + \frac{3}{90} = \frac{24}{30} + \frac{1}{30} = \frac{25}{30} = \frac{5}{6}$$

$$0.208\dot{3} = \frac{2}{10} + \frac{8}{1,000} + \frac{3}{9,000} = \frac{600}{3,000} + \frac{24}{3,000} + \frac{1}{3,000}$$

$$= \frac{625}{3,000} = \frac{125}{600} = \frac{25}{120} = \frac{5}{24}$$

Check $\quad \dfrac{5}{4 \times 6} \quad \Rightarrow \quad$
$$6 \overline{\smash{)}\,1.250000}^{\,0.208\dot{3}33}$$
$$4 \overline{\smash{)}\,5.00}$$

Where more than one digit recurs

In this case, you just use as many 9s as digits that repeat. Then, cancel down to the simplest form of fraction and check your result by dividing numerator by denominator to see if you get your original recurring decimal. Some of this technique is explained further after the next question and problem section.

WHEN THERE ARE MORE THAN ONE RECURRING DECIMALS, USE A 9 FOR EACH IN ITS PROPER PLACE!

$$0.\dot{4}\dot{5} = \frac{45}{99} = \frac{5 \times 9}{11 \times 9} = \frac{5}{11}$$

$$0.\dot{4}\dot{7} = \frac{47}{99}$$

Check

$$\begin{array}{r} 99 \\ \underline{4} \\ 396 \end{array} \qquad \begin{array}{r} 99 \\ \underline{7} \\ 693 \end{array}$$

FIGURE
REPEATS

$$99 \overline{\smash{)}\,47.00}^{\,0.\dot{4}7\dot{4}}$$
$$\underline{39.6}$$
$$7.40$$
$$\underline{6.93}$$
$$470$$

$$0.\dot{7}2\dot{9} = \frac{729}{999} = \frac{81}{111} = \frac{27}{37}$$

DIVIDE TOP
AND BOTTOM
BY 9   BY 3

Check

```
                          0.7̇29̇7
                    99 / 27.0000
   37          37        25.9
    7   37      2         1.10
  259    9     74          .74
       333              360
             FIGURE      333
             REPEATS     270
```

Questions and problems

1. Make the following divisions:

    (a) 7)343   (b) 9)729   (c) 8)4928   (d) 5)3265   (e) 3)6243

    (f) 2)7862   (g) 4)3936   (h) 6)3924   (j) 13)3081

    (k) 11)16324   (l) 17)6443   (m) 19)8341   (n) 23)28382

Check your answers by multiplication.

2. Make the following divisions by successive (division by factors) and long division; if your answers do not agree, check them with long multiplication:

    (a) 15)3600   (b) 21)15813   (c) 25)73625   (d) 28)10136

3. Make the following divisions and write the remainder as a fraction, using whichever method of conversion you like best:

    (a) 7)3459   (b) 8)23431   (c) 9)13263   (d) 3)14373

    (e) 6)29336   (f) 17)8239   (g) 28)34343   (h) 29)92929

4. A profit of $14,000,000 has to be shared among holders of 2,800,000 shares of stock. What is the profit per share?

5. Total operating cost for an airline flight between two cities is estimated as $8,415. What fare should be charged so that a flight with 55 passengers just meets operating cost?

6. A part needs a special tool that costs $5,000. With this tool, the machine makes parts for 25 cents each. But the price of parts must also pay for the tool. If the tool cost is to be paid for out of the first 10,000 parts made, what will be the cost of each part?

7. A freight car carries 58 tons (1 ton = 2000 pounds), including its own weight, and runs on 8 wheels. Its suspension distributes the weight equally among the wheels. What is the weight on each wheel?

8. A man makes 1,200 of a certain part in 8 hours. How much time is spent for making each part?

9. A package of 10,000 small parts weighs 1,565 pounds. The empty package weighs 2.5 pounds. How much does each part weigh (hint, convert the pounds to ounces)?

10. Another package weighs 2,960 pounds full and 5 pounds empty. One part weighs 3 ounces. How many parts are in the full package?

11. A narrow strip of land, 1 mile long, is to be divided into lots of equal width. How wide is each lot?

12. On a test run, a car travels 462 miles on 22 gallons of gas. Assuming performance is uniform, how far does it go on each gallon of gas?

13. A particular mixture is usually made up 160 gallons at a time. It uses 75 gallons of ingredient 1, 50 gallons of ingredient 2, 25 gallons of ingredient 3, and 10 gallons of ingredient 4. If only 1 gallon is required, what amounts of each ingredient should be used?

14. In question 6, how many parts would have to be made to bring the cost down to 35 cents apiece?

15. Find the simplest fractional equivalent of the following decimals:

(a) 0.875   (b) 0.6   (c) 0.5625   (d) 0.741   (e) 0.128

16. In the following decimals, all the digits after the decimal point repeat. Find their fractional equivalents.

(a) 0.416   (b) 0.21   (c) 0.189   (d) 0.571428   (e) 0.909

Check each by dividing back to decimal form again.

5

CHAPTER

Fractions

Different fractions with the same value

In the previous chapter, you did some things that you might not yet understand, or had difficulty doing. In this chapter, you can "catch up." Seeing a fraction as a piece of pie helps. Notice that the simple fraction 1/4 can be cut in smaller pieces without changing its value as part of the whole.

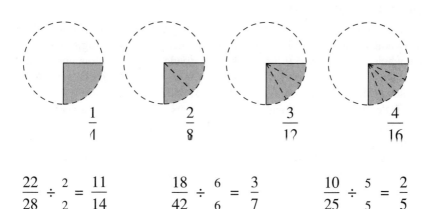

$$\frac{1}{4} \qquad \frac{2}{8} \qquad \frac{3}{12} \qquad \frac{4}{16}$$

$$\frac{22}{28} \div \frac{2}{2} = \frac{11}{14} \qquad \frac{18}{42} \div \frac{6}{6} = \frac{3}{7} \qquad \frac{10}{25} \div \frac{5}{5} = \frac{2}{5}$$

Factors help find the simplest form—cancellation

A fraction might have large numbers for both the numerator and denominator. See whether the fraction can be reduced to a simpler form. A calculator that handles fractions will automatically find and present them in their simplest form. If your calculator can not handle fractions, you must know how to calculate them yourself.

Always Find the Factors in a Fraction

$$\frac{160}{1,280} = \frac{16}{128} = \frac{8}{64} = \frac{1}{8} \quad \text{Simplest Form}$$

$$\frac{455}{462} = \frac{5 \times 91}{2 \times 231} = \frac{5 \times 7 \times 13}{2 \times 3 \times 7 \times 11}$$

by canceling 7's $\quad \dfrac{5 \times \cancel{7} \times 13}{2 \times 3 \times \cancel{7} \times 11} = \dfrac{5 \times 13}{2 \times 3 \times 11} = \dfrac{65}{66}$

If the numerator and denominator both end in zeros, you can strike off the same number of zeros from each. If both are even, divide both by 2.

Sometimes the factors aren't so obvious. Here, for the fraction 455/462, is one way. After checking the factors of each, they both contain a 7. If 7 is a factor of both the numerator and denominator, divide both by 7.

Spotting the factors

Here are rules for spotting factors. If the last digit of a number divides by 2, then the whole number does. This rule can go further. If the last 2 digits divide by 4, the whole number does. If the last 3 digits divide by 8, the whole number does. That leads to a whole series of checks for powers of 2: 4, 8, 16, 32, etc. A similar set works for powers of 5: 25, 125, 625, 3125, etc.

Rules also exist for 3s and 9s. Add the digits together. If the sum of the digits divides by 3, the whole number does. If the sum of digits divides by 9, the whole number does.

The check for dividing by 11 is more complicated. Add alternate digits in two sets. If the sums are identical, differ by 11, or differ by a multiple of 11, the whole original number divides by 11.

FINDING FACTORS

51,756 $\longrightarrow$ **IF THESE TWO FIGURES DIVIDE EXACTLY BY 4, THE WHOLE NUMBER DOES**

23,128 $\longrightarrow$ **IF THESE THREE FIGURES DIVIDE EXACTLY BY 8, THE WHOLE NUMBER DOES**

138 $\quad 1 + 3 + 8 = 12 \quad \dfrac{12}{3} = 4 \qquad 3 \overline{)138}^{\,46 \text{ exactly}} \qquad$ **CHECK FOR 3**

135 $1 + 3 + 5 = 9$

$$9 \overline{)\, 135} \quad \substack{15 \text{ exactly}}$$

CHECK FOR 9

738 $7 + 3 + 8 = 18 \quad \dfrac{18}{9} = 2$

$$9 \overline{)\, 738} \quad \substack{82 \text{ exactly}}$$

28,347 $2 + 3 + 7 = 12$

$8 + 4 = 12$

$$11 \overline{)\, 28,347} \quad \substack{2,577 \text{ exactly}}$$

CHECK FOR 11

869 $8 + 9 = 17$

$17 - 6 = 11$

6

$$11 \overline{)\, 869} \quad \substack{79 \text{ exactly}}$$

Rules for finding factors

That covers finding simple numbers as factors: 2, 3, 4, 5, 8, 9, 11, etc. If 2 and 3 both "go," then 6 is a factor. No easy check exists for 7.

When finding factors, try *primes*—numbers that won't factorize into smaller numbers.

This table shows factors of numbers and identifies primes up to the number 20. We have to try each prime when looking for factors. Isn't there an easier way?

HOW FAR SHOULD YOU TRY FACTORS?

Factors of 139?

$2 \times ?$

$3 \times ?$

$5 \times ?$

$7 \times ?$

$11 \times ?$

$12 \times 12 = 144$ – too big

2: 3: $1 + 3 + 9 = 13$

11: $1 + 9 = 10$

$$\dfrac{3}{7}$$

5: $7 \overline{)\, 139} \quad \substack{19\frac{6}{7}}$

So 139 is a prime number

Factors of 493?

$2 \times ?$

$3 \times ?$

$5 \times ?$

$7 \times ?$

$11 \times ?$

$13 \times ?$

$17 \times ?$

$19 \times ?$

$23 \times 23 = 512$ – too big

2: 3: $4 + 9 + 3 = 16$

11: $\begin{array}{l} 4 + 3 = 7 \\ 9 - 7 = 2 \end{array}$

$$17 \overline{)\, 493} \quad \substack{29 \text{ exactly}}$$

5: $7 \overline{)\, 493} \quad \substack{70\frac{3}{7}}$

$$13 \overline{)\, 493} \quad \substack{37\frac{12}{13}}$$

493 = 17 × 29

How far to try

A useful principle can save you from wasting time when looking for factors. Remember the multiplication table—that line of squares down the diagonal? It goes far beyond the numbers in the table. Look at it this way: If a number does not have at least two factors, it must be prime. So, if it has any factors at all, at least one of them must be smaller than the nearest square.

Look at these two examples. The nearest square to 139 is 144. If 139 has any factors, at least one of them must be less than 12. After trying up to 11, none of them work, so 139 must be prime.

The square of 23 is larger than 493. Try squaring each number before you try it as a factor. 23 squared is 529. Try all the primes up to 23. When you reach 17, you will find that 17 divides into 483 29 times. So, the factors of 493 are 17 and 29.

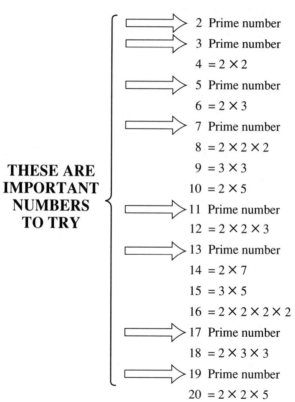

THESE ARE
IMPORTANT
NUMBERS
TO TRY

2 Prime number
3 Prime number
4 $= 2 \times 2$
5 Prime number
6 $= 2 \times 3$
7 Prime number
8 $= 2 \times 2 \times 2$
9 $= 3 \times 3$
10 $= 2 \times 5$
11 Prime number
12 $= 2 \times 2 \times 3$
13 Prime number
14 $= 2 \times 7$
15 $= 3 \times 5$
16 $= 2 \times 2 \times 2 \times 2$
17 Prime number
18 $= 2 \times 3 \times 3$
19 Prime number
20 $= 2 \times 2 \times 5$

Squares and primes

Suppose you want the factors of 8,249. The table of squares shows that the square of 91 is 8,281. Try primes up to 89, the last one before 91. You will find that 73 and 113 are factors. If you had reached 89 with no factor, the original number (8,249) would have been prime.

When making a table of squares, look for patterns, like you did with the much simpler multiplication table. Look at the middle column. The last two digits run the squares of 9 down to 0 and then up to 9 again for the square of 59. This occurs because the square of 50 ends in double zero and twice 5 (the first digit) is 10. A similar sequence begins at 91 and runs up to 109.

PRIME
NUMBERS,
1–100

TABLE OF SQUARES, 1–100

NO.	SQ.	NO.	SQ.	NO.	SQ.	NO.	SQ.	NO.	SQ.
1	1	21	441	41	1,681	61	3,721	81	6,561
2	4	22	484	42	1,764	62	3,844	82	6,724
3	9	23	529	43	1,849	63	3,969	83	6,889
4	16	24	576	44	1,936	64	4,096	84	7,056
5	25	25	625	45	2,025	65	4,225	85	7,225
6	36	26	676	46	2,116	66	4,356	86	7,396
7	49	27	729	47	2,209	67	4,489	87	7,569
8	64	28	784	48	2,304	68	4,624	88	7,744
9	81	29	841	49	2,401	69	4,761	89	7,921
10	100	30	900	50	2,500	70	4,900	90	8,100
11	121	31	961	51	2,601	71	5,041	91	8,281
12	144	32	1,024	52	2,704	72	5,184	92	8,464
13	169	33	1,089	53	2,809	73	5,329	93	8,649
14	196	34	1,156	64	2,916	74	5,476	94	8,836
15	225	35	1,225	55	3,025	75	5,625	95	9,025
16	256	36	1,296	56	3,136	76	5,776	96	9,216
17	289	37	1,309	37	3,249	77	5,929	97	9,409
18	324	38	1,444	58	3,364	78	6,084	98	9,604
19	361	39	1,521	59	3,481	79	6,241	99	9,801
20	400	40	1,600	60	3,600	80	6,400	100	10,000

2
3
5
7
11
13
17
19
23
29
31
37
41
43
47
53
59
61
67
71
73
79
83
89
97

Factoring with a calculator

Of course, a calculator can help you make such a table. It can also help you find factors. The easy way is to put the number for which you want factors in memory, then keep withdrawing it with the *MR* (memory recall) key to try the next prime. If it displays a number with a decimal, you don't have a factor. When you hit a factor, the readout is a whole number.

Thus, in the example on the next page, when you get to 73, the calculator reads 113. Here's how it goes:

Enter 8249 Press *M* in or *M* + (making sure that the memory is empty)
Enter divide by   7   and = reads 1178.428571   Press *MR* and

"	"	11	"	79.9090909	"	"
"	"	13	"	634.5384615	"	"
"	"	17	"	485.2352941	"	"
"	"	19	"	434.1578947	"	"
"	"	23	"	358.6521739	"	"
"	"	29	"	284.4482759	"	"
"	"	31	"	266.0967742	"	"
"	"	37	"	222.9459459	"	"
"	"	41	"	201.1951220	"	"
"	"	43	"	191.8372093	"	"
"	"	47	"	175.5106383	"	"
"	"	53	"	155.6415094	"	"
"	"	59	"	139.8135593	"	"
"	"	61	"	135.2295082	"	"
"	"	67	"	123.1194030	"	"
"	"	71	"	116.1830986	"	"
"	"	73	"	113	That's it!	

Adding and subtracting fractions

If your calculator handles fractions, it keeps finding the simplest form for you. But it helps if you know what it's doing. If your calculator doesn't handle fractions, it will calculate everything in decimals. This system is difficult to verify, unless you do it yourself, the old way.

Remember: in adding or subtracting fractions, they must have the same denominator. For instance, to add 1/2, 2/3, and 5/12, both 1/2 and 1/3 can be changed to 12ths. 1/2 is 6/12 and 2/3 is 8/12. Now, just add numerators, because they are all 12ths: 6 + 8 + 5 = 19. 19/12 is more than 1. Subtract 12/12 for 1. 1 7/12 is the answer.

$$\text{ADD} \quad \frac{1}{2} + \frac{2}{3} + \frac{5}{12}$$

$$\frac{1}{2} = \frac{6}{12} \qquad \text{MAKE EACH}$$

$$\frac{2}{3} = \frac{8}{12} \qquad \begin{array}{l}\text{DENOMINATOR}\\ \text{THE SAME}\end{array}$$

$$\frac{1}{2} + \frac{2}{3} + \frac{5}{12}$$

$$\frac{6}{12} + \frac{8}{12} + \frac{5}{12} = \frac{19}{12} = 1\frac{7}{12}$$

Suppose you must subtract 3 3/5 from 7 5/12. 60 is the common denominator. 3/5 is 36/60 and 5/12 is 25/60. You cannot subtract 36 from 25. So you borrow 1, converting it to 60ths. Now, subtract 3 36/60 from 6 85/60. 36 from 85 is 49. 3 from 6 is 3 and the complete answer is 3 49/60.

To Add Fractions, Each One Must Have the SAME DENOMINATOR

Subtract $3\dfrac{3}{5}$ from $7\dfrac{5}{12}$

1. Common denominator = 60

2. $\dfrac{3}{5} = \dfrac{3 \times 12}{5 \times 12} = \dfrac{36}{60}$ and $\dfrac{5}{12} = \dfrac{5 \times 5}{12 \times 5} = \dfrac{25}{60}$

3. Because $\dfrac{36}{60}$ is bigger than $\dfrac{25}{60}$, change $7\dfrac{25}{60}$ to $6 + 1\dfrac{25}{60} = 6\dfrac{60 + 25}{60} = 6\dfrac{85}{60}$

4. $6\dfrac{85}{60} - 3\dfrac{36}{60} = 3\dfrac{49}{60}$ OR $7\dfrac{5}{12} - 3\dfrac{3}{5} = 3\dfrac{49}{60}$

Finding the common denominator

How do you find a common denominator? Sometimes it's easy. It's not always so obvious. Older textbooks had a routine for the job, but it wasn't easy to understand. Here's a way you can understand. See illustration at top of page 76.

Suppose you have:

$$1/4 + 1/3 + 2/5 + 1/6 + 5/12 + 3/10 + 7/30 + 4/15.$$

Find the factors of each denominator: $4 = 2 \times 2$; 3 and 5 are both prime; 6 is 2×3; 12 is $2 \times 2 \times 3$; 10 is 2×3; 30 is $2 \times 3 \times 5$; and 15 is 3×5.

Our common denominator must contain every factor that is in any denominator. The factors are 2, 3, and 5. The common denominator is $2 \times 2 \times 3 \times 5 = 60$. Now convert all the fractions to 60ths. The numerators are: $15 + 20 + 24 + 10 + 25 + 18 + 14 + 16$, which adds up to 142. Reduce the final form, because both divide by 2, yielding 71/30. Taking out 30/30 for a whole 1, twice, the answer is 2 11/30.

$$\overbrace{\frac{1}{4} + \frac{1}{3} + \frac{2}{5} + \frac{1}{6} + \frac{5}{12} + \frac{3}{10} + \frac{7}{30} + \frac{4}{15}}^{\textbf{Add together}}$$

$$\frac{1}{4} = \frac{1}{2 \times 2} \qquad \frac{1}{6} = \frac{1}{2 \times 3} \qquad \frac{5}{12} = \frac{5}{2 \times 2 \times 3}$$

$$\frac{3}{10} = \frac{3}{2 \times 5} \qquad \frac{7}{30} = \frac{7}{2 \times 3 \times 5}$$

$$\frac{4}{15} = \frac{2 \times 2}{3 \times 5}$$

$$15 + 20 + (2 \times 12 = 24) + 10 + (5 \times 5 = 25) + (3 \times 6 = 18) + (7 \times 2 = 14) + (4 \times 4 = 16)$$

$$2 \times 2 \times 3 \times 5 = 60$$

$$\frac{142}{60} = \frac{71}{30} = 2\frac{11}{30}$$

same answer

$$\text{or} \quad \frac{142}{60} = 2\frac{22}{60} = 2\frac{11}{30}$$

Sometimes the working denominator is not so obvious!

Calculators that "do" fractions

Some pocket calculators "do" fractions. Knowing how to use fractions and knowing calculator limitations can help you understand more about fractions and decimals. Individual calculators might use different ways of keying the problem in, but the way it works is similar. The readout might also differ. See the example at the top of the facing page.

Enter 4 and 2/7. Other keys change the readout first to 30/7, called in the old parlance an *improper fraction*, meaning that its value is more than 1; then to 4.28571428 . . ., its decimal form.

Maybe calculators will soon be able to convert a decimal to a fraction. To do that, it must have circuits that receive its recurring nature, then use the information properly. Without that, your calculator has now "lost" the fractional form.

Now, multiply by 7. The calculator will read 30. Divide by 7. It does not give you 4 and 2/7 again, but the decimal form, 4.28571428 . . . Do some of the things, adding, subtracting, multiplying, or dividing fractions that this book has shown you. The calculator will give the same answers, always presenting them in their simplest form.

These methods will help you understand just what your calculator does and how it does it. In turn, this knowledge will help you understand the whole process.

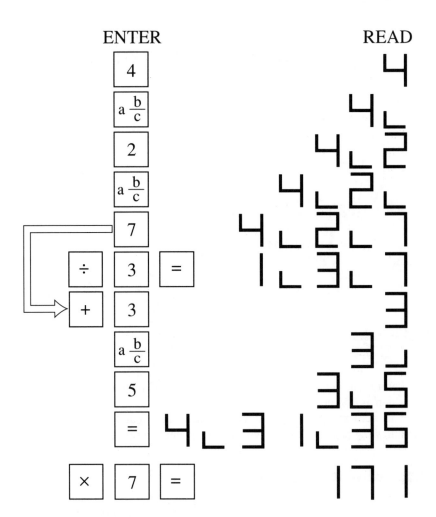

Significant figures

Just how accurate is the figure you look at? Ask yourself how reliable the number is that you look at. Devices that give you numerical information might be either analog or digital. Some think that digital is much more accurate. However, numbers can be deceptive.

Suppose you talk about a 150-pound man. Does he weigh exactly 150 pounds—not an ounce more or less? Is he between 149.5 and 150.5 pounds, or is he between 145 and 155 pounds? The answer depends on what figures in that number 150 are "significant."

If the 0 is just a placeholder, then if he weighed less than 145 or more than 155, the number would be stated as 140 or 160. If the 0 is "significant," then 150 means that he weighs from 149.5 to 150.5 pounds. If he weighed less than 149.5 or more than 150.5, the number would be 149 or 151.

ACCURACY IS RELATIVE . . . IT DEPENDS ON WHAT YOU ARE FIGURING

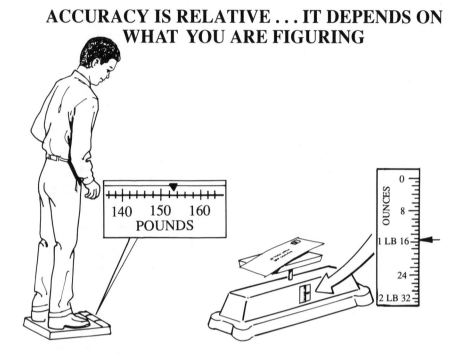

For it to mean not an ounce more or less than 150 pounds, the weight should be written as 150 pounds, 0 ounces.

The word *significant* should tell you that other figures are not significant. The following pages will show why this point can be important.

Approximate long division: why use it

Long division is something you can't "tell" a calculator. Most students had trouble with it because they didn't understand it. Suppose you divide 150 by 7. Your calculator gives you something like 21.4285714 . . . You are tempted to believe all those figures. Now apply what the previous page said. If 150 means more than 145 and less than 155, which means only two figures are significant (the 1 and 5), then dividing by 7 could yield something between 145/7 and 155/7, which read out as 20.71527571 . . . and 22.14285714 . . . The only figure that doesn't change is the 2 of 20!

The other figures, in any of these answers, can't be reliable. You could say the possible "spread" is from 20.7 to 22.1. Now, if the 0 of 150 was significant, the result can be from 149 5/7 to 150 5/7, which read out as 21.3571428 and 21.5 (surprise, 150.5 divides by 7 exactly!) Now the "spread" is reduced to between 21.357 and 21.50.

Suppose you have 153 (an extra significant figure) to divide by 7. Examine the possibilities here.

APPROXIMATE LONG DIVISION

DIVIDE 153 by 7:

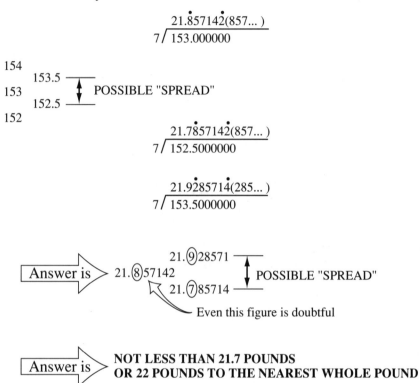

Answer is > NOT LESS THAN 21.7 POUNDS
OR 22 POUNDS TO THE NEAREST WHOLE POUND

Longhand procedure

Suppose that you have to divide 23,500 by 291 and you believe those end zeros in the dividend aren't significant. You could assume it was between 23,450 and 23,550, perform both divisions, and then decide what was significant. But longhand, that's a lot of work! The practice was to draw a vertical line where figures begin to be progressively more doubtful.

You could divide between the "limiting values" as the possible errors, because only so many figures are significant, then you could guess at the most probable value. This procedure is illustrated at the top of page 80.

Using a calculator to find significance

Exploring shows what you can do with a calculator that wouldn't have been practical longhand. Take values that represent the biggest variation on either side of the stated value, then deduce how accurate the answer will be. Notice that in

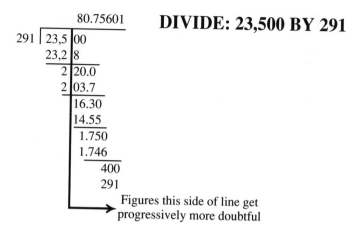

DIVIDE: 23,500 BY 291

Figures this side of line get progressively more doubtful

APPROXIMATE METHOD:

```
           80.76
    291 | 23,500
           23,28
           ─────
            220
7 × 29 ──→  203
           ─────
            17
5 × 3 = 15
6 × 3 = 18 ──→ nearest
```

Nearest third
significant figure
80.8

Finding the limits of accuracy

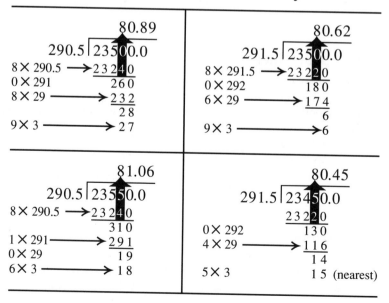

division, both numbers (the dividend and the divisor) can have a number of significant figures. An answer cannot have more significant figures than the number with the least number of significant figures used as "input" for the problem.

Approximate long multiplication

The same methods for long division work for multiplication, as shown here for multiplying 5.32 by 2.91. It also applies to addition or subtraction. Suppose you add 55 to 1,000,000. What does that million mean? It could mean "give or take" 500,000. More likely, it would mean "give or take" 50,000. Even at that figure, adding 55 is unlikely to make a difference. That is an extreme example. The result of any operation can only be as accurate as the "weakest link" which in that case would be the million figure. Only in rare instances would the number 1,000,000 mean exactly that number, not 999,999 or 1,000,001.

Multiply 2.91 by 5.32

$$
\begin{array}{r}
2.91 \\
5.32 \\
\hline
14.55 \\
.873 \\
582 \\
\hline
15.4812
\end{array}
$$

These figures are doubtful

$$
\begin{array}{r}
5.32 \\
2.91 \\
\hline
\end{array}
$$

2 × 5.32 ⟹ 10.64
0.9 × 5.3 ⟹ 4.79
0.01 × 5 ⟹ 5
$$\overline{15.48}$$ Nearest third figure 15.5

Questions and problems

1. Arrange the following fractions in groups that have the same value:

   1/2, 1/3, 2/5, 2/3, 3/4, 3/6, 4/6, 4/8, 3/9, 4/10, 4/12,
   8/12, 9/12, 5/15, 6/15, 6/18, 9/18, 8/20, 10/20, 15/20, 7/21

2. Reduce each of the following fractions to its simplest form:

   7/14, 26/91, 21/91, 52/78, 39/65, 22/30, 39/51, 52/64,
   34/51, 27/81, 18/45, 57/69

3. Without actually performing the divisions, indicate which of the following numbers divide exactly by 3, 4, 8, 9, or 11:

   (a) 10,452   (b) 2,088   (c) 5,841   (d) 41,613
   (e) 64,572   (f) 37,848

4. Find the factors of the following:

   (a) 1,829 (b) 1,517 (c) 7,387
   (d) 7,031 (e) 2,059 (f) 2,491

5. Add together the following groups of fractions:

   (a) 1/5 + 1/6 + 4/15 + 3/10 + 2/3
   (b) 1/8 + 1/3 + 5/18 + 7/12 + 4/9
   (c) 1/4 + 1/5 + 1/6 + 1/10 + 1/12
   (d) 4/7 + 3/4 + 7/12 + 8/21 + 5/6

and reduce each to its simplest form.

6. Multiply the following pairs of quantities:

   (a) 3/4 × 4/5   (b) 37/73 × 5/6   (c) 51/57 × 19/26
   (d) 55/111 × 37/44 (e) 19/119 × 17/57

7. Multiply the following pairs of quantities:

   (a) 2 3/5 × 1 7/13 (b) 3 1/6 × 2 4/19 (c) 7 4/6 × 8 5/8
   (d) 6 1/8 × 7 1/7   (e) 1 2/3 × 3 4/5

8. Divide the following pairs of quantities:

   (a) 2 1/4 divided by 3   (b) 3 1/3 divided by 5 (c) 7 1/7 divided by 5/7
   (d) 3/4 divided by 2 1/4   (e) 30 divided by 4 2/7

9. Find the simplest fractional equivalent for the following decimals:

   (a) 0.875 (b) 0.6 (c) 0.5625 (d) 0.741 (e) 0.128

10. Find the decimal equivalent of the following fractions:

    (a) 2/3 (b) 3/4 (c) 4/5 (d) 5/6
    (e) 6/7 (f) 7/8 (g) 8/9

11. Find the decimal equivalent of the following fractions and thus check question 10s answer, by adding pairs that should make 1:

    (a) 1/3 (b) 1/4 (c) 1/5 (d) 1/6
    (e) 1/7 (f) 1/8 (g) 1/9

12. Find the fraction equivalent to the following recurring decimals:

    (a) 0.416   (b) 0.21   (c) 0.189
    (d) 0.571428   (e) 0.909   (f) 0.090

NOTE: don't confuse the last two with fractions that have 11 as a denominator.

13. The following interesting number sequence is sometimes shown as "mathematical magic:"

142,857 × 2 = 285,714   142,857 × 3 = 428,571
142,857 × 4 = 571,428   142,857 × 5 = 714,285
142,857 × 6 = 857,142

Each result uses the same numbers, in the same order, but starting in a different place. But 142,857 × 7 = 999,999, breaking the magic. Can you explain this?

14. What is meant by significant figures? To illustrate, show the limits of possible meaning for measurements given as 158 feet and 857 feet.

15. Using the approximate method, divide 932 by 173. Then by dividing (a) 932.5 by 172.5 and (b) 931.5 by 173.5, show how many of your figures are justified. Noting that 932 and 173 each have three significant figures, what conclusion would you draw from these calculations?

16. Divide 93,700 by 857 using an approximate method. Then by dividing 93,750 by 856.5 and 93,650 by 857.5, show how many of your figures are justified. Can you shorten your method still further to avoid writing down meaningless figures?

6
CHAPTER

Area:
the second dimension

Scales of length: units and measurement

So far, this book dealt with counting various things, such as money, weights, measures, etc. The things have been placed in rows, in squares, or stacked into cubes to conveniently visualize the count.

Now, see how math can help to relate different measures together. Suppose

You can't multiply ORANGES by PEARS...
or inches by gallons, or tons by miles

someone asked you to multiply 17 oranges by 23 pears, what would you do? You can multiply 17 by 23. But the answer is 391 what? It doesn't make sense! However, multiplying a length by a length can make sense. The second dimension is a result of these multiplications, as you shall see.

Length times length is area

You probably already guessed that multiplying a length by a length produces an area. If a piece of wallpaper 27 inches wide is 108 inches long, its area is 27 inches times 108 inches. The answer is in square inches. The answer is deduced from the way you began counting: laying articles in rows and thus building squares or areas of other shapes.

If you lay 27 square inches in a row, then line 108 rows of 27 square inches, the total area (the wallpaper in the question) is 27 times 108 square inches. If you count the little squares, you will find 2,916 square inches.

But you can multiply INCHES by INCHES.

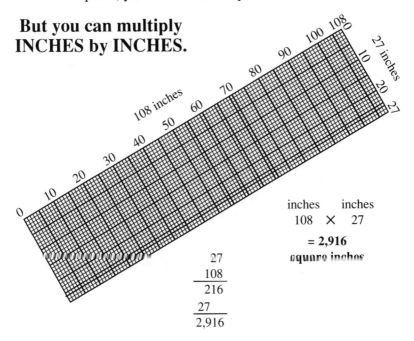

inches   inches
108 ✕   27

= 2,916
square inches

      27
    108
    216
    27
   2,916

What is square?

We are so used to the shape we call square, that we've probably never bothered to define it. Counting those square inches, you probably thought of them as measuring 1 inch "each way." But if you measure 1 inch along four edges, you might not end up with a square. What makes it a square?

The fourth side must end where the first side began. Also, opposite sides must be parallel. Even then, the figure still might not be square. The angles must be what we call "right" angles. See the example at the top of page 86.

What is Square?

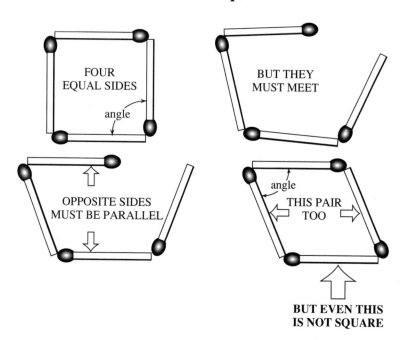

FOUR
EQUAL SIDES

angle

BUT THEY
MUST MEET

OPPOSITE SIDES
MUST BE PARALLEL

angle

THIS PAIR
TOO

BUT EVEN THIS
IS NOT SQUARE

The right angle

What is a right angle? Look at what the phrase originally meant.

Suppose a carpenter makes a table. He must attach its four legs to the table top. All four legs should be attached at one particular angle so that the table stands securely (unless you use some fancy means to hold the table). This was called the *right angle*, from which the word originated. Other angles are wrong angles, without some extra construction to strengthen them.

Diamonds do not have the RIGHT angles.

So they will not always fit.

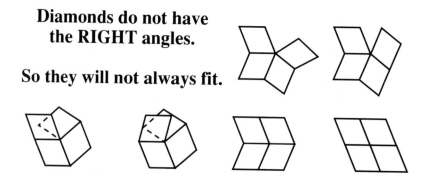

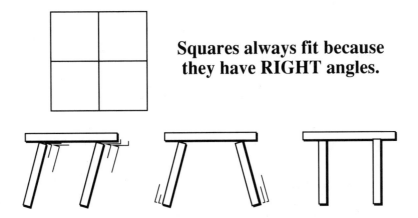

Squares always fit because they have RIGHT angles.

Different shapes with the same area

Take a piece of drawing paper 22 inches by 30 inches, cut it across, and rejoin it to be 44 inches by 15 inches. The area is the same both times because it's the same paper, just rearranged. 22 × 30 is 660 square inches. 44 × 15 also is 660 square inches.

The same number of square inches could be rearranged into almost any number of different shapes.

All these shapes are 660 square inches.

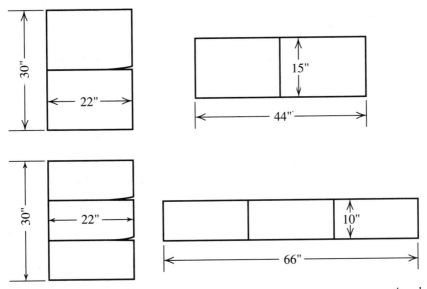

continued

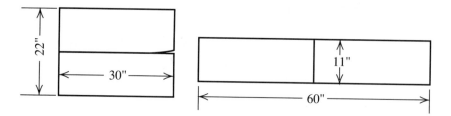

Square measure

In linear measure—measuring along a line in a single direction—each foot has 12 inches and each yard has 3 feet. But how many square inches are in a square foot or square feet are in a square yard?

That's easy to figure out. When we were counting, 10 rows of 10 is 100, which is 10 times 10. So, 12 times 12 or 144 square inches are in a square foot.

Notice one more thing: 6 square inches are different from a 6-inch square. A 6-inch square is a square, each side of which is 6 inches long, so it contains 6 times 6, or 36 square inches. 6 square inches could be an area 6 inches long by 1 inch wide, or 3 inches long by 2 inches wide, or even 4 inches long by 1 1/2 inches wide, each of which multiplies out to 6 square inches.

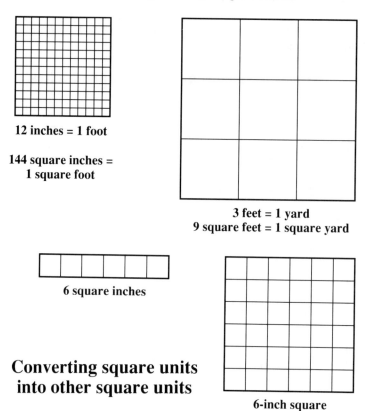

12 inches = 1 foot

144 square inches = 1 square foot

3 feet = 1 yard
9 square feet = 1 square yard

6 square inches

Converting square units into other square units

6-inch square

From oblongs to triangles

All the areas that have been considered so far have had four sides with right angles between the sides. Mathematicians call such shapes *rectangles*, a word of Latin origin, which means "having right angles." An easy way to understand areas of triangles is to think of a triangle as a rectangle (oblong) cut in half diagonally.

Notice that this is a special kind of triangle with a square corner. Other kinds of triangles will be described later on.

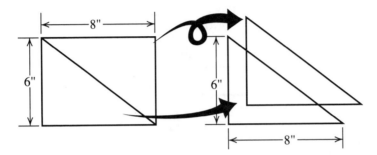

OBLONG

AREA = LENGTH TIMES BREADTH
= 8 inches × 6 inches
= 48 Square Inches

SQUARE-CORNERED TRIANGLE

AREA = HALF LENGTH TIMES BREADTH
$$= \frac{1}{2} \times 8 \text{ inches} \times 6 \text{ inches}$$
= 24 Square Inches

Parallelograms

Going back to four-sided shapes. So far, squares and rectangles have been covered. What about other four-sided shapes that have parallel sides, but not right angles? Geometry distinguishes two kinds, just as squares and rectangles have right-angle corners. If the four sides are equal, it's called a *rhombus*, or in common terms, a *diamond*. If the sides are unequal, it can be called a *rhomboid* or a *parallelogram*.

The illustration at the top of the next page shows that, if the sides of a rectangle are kept the same, but the angles are changed (it's now a parallelogram) its area decreases, until eventually it disappears, when it "squashes" into a straight line.

PARALLELOGRAMS

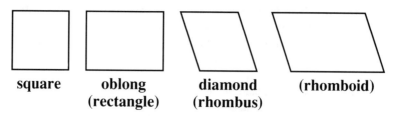

| square | oblong (rectangle) | diamond (rhombus) | (rhomboid) |

AS A RECTANGLE IS CHANGED INTO A RHOMBOID,
ITS AREA DECREASES

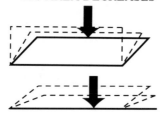

Area of parallelograms

One way to find the area of a parallelogram is to take it from a different rectangle—one that has one pair of sides the same as the parallelogram, but the other two sides are shorter. The straight-across distance between the first two sides is the same as the distance between those two sides in the parallelogram.

1. Sides 10" and 15"
Distance squarely
between 15" sides is 8"

2. Same parallelogram
Sides 10" and 15" Distance
between 10" sides is 12"

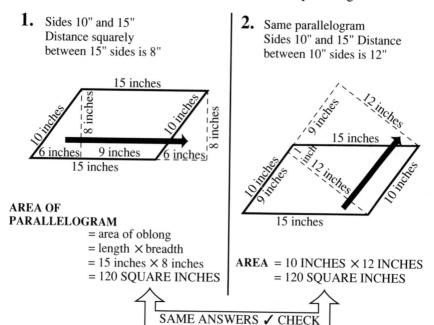

**AREA OF
PARALLELOGRAM**
= area of oblong
= length × breadth
= 15 inches × 8 inches
= 120 SQUARE INCHES

AREA = 10 INCHES × 12 INCHES
= 120 SQUARE INCHES

SAME ANSWERS ✓ CHECK

A way of seeing that the rectangle has the same area as the parallelogram is to note that the parallelogram turns into the rectangle by moving the same size of triangle from one end to the other.

Area of acute triangles

Any triangle that is not square cornered, can consist of two square-cornered triangles, each of which is half of a corresponding rectangle. The two rectangles, put together, form one larger rectangle. And the two square-cornered triangles form one larger triangle.

The dimension of the rectangle side that is between the two smaller rectangles also becomes the vertical height of the triangle, if the two sides joined together are seen as the base. In the example on this page, the triangle breaks into two square-cornered triangles, each with one side 12 inches and two other sides are 5 inches and 9 inches, respectively. The whole side, the *base* of the triangle, is $5 + 9 = 14$ inches. 12 inches is its vertical height.

The entire rectangle contains $12 \times 14 = 168$ square inches. So, the area of the triangle is half of the rectangle, 84 square inches. A rule begins to emerge. The area of a triangle is half of the base times the vertical height.

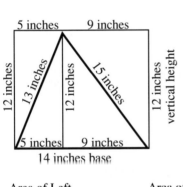

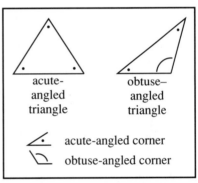

Area of Left Oblong = 5" × 12"	Area of Right Oblong = 9" × 12"	Area of Combined Oblong = 14" × 12"
= 60 Square Inches	PLUS = 108 Square Inches	= 168 Square Inches
Square–Cornered Triangle = $\frac{1}{2} \times 60$	Square–Cornered Triangle = $\frac{1}{2} \times 108$	Triangle = $\frac{1}{2} \times 168$
= 30 Square Inches	PLUS = 54 Square Inches	= 84 Square Inches

HALF BASE TIMES VERTICAL HEIGHT

Area of obtuse triangles

An obtuse triangle has one angle wider than a right angle. At the way it is on this page, the vertical height "overhangs" the base. The main triangle is a larger right triangle with a smaller one taken away from it. So, each triangle is half the corresponding rectangle and the formula holds good, although the vertical height is measured outside the base of the main triangle.

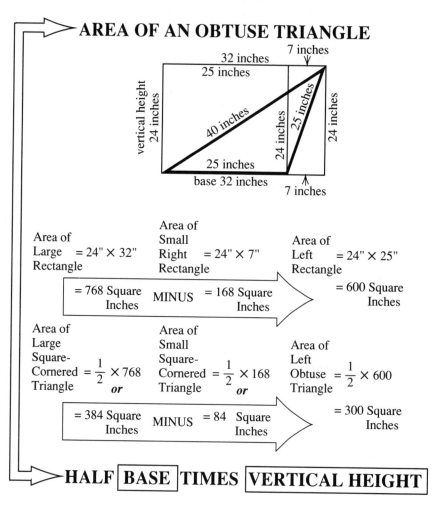

AREA OF AN OBTUSE TRIANGLE

Area of Large Rectangle = 24" × 32"

Area of Small Right Rectangle = 24" × 7"

Area of Left Rectangle = 24" × 25"

= 768 Square Inches   MINUS   = 168 Square Inches   = 600 Square Inches

Area of Large Square-Cornered Triangle = $\frac{1}{2}$ × 768 *or*

Area of Small Square-Cornered Triangle = $\frac{1}{2}$ × 168 *or*

Area of Left Obtuse Triangle = $\frac{1}{2}$ × 600

= 384 Square Inches   MINUS   = 84 Square Inches   = 300 Square Inches

HALF | BASE | TIMES | VERTICAL HEIGHT

Area of triangles

You can turn that same triangle around, using its longest side as base, instead of one of the shorter sides. Now, find its area in the same way that you did for the acute triangle. Notice that so long as you take the vertical height from the base to the remaining angle (corner), the formula is true and gives the same answer for the same triangle.

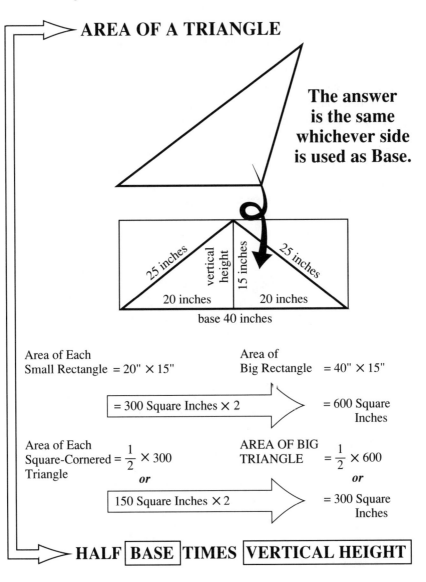

AREA OF A TRIANGLE

The answer is the same whichever side is used as Base.

25 inches
vertical height
15 inches
25 inches
20 inches
20 inches
base 40 inches

Area of Each Small Rectangle $= 20'' \times 15''$

Area of Big Rectangle $= 40'' \times 15''$

$= 300$ Square Inches $\times 2$

$= 600$ Square Inches

Area of Each Square-Cornered Triangle $= \frac{1}{2} \times 300$

or

AREA OF BIG TRIANGLE $= \frac{1}{2} \times 600$

or

150 Square Inches $\times 2$

$= 300$ Square Inches

HALF **BASE** **TIMES** **VERTICAL HEIGHT**

Metric measure

For many years, inconsistency in systems of measurement made learning arithmetic difficult. 12 inches to the foot, 3 feet to the yard, 220 yards to the furlong, and 8 furlongs to the mile. Then, 4 gills to the pint, 2 pints to the quart, and 4 quarts to the gallon in liquid measure.

Several measures of weight are used. The common one is *avoirdupois*: 16 drams to the ounce, 16 ounces to the pound, 14 pounds to the stone, 2 stones to the quarter, 4 quarters to the hundredweight, and 20 hundredweights to the ton. Some of these measures have not been used in recent years.

Troy weight, used for jewelry, has 24 grains to the pennyweight, 20 pennyweights to the ounce, and 12 ounces to the pound. Isn't it confusing?

Some countries have adopted the metric system, in which every measure is based on 10s. This system is certainly easier to learn. The problem is that so many people already learned to use the old systems, so the new ones, though simpler to learn, are strange to them.

Such a change involves making practical changes. For example, plywood is made in sheets 4 feet by 8 feet. What is that in meters? It comes out to an awkward fraction. What do we do? Change the standard size for plywood sheets or describe the present size with rather awkward numbers?

Metric Measure

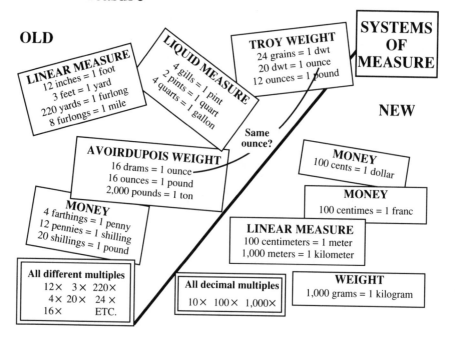

The metric system

Metric has advantages in scientific work. This book won't have to get into that. To use it, for people accustomed to older systems, requires conversions. Here are the conversions for linear measure. As you go through, I'll let you know about metric units in other measures.

CONVERSION between METRIC and ENGLISH UNITS

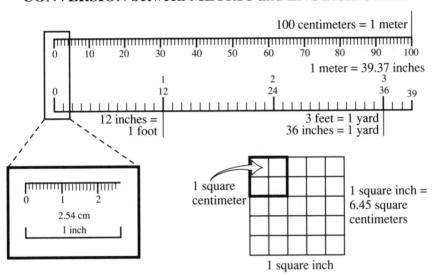

100 centimeters = 1 meter

1 meter = 39.37 inches

12 inches = 1 foot

3 feet = 1 yard
36 inches = 1 yard

2.54 cm
1 inch

1 square centimeter

1 square inch = 6.45 square centimeters

1 square inch

Area problems

Simple area problems are concerned with an amount of area, using whatever square units are appropriate. However, sometimes these problems are a matter of "fitting" something into a space of stated dimensions. Areas are not always simple in shape. So, you need ways to figure more complicated shapes.

If the shapes are square-cornered, but more than a simple rectangle, several ways might be used to figure them. Shapes other than *rectilinear* (having straight sides) are covered later in this book.

Problems, such as papering a wall can, in turn be complicated by a pattern, which must be matched on adjoining strips. Tiles or material that comes in standard square or oblong pieces give another kind of "matching" problem: how do you cut pieces to fill the space so as to minimize wasted pieces? Many of these problems can be worked out by making detailed trial calculations. Examples are shown on the next two pages.

Find the Area of this Floor

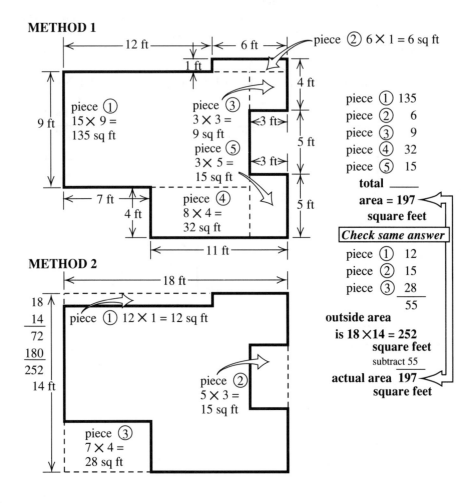

METHOD 1

piece ② 6 × 1 = 6 sq ft

12 ft · 6 ft

1 ft

4 ft

piece ① 15 × 9 = 135 sq ft

9 ft

piece ③ 3 × 3 = 9 sq ft

<3 ft>

piece ⑤ 3 × 5 = 15 sq ft

<3 ft>

5 ft

7 ft

4 ft

piece ④ 8 × 4 = 32 sq ft

5 ft

11 ft

piece ① 135
piece ② 6
piece ③ 9
piece ④ 32
piece ⑤ 15
total _____
area = 197
square feet

Check same answer

piece ① 12
piece ② 15
piece ③ 28
55

outside area
is 18 × 14 = 252
square feet
subtract 55
actual area 197
square feet

METHOD 2

18 ft

18
14
72
180
252
14 ft

piece ① 12 × 1 = 12 sq ft

piece ② 5 × 3 = 15 sq ft

piece ③ 7 × 4 = 28 sq ft

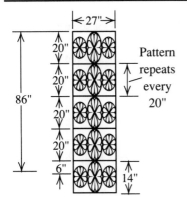

Papering a Wall ...

Five panels must be allowed. Each 86"
long. Total 430" or 35 feet 10 inches
Avoid joins in panels. If 30-ft rolls are
used (360"), this will do 4 panels: 344".
Last panel must start new roll.

... when there's a pattern

If pattern has to match, length must be allowed
to next larger complete pattern length. 80" is
not enough. Needs 100" so 30-ft roll will only
do 3 panels (300").

Remaining 60" is waste.

Tiling a Floor ... with square tiles

All pieces must be counted as well as whole
tiles. $8 \times 4 = 32$ tiles needed.

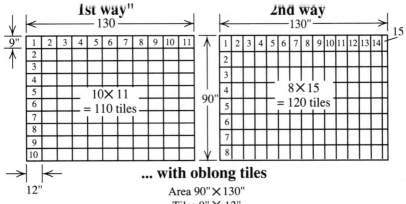

... with oblong tiles

Area 90" × 130"
Tiles 9" × 12"
1st way saves 10 tiles
over 2nd way.

Questions and problems

1. Find the area of the following rectangles:
   - (a) 54 inches by 78 inches
   - (b) 13 feet by 17 feet
   - (c) 250 yards by 350 yards
   - (d) 3 miles by 7 miles
   - (e) 17 inches by 5 feet
   - (f) 340 yards by 1 mile

2. What is a right angle? Why is it so named?

3. Starting with a piece of paper 36 by 25 inches, cut it into three pieces that can be arranged to prove that 37 1/2 by 24 inches has the same area.

4. How many square feet are in (a) 5 square feet, (b) a 5-foot square?

5. Find the area of square-cornered triangles, where the sides by the square corner have the following dimensions:
   - (a) 5 inches by 6 inches
   - (b) 12 feet by 13 feet
   - (c) 20 yards by 30 yards
   - (d) 3 miles by 4 miles
   - (e) 20 inches by 2 feet
   - (f) 750 yards by 1 mile

6. A field was thought to have four straight sides. Opposite pairs of sides measure 220 yards and 150 yards respectively. But the field does not have square corners. A measurement between the opposite 220-yard sides finds that the straight-across distance is 110 yards. Find the area of the field in acres (an acre is 4,840 square yards).

7. A parallelogram has sides 20 inches and 15 inches long. The straight-across distance between the 20-inch sides is 12 inches. Calculate the straight-across distance between the 15-inch sides. (HINT: use the fact that the area can be calculated in two ways.)

8. Find the area of the following triangles:
   - (a) base 11 inches, height 16 inches
   - (b) base 31 inches, height 43 inches
   - (c) base 27 inches, height 37 inches

9. Two sides of a triangle are 39 inches and 52 inches. When the 39-inch side is used as the base, the vertical height is 48 inches. What is the vertical height when the 52-inch side is used as the base?

10. A piece of property has two square corners. The side that joins these square corners is 200 yards long. Measuring from each square corner, the two adjoining sides are 106.5 yards and 256.5 yards. The fourth side, joining the ends of these sides is 250 yards long. What is the acreage of this property? (HINT: treat this area in two parts, an oblong and a square-cornered triangle.)

11. A piece of property measures 300 yards by 440 yards. The owner wants to keep a smaller piece inside that piece that measures 110 yards by 44 yards and sell the rest. What is the area he wants to sell?

12. A farmer has 60 6-foot lengths of portable fencing. What is the largest area he can enclose? (HINT: try different-shaped oblongs that exactly use the 360 feet of fencing he has, and see which gives the biggest area.)

13. Find the area of the floor in the drawing.

ROOM DIMENSIONS

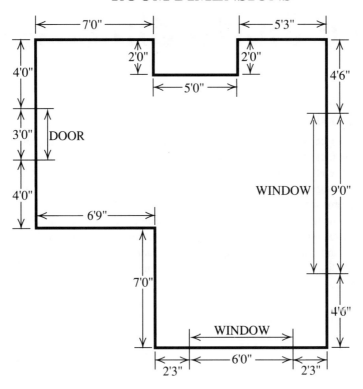

14. The walls of the room in question 13 are 7-feet high. Doors and windows at the positions shown, run from floor to ceiling. How much wallpaper, in 30-foot rolls that are 27 inches wide, will be needed to paper the room if the pattern on the paper repeats every 22 inches? Don't allow for a strip of wallpaper to turn the corner.

15. How many tiles will be needed to cover the ceiling of the same room (using the most economical way) with 9″ × 12″ tiles? How many tiles can you save over the less-economical way?

7
CHAPTER

Time:
the fourth dimension

What is dimension?

Things can be measured with different scales. You might use inches or centimeters to measure length, for instance. But whichever system you use, it still measures length. That is the first dimension. Breadth, depth, height, and width can be measured with the same units, inches or centimeters. If you measure only one of them, that is still the first dimension.

Measuring area is the second dimension. Two dimensions multiplied together are an area. Using a third dimension changes an area into a volume. Just as the second dimension is in square units (square inches or square centimeters), the third dimension, volume, is in cubic units (cubic inches or cubic centimeters).

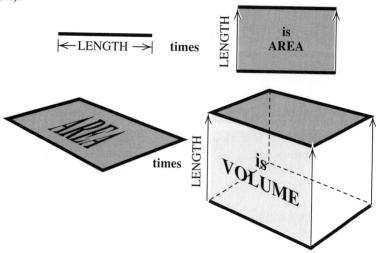

Later, volume is related to measure and weight. For now, the fourth dimension needs our attention: time.

The fourth dimension: time

How can time be a dimension? Think about it. We use a variety of ways to measure it. The ancients used sand in an hour glass. Until recently, we used spring-wound or weight-driven clocks or watches. Quartz electronic instruments use frequency counted off digitally, as a measure of time. More than measuring time, here we relate it to other measures, making it a fourth dimension.

A measure is "in" a dimension. The length (or width) of a desk is a specific number of inches, though the measure goes on beyond the ends of your desk. Just so, when we measure time, we measure a particular duration of time, but time goes on, beyond the piece we measure. Time is a dimension.

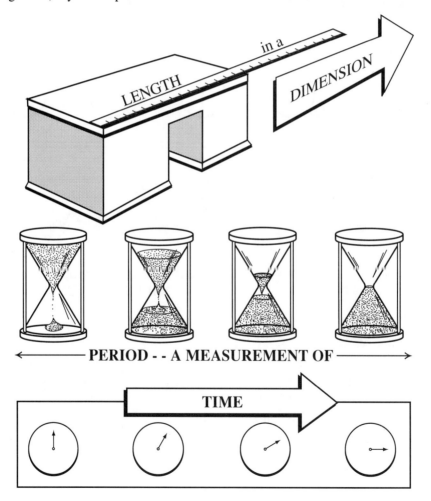

LENGTH in a DIMENSION

←——————— PERIOD - - A MEASUREMENT OF ———————→

TIME

Using time to build more dimensions

Combining length one way with length another way forms an area. Similarly, combining time with length or distance, forms speed. Suppose you walk at a steady rate and an even pace to go a mile in 15 minutes. If you don't change that pace, you will go another mile in the next 15 minutes. In an hour (60 minutes) you'll go 4 miles. Our speed is 4 miles per hour (mph).

Keep it up, and you'll do 4 more miles in the next hour, and so on. To find distance travelled, multiply speed by time. The same is true of driving. If you drive 60 miles per hour, you go a mile every minute. In 60 minutes (1 hour), you will go 60 miles.

To know speed, you must know how far in how much time. If you travel 300 miles in 6 hours at a steady speed, you must be doing 50 miles every hour. Six hours at 50 miles each hour is 300 miles. Speed is distance (a length dimension) divided by time.

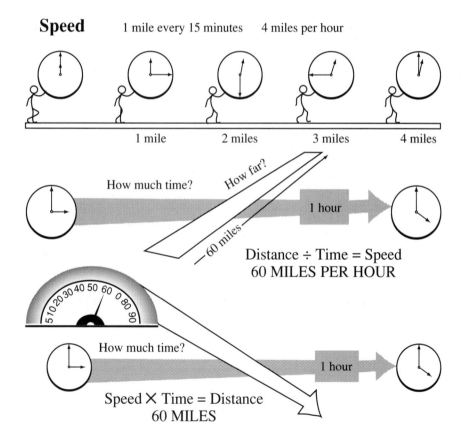

Speed   1 mile every 15 minutes   4 miles per hour

1 mile   2 miles   3 miles   4 miles

How much time?   How far?

60 miles

1 hour

Distance ÷ Time = Speed
60 MILES PER HOUR

How much time?

1 hour

Speed ✕ Time = Distance
60 MILES

Average speed

In the last section, it was assumed that speed was steady. It isn't always. On a 300-mile journey, you might go 30 miles the first hour, 45 miles the second, 60 miles the third, 55 miles the fourth, fifth and sixth hours. A total of 300 miles are travelled in 6 hours. So, the average speed is 300/6 = 50 miles an hour. You probably didn't travel at a steady speed all the way. Sometimes you went faster, sometimes slower. This measure is called *average speed*. It is the steady speed necessary to cover the distance in the same time.

Finding Average Speed

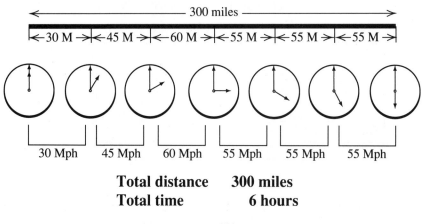

Total distance   **300 miles**
Total time   **6 hours**

Average Speed $\dfrac{300}{6}$ = 50 MPH

FIGURING SPEED FOR A SPECIFIC DISTANCE

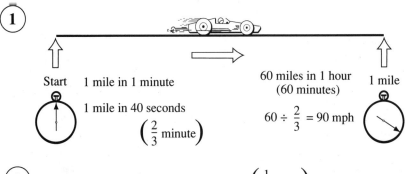

1

Start   1 mile in 1 minute   60 miles in 1 hour   1 mile
(60 minutes)

1 mile in 40 seconds

$\left(\dfrac{2}{3} \text{ minute}\right)$ $60 \div \dfrac{2}{3}$ = 90 mph

2   Lap is 3 miles;   time is 2 minutes $\left(\dfrac{1}{30} \text{ hour}\right)$

Speed is $3 \div \dfrac{1}{30}$ = 90 mph

Watch your car's speedometer. You will find that you can seldom go at a steady speed for a whole hour, or even for a few minutes. Measuring the distance covered every hour shows the average speed during that hour.

Instead of noting the distance travelled every hour, you might notice your speed on the speedometer every few minutes. The speedometer will show how your speed varies during each hour. Perhaps you have to brake to a stop during the hour. What does that do to the average? Later, calculus will help you answer that problem.

A racing driver watches his speed closely. Each lap of track might be 1 mile long. Speed for the lap can be figured by timing that lap. If the lap is made in exactly 1 minute, the speed is 60 miles per hour. If a mile requires 40 seconds, he would drive 1 1/2 miles in 60 seconds, which would be 90 miles per hour.

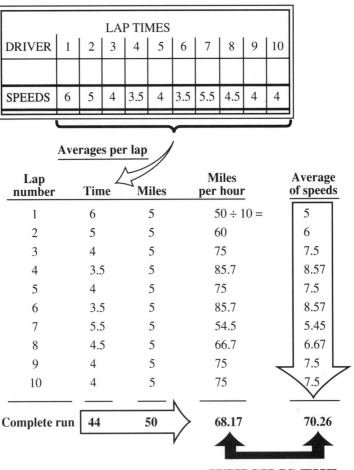

LAP TIMES										
DRIVER	1	2	3	4	5	6	7	8	9	10
SPEEDS	6	5	4	3.5	4	3.5	5.5	4.5	4	4

Averages per lap

Lap number	Time	Miles	Miles per hour	Average of speeds
1	6	5	50 ÷ 10 =	5
2	5	5	60	6
3	4	5	75	7.5
4	3.5	5	85.7	8.57
5	4	5	75	7.5
6	3.5	5	85.7	8.57
7	5.5	5	54.5	5.45
8	4.5	5	66.7	6.67
9	4	5	75	7.5
10	4	5	75	7.5
Complete run	44	50	68.17	70.26

WHICH IS THE RIGHT AVERAGE?

A lap is 3 miles and the time is 2 minutes (1/30 of an hour). Dividing 3 miles by 1/30, which is the same as multiplying by 30, gives the speed as 90 miles per hour.

Suppose you time a 50-mile race, made by driving 10 laps on a 5-mile course. The times made by one driver for the 10 laps were 6, 5, 4, 3.5, 4, 3.5, 5.5, 4.5, 4, and 4 minutes. The total time for 50 miles was 44 minutes, an average speed of 68.17 mph.

You could calculate the average speed for each lap and then average the speed over 10 laps. Each represents the speed for 1/10 of the total distance. So, shouldn't the average be found by adding together the speed for each lap and dividing by 10? Doing that gives an average of 70.26 mph, a different figure. Which is right?

The reference quantity

Speed (movement) involves two dimensions: distance and time. Which is more important? Usually, time is. You want to get somewhere, but you want to know when. Time will pass whether you go fast or slow. So, the distance is measured against time, miles per hour. Miles are referred to the time required to travel.

Lap number	Time	Miles	MINUTES PER MILE	Average of minutes per mile
1	6	5	1.2 ÷ 10 =	0.12
2	5	5	1	0.10
3	4	5	0.8	0.08
4	3.5	5	0.7	0.07
5	4	5	0.8	0.08
6	3.5	5	0.7	0.07
7	5.5	5	1.1	0.11
8	4.5	5	0.9	0.09
9	4	5	0.8	0.08
10	4	5	0.8	0.08
Complete Run	44	50	**68.17**	0.88

0.88 minutes per mile
= 0.88 × 50 = 44 minutes for 50 miles.

THE REFERENCE QUANTITY:	MILES PER HOUR (TIME) or MINUTES PER MILE (DISTANCE)

In the previous section, distance was the reference used: time was measured every 5 miles, rather than distance every so many minutes. That is why the discrepancy occurs.

Look at slowness instead of speed: minutes per mile, instead of miles per minute. We don't usually measure in minutes per mile. The figures would be 1.2, 1, 0.8, 0.7, 0.8, 0.7, 1.1, 0.9, 0.8, and 0.8 minutes per mile. Adding the figures produces 8.8 minutes. As each was taken over 5 miles, not 1, 50 miles will require 5 times 8.8 (44 minutes), an average of 68.17 miles per hour.

HOW PROPORTION
OF TIME AFFECTS AVERAGE

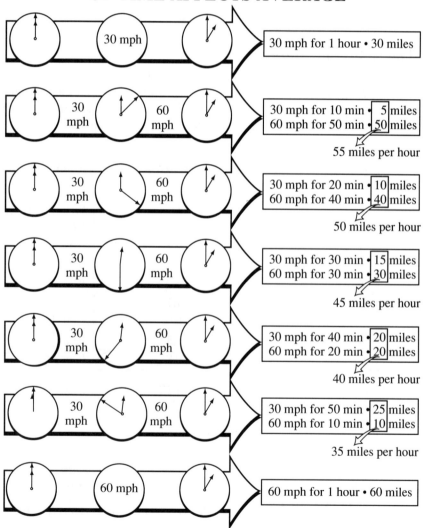

30 mph for 1 hour • 30 miles

30 mph for 10 min • 5 miles
60 mph for 50 min • 50 miles

55 miles per hour

30 mph for 20 min • 10 miles
60 mph for 40 min • 40 miles

50 miles per hour

30 mph for 30 min • 15 miles
60 mph for 30 min • 30 miles

45 miles per hour

30 mph for 40 min • 20 miles
60 mph for 20 min • 20 miles

40 miles per hour

30 mph for 50 min • 25 miles
60 mph for 10 min • 10 miles

35 miles per hour

60 mph for 1 hour • 60 miles

Changing the average

Suppose you travel for 1 hour. Assume that you went 30 mph for the whole hour: you will have gone 30 miles. Now suppose that for the last 10 minutes you go 60 mph: you went 25 miles at 30 mph and 10 miles at 60 mph, a total of 35 miles. The average is now 35 mph. The rate increases as more time is travelled at the higher speed.

Making up time

Now suppose you have 45 miles to go and an hour to do it in. At a steady speed that would be 45 mph. But suppose you go at 30 mph for 10 minutes—that's 5 miles travelled. Now you need to go 40 miles in 50 minutes—48 mph. That's just 3 mph faster than going steady all the way.

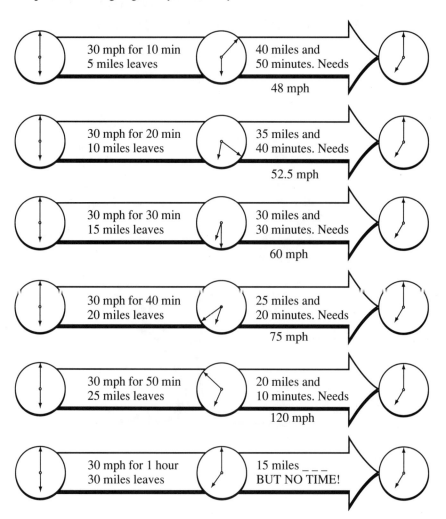

30 mph for 10 min
5 miles leaves

40 miles and
50 minutes. Needs

48 mph

30 mph for 20 min
10 miles leaves

35 miles and
40 minutes. Needs

52.5 mph

30 mph for 30 min
15 miles leaves

30 miles and
30 minutes. Needs

60 mph

30 mph for 40 min
20 miles leaves

25 miles and
20 minutes. Needs

75 mph

30 mph for 50 min
25 miles leaves

20 miles and
10 minutes. Needs

120 mph

30 mph for 1 hour
30 miles leaves

15 miles _ _ _
BUT NO TIME!

Rate of growth

Speed and rate of growth are similar ideas, rather like comparing the hare and the tortoise method of travel. In terms of minutes or hours, growth might be imperceptible. You can measure time in days or weeks. You can measure growth in inches or feet.

The rate of growth raises another question of reference quantity. If a seedling is 2 inches high today, and tomorrow it's 10 inches high, that sounds like fast growth. However, if a 30-foot tree grows to 30 feet 8 inches by tomorrow, you'd have to look twice to see if it had grown overnight at all. When you add 8 inches to 2 inches, that's a big growth. But added to 30 feet, 8 inches doesn't seem like much.

BOTH GROW 8 INCHES, BUT ONE DOESN'T SEEM AS SIGNIFICANT AS THE OTHER

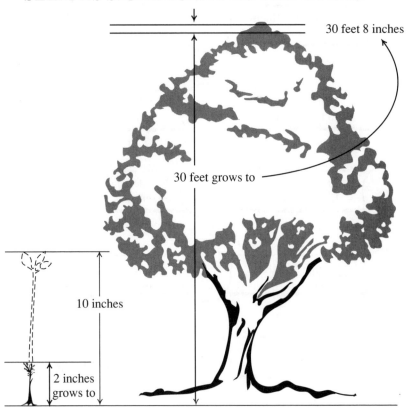

30 feet 8 inches

30 feet grows to

10 inches

2 inches grows to

Fractional increase

The seedling grew from 2 inches to 10 inches, increasing 4 times yesterday's height. The 30-foot tree added 2/3 foot (8 inches), which is only 2/3 divided by 30, as a fraction of the height of the tree: 2/90 or 1/45.

Considered as a fractional increase, the seedling adds 4 times its height, the tree only 1/45, although both actual measurements are 8 inches. Using fractional increase as a reference, the seedling grows 180 times as fast as the tree.

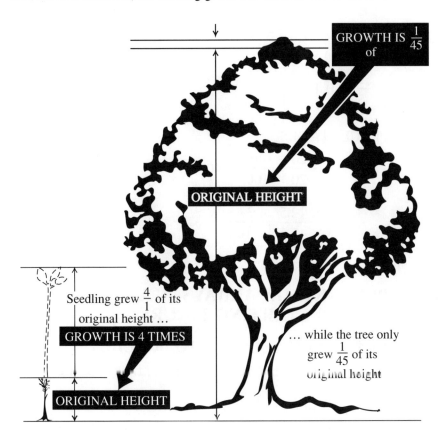

GROWTH IS $\frac{1}{45}$ of

ORIGINAL HEIGHT

Seedling grew $\frac{4}{1}$ of its original height ...

GROWTH IS 4 TIMES

... while the tree only grew $\frac{1}{45}$ of its original height

ORIGINAL HEIGHT

Percentages

Percentages are a standard way to express things as a fractional reference. Percentages were developed before decimals to make working in fractions easier. Decimals might be easier to understand directly, but percentages were used so long that they became a habit for many purposes.

Percentages began because fractions are clumsy. If you were asked which is the bigger fraction, 2/5 or 3/8, could you answer just by looking at them? Converting to decimals makes it easy: 2/5 is 0.4 and 3/8 is 0.375. So, 2/5 is larger.

Percentages use 100 as the denominator, so the numerator can often be a

WHICH IS BIGGER? $\frac{2}{5}$ or $\frac{3}{8}$

1. BY FRACTIONS   COMMON DENOMINATOR IS $5 \times 8 = 40$

$$\frac{2}{5} = \frac{2 \times 8}{5 \times 8} = \frac{16}{40} \qquad \frac{3}{8} = \frac{3 \times 5}{8 \times 5} = \frac{15}{40}$$

THIS IS BIGGER THAN THIS

2. BY DECIMALS

$$\frac{2}{5} = 0.4 \qquad \frac{3}{8} = 0.375 \qquad 5\overline{\smash)2.0}^{\,0.4} \qquad 8\overline{\smash)3.000}^{\,0.375}$$

3. BY PERCENTAGES

$$\frac{2}{5} = \frac{2 \times 20}{5 \times 20} = \frac{40}{100} = 40\%$$

$$\frac{3}{8} = \frac{3 \times 12.5}{8 \times 12.5} = \frac{37.5}{100} = 37.5\%$$

PERCENTAGE ALWAYS RELATES TO STARTING FIGURE

STARTS AT 2 INCHES; GROWS TO 10 INCHES

Growth is 8 inches which is 4 times starting height

4 times is $\dfrac{400}{100}$ or **400%**

STARTS AT 30 FEET; GROWS 8 INCHES MORE $\left(\frac{2}{3} \text{ FOOT}\right)$

Growth is $\dfrac{2}{3} \div 30 = \dfrac{2}{90} = \dfrac{1}{45}$

$$\frac{\frac{100}{45}}{100} = \frac{\frac{20}{9}}{100} = \frac{2\frac{2}{9}}{100} = 2\frac{2}{9}\%$$

simple number. You can see that percentages, like decimals, make it easier to compare fractions at a glance. 40% is obviously 2.5% more than 37.5%.

Another reason for using percentages is that they always refer to the starting size or number. If you say something grew 8 inches, you don't know whether to think if that's fast or slow, unless you know how big it was at the beginning. For the seedling, it is 400%. For the tree, it is 2/45 × 100 = 200/45 = 2 2/9% or 2.222%.

A percentage is a number divided by 100, taken as a fraction of the number that you started with.

Percentages with money

When dealing with money, you often hear about percentages. If railway or airline fares increase, it's usually figured as a percentage. Dividends are paid as a percentage, so that everything can be divided fairly.

Flat rate increase:

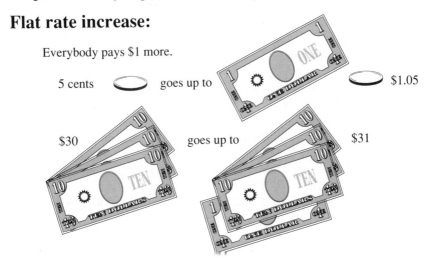

Everybody pays $1 more.

5 cents   goes up to $1.05

$30   goes up to $31

Percentage increase:

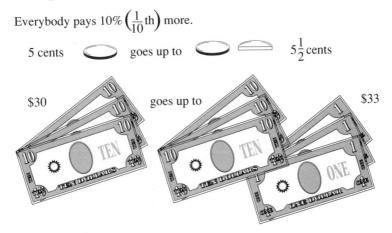

Everybody pays 10% $\left(\frac{1}{10}\text{th}\right)$ more.

5 cents   goes up to $5\frac{1}{2}$ cents

$30   goes up to $33

A railway company has to raise fares because costs have increased. It would not be fair to charge everyone $1 more. That would raise a 5-cent fare to $1.05 and a $30 fare to only $31.

If the profits for a company are $10,000, $50 would be given to each of the 200 stockholders. One stockholder might have invested $1, another $100,000. Would it be fair for each of them to get $50 of the profit?

Such things are worked out on a percentage basis. If the railway's costs rise from $100,000,000 to $110,000,000, that is a 10% increase. To get this money back in fares charged, each should increase by 10%. Thus, the 5-cent fare would then only increase to 5.5 cents (probably 6 cents), and the $30 fare would go to $33.

Similarly, if profits are $10,000 on a total investment of $2,000,000, the rate is 5%. The stockholder who invested $1 gets 5 cents. The one who invested $100,000 gets $5,000 dividend.

Percentages up and down

One thing to watch in percentages: always use the starting figure of a transaction or calculation as the 100% point.

Suppose a man buys property for $100,000 and its value increases, so he sells it for $125,000. He's made 25% profit on the deal: it cost him $100,000, he recovered his $100,000, and made $25,000 more. The profit is $25 for every $100 of the starting price.

The value decreases, so the second man sells it for its original price of $100,000. Being back to its original price, after having gone up 25%, you might think it dropped 25%. But it hasn't. The second man paid $125,000. Of his original investment, he gets $100,000.

The loss is still $25,000, but now it's a fraction or percentage of $125,000, not of $100,000. The loss is 20%. He lost $20 for every $100 of his purchase price. $20 × 1250 = $25,000.

FIRST MAN

Buys   for $100,000

Sells it for $125,000

Profit $25,000 $\dfrac{25,000}{100,000} = 25\%$

SECOND MAN

Buys   for $125,000

Sells it for $100,000

Loss $25,000 $\dfrac{25,000}{125,000} = 20\%$

A 25% increase is the same fractional change, reversed, as 20% decrease. Smaller percentages come nearer to being the same, either way. For larger percentages, the difference is larger. An increase of 100% is doubling, but the reverse, which is halving, is a decrease of only 50%.

Graphical representation of facts

Visual presentation of statistics is common these days. Commercials use it all the time—even if their "facts" are questionable! Visual comparison conveys an impression more quickly and effectively than numbers can.

So, graphic presentation is widely used. Lengths replace numbers. Suppose you want to show your club's growth in membership. The figures for successive years are: 47, 52, 65, 73, 76, 77, 85, 96, 110. To show this growth, you draw

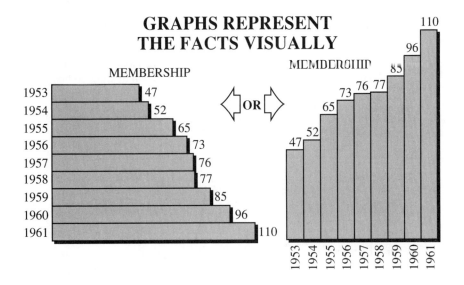

GRAPHS REPRESENT THE FACTS VISUALLY

lines or blocks to represent the number of members for each year. If you draw each member at 1/16 inch, 100 members are 6 7/8″ high. Having drawn the lines, equally spaced, to represent 1-year intervals, you have a visual picture of membership growth. You can place the lines either horizontally or vertically.

Graphs

Visual presentations can help in many ways. A club might want to show how many members belong to certain groups: engineers, doctors, lawyers, salesmen, factory workers, shop assistants, truck drivers, musicians, etc. They could list these occupations as percentages or a number of lines could be placed side by side, as on the previous page. But here, the idea is to show how much each is, of the whole. So, a square of suitable width is marked off in 100 units and each group is given a space, which represents that percentage of the total. Since the total is 100%, all the widths together must fill the box.

MEMBERSHIP COMPOSITION

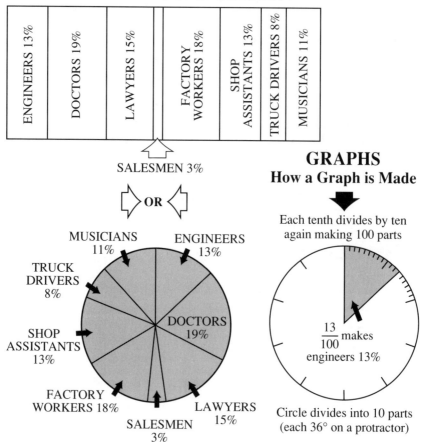

ENGINEERS 13%   DOCTORS 19%   LAWYERS 15%   FACTORY WORKERS 18%   SHOP ASSISTANTS 13%   TRUCK DRIVERS 8%   MUSICIANS 11%

SALESMEN 3%

OR

MUSICIANS 11%   ENGINEERS 13%
TRUCK DRIVERS 8%
SHOP ASSISTANTS 13%
DOCTORS 19%
FACTORY WORKERS 18%
LAWYERS 15%
SALESMEN 3%

GRAPHS
How a Graph is Made

Each tenth divides by ten again making 100 parts

$\frac{13}{100}$ makes engineers 13%

Circle divides into 10 parts
(each 36° on a protractor)

Another way that is sometimes favored, is to divide a circle up in the same way. Its circumference is divided into 100 parts (10 are shown here to make it clearer). Then, divide the circumference according to the percentages in the groups and draw lines from each marker to the center.

Graphs are also used to help make calculations. Suppose tests on an electric motor relate electrical wattage input to horsepower output. Results are tabulated—from electrical power running "light" with no mechanical output, to electrical power inputs for various horsepower outputs.

Suppose you have a job that requires an unlisted amount of horsepower. How do you calculate the electrical power it needs? A graph makes the job easier. Mark points on squared paper to show all the figures in the table, then join the dots made. You need not know anything about electricity to do this graph.

Find the amount of electrical power that is needed for the new job by reading it on your graph. This process is called *interpolation*.

Graphs can show a lot of things that are not obvious from the figures used to make them. They also are useful to check figures. In making the tests tabulated, you read meters to write down numbers. A meter might have numbers on its scale at 20 and 30, with 9 unmarked lines in between. The reading should be 23, but maybe you wrote 27 by mistake. All the other readings are correct.

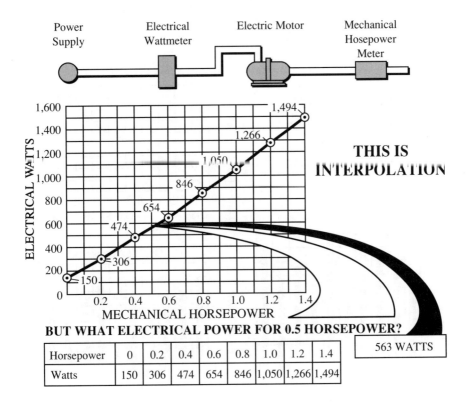

Horsepower	0	0.2	0.4	0.6	0.8	1.0	1.2	1.4
Watts	150	306	474	654	846	1,050	1,266	1,494

GRAPHS HELP FIND MISTAKES

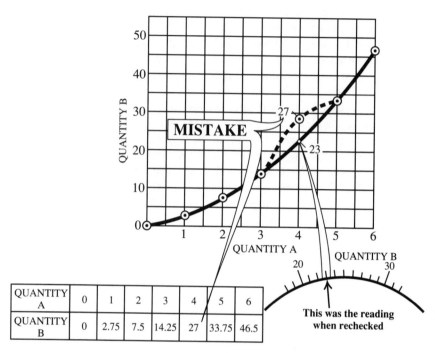

MISTAKE

QUANTITY A	0	1	2	3	4	5	6
QUANTITY B	0	2.75	7.5	14.25	27	33.75	46.5

This was the reading when rechecked

The same basic information— presented different ways

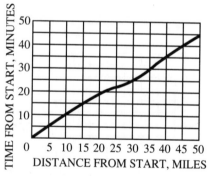

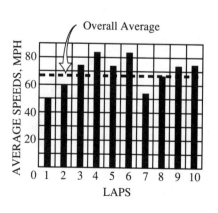

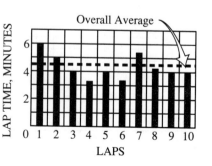

You plot your graph using your figures. All the points line up nicely, except the one that you wrongly copied as 27. This point immediately gives you a clue where you made the mistake. So, you take that reading again and find that you misread the meter.

Modern meters have digital readings. You could copy the numbers down wrong. Nobody is exempt from making occasional mistakes.

The same basic information can often be presented in a variety of different ways. Going back to the race track situation from earlier in this chapter, here are some ways to show the driver's performance visually. Here *visual* can have two meanings: watching him go around the track or using graphs that analyze his performance visually.

Questions and problems

1. How far will a car go at 35 mph for 36 minutes? (HINT: find what fraction 36 minutes is of one hour.)

2. A riverboat makes a water speed (at which its motor propels it through the water) of 10 mph. The river has a downstream current flowing at 2 mph. How fast does the boat go (a) upstream? (b) downstream?

3. How long will the riverboat in question 2 take to make a journey of 96 miles (a) upstream? (b) downstream?

4. Traveling at a water speed of 10 mph, the riverboat burns half a ton (1 ton is 2000 pounds) of fuel per hour. How much fuel does the boat use on its 96-mile journey (a) upstream? (b) downstream?

5. If the boat slows down to make the downstream journey in the same time as the upstream one, and if that reduces fuel burned in proportion to speed reduction, (a) Will reducing the downstream speed save fuel? (b) How much? (c) What would be the percentage saving (on the round trip)?

6. If the boat reduces its water speed on the upstream run, will it save fuel or use more? How much? What percentage?

7. A man invests $50,000 in stock. For the first year, it pays a dividend of 5% on his investment. At the same time, the value of his stock rises to $60,000. If he sells the stock, how much profit will he make (a) in cash? (b) as a percentage?

8. During early weeks of growth, a tree's height is recorded every week. For 5 successive weeks, the heights are 16″, 24″, 36″, 54″, and 81″. What is the percentage growth per week for each of the 4 weeks? What is the percentage for the whole month?

9. Make a graph of the tree's growth for the month. From the graph, estimate the height of the tree in the middle of the second week.

10. A race track has an 8-mile lap that consists of 5 miles with many hairpins, corners, and grades. The remaining 3 miles has straights and banked curves. The best time any car can make over the 5-mile part is 6 minutes, but the 3-mile section lets drivers "open up." Two drivers tie for the best on the 5-mile part, but one averages 90 mph on the 3-mile part, and the other averages 120 mph. Find the average speed for each on the whole lap.

11. A car is checked for mileage per gallon and is found to give 32 mpg on straight turnpike driving. How far will it go on a tankful if the tank holds 18 gallons?

12. A company needs printed circuit boards for which two processes are available. The first needs a tool that costs $2,000, then makes boards for 15 cents apiece. The other uses a procedure that initially costs $200, then makes boards for 65 cents each. Find the cost per board, assuming the total quantity ordered is: 100, 500, 1,000, 2,000, 5,000, and 10,000 units, by each process.

13. Plot a graph of the cost per board by the two processes (question 12) for quantities from 1,000 to 10,000 boards. For how many boards would the cost of both processes be the same?

14. Driving a car at a steady 40 mph produces a mileage of 28 mpg. Driving the same car at 60 mph reduces the mileage to 24 mpg. On a journey of 594 miles, how much gas will be saved by driving at the slower speed and how much longer will the journey take?

15. A man pays $200,000 for some property. After a year its value rises 25%, but he does not sell. During the next year its value drops 10%, after which he sells. What profit did he make on his original investment (a) in cash? (b) percent? Why was it not 25 − 10 = 15% profit?

16. After all allowances and deductions have been made, a man's taxable income is $120,000. How much tax will he pay at 20% on the first $30,000 and 22% on the rest?

17. A square-cornered triangle with 12-foot and 16-foot sides against the square corner has the same area as a parallelogram with opposite pairs of sides that are 10 feet and 16 feet long. What is the distance between the 16-foot sides?

18. An aircraft gains altitude at 1,000 feet per minute. How long does it take to climb to its flying altitude of 22,000 feet? If its forward speed while climbing is 360 mph, how far does it travel while climbing?

19. An aircraft in level flight has a speed of 420 mph, but it comsumes fuel at half the rate (per minute) compared with climbing. How far can the plane fly level, using the same amount of fuel used in climbing to 22,000 feet?

20. How much further would the plane of the previous two questions fly on the same total fuel if it leveled off at 11,000 feet, instead of 22,000 feet? It is assumed that altitude does not affect speed or fuel consumption and the wind effects are ignored.

Part 2

Introducing algebra, geometry, and trigonometry as ways of thinking in mathematics

$$8$$

CHAPTER

First notions
leading into algebra

Shorter methods for longer problems

As problems get more complicated, so does the arithmetic to solve them. People invented multiplication and division to shortcut the repeating of addition and subtraction. Algebra began to handle more involved problems, where older short cuts didn't help much. Even that has become more complicated now that computers and calculators use algebra. Taking these problems one step at a time makes it easier.

This page shows a relatively simple problem that algebra helps: fencing for a double-fenced enclosure. Two things are fixed: the length of the inner enclosure must be 3 times its width and the spacing between the two fences must always be 3 feet.

This can set two kinds of problems. Given enclosure size, how much fencing is needed? Given a length of fencing, how big an enclosure will it make? The first question can be answered by the arithmetic tabulated; no algebra is needed. The information at the top of the next page shows the arithmetic that is needed for this part. The second isn't so simple.

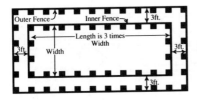

Outer Fence — Inner Fence → 3ft.
—Length is 3 times—
Width
3ft. Width 3ft.
3ft.

PROBLEM

① *How much fencing?*

② How large is the enclosed area, with a given amount of fencing?

123

1	2	3	4	5	6	7	8	9	10	11	12	13
		←		Inner Fence		→	←		Outer Fence		→	TOTAL
width	length	2 pieces (width)	take	2 pieces (length)	take	TOTAL	2 pieces (width)	take	2 pieces (length)	take	TOTAL	BOTH FENCES
10	30	10	20	30	60	80	16	32	36	72	104	184
15	45	15	30	45	90	120	21	42	51	102	144	264
20	60	20	40	60	120	160	26	52	66	132	184	344
25	75	25	50	75	150	200	31	62	81	162	224	424
30	90	30	60	90	180	240	36	72	96	192	264	504
35	105	35	70	105	210	280	41	82	111	222	304	584
40	120	40	80	120	240	320	46	92	126	252	344	664

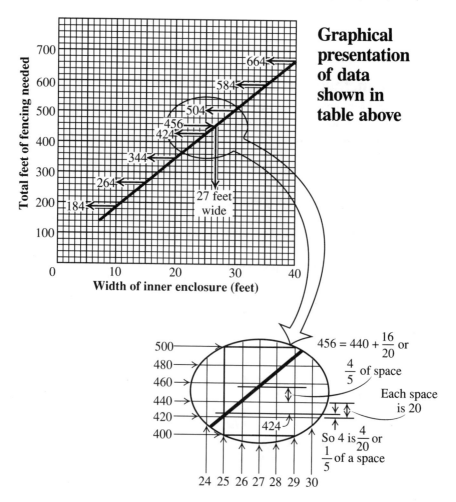

Graphical presentation of data shown in table above

Total feet of fencing needed

Width of inner enclosure (feet)

27 feet wide

$456 = 440 + \frac{16}{20}$ or $\frac{4}{5}$ of space

Each space is 20

So 4 is $\frac{4}{20}$ or $\frac{1}{5}$ of a space

Graphs check arithmetic and find solutions

With all the steps in the calculations, you might make a mistake. By graphing your results you can find mistakes. The results should be on a straight line or on a smooth curve (according to the type of problem). A point that isn't on the line or curve, alerts you to a mistake.

Make the graph with squared paper. Choose a scale that uses as much paper as possible without making the values awkward to read. Here, each small space represents 20 feet of fencing. The problem says you have 456 feet of fencing, so the graph shows that the width of the inner enclosure can be 27 feet.

Algebra: a more direct way

The methods of the last two pages are quite long. They involve making the same series of calculations several times with different numbers. A pocket calculator makes these problems easier than they used to be. However, it's still a long way around, especially if you only want one answer. See example on top of page 126.

Without actually using algebra, first write all the pieces, in terms of either width or feet, and add them. The total length of fencing is 16 widths + 24 feet. You calculate each answer in the table with this formula, using only one multiplication (× 16) and one addition (+ 24). Or, turn everything around with a process you will become more familiar with in algebraic terms.

This is an *equation*:

Total length of fencing = 16 widths + 24 feet.

The words on either side of the equals sign have the same value; they are different ways of naming the same amount. Next, if you subtract 24 feet from each side of that equals sign, the statement or equation will still be true. Each side will be 24 feet less. Now you have:

Total length of fencing: 24 feet = 16 widths.

Divide both sides by 16 and it will still be true. Each side will be 1/16 what it was before.

Length of fencing: 24 feet, all divided by 16 = width.

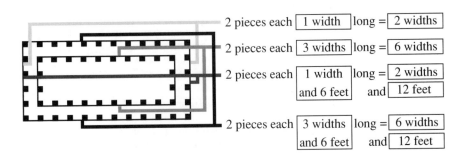

2 pieces each	1 width	long =	2 widths
2 pieces each	3 widths	long =	6 widths
2 pieces each	1 width and 6 feet	long =	2 widths and 12 feet
2 pieces each	3 widths and 6 feet	long =	6 widths and 12 feet

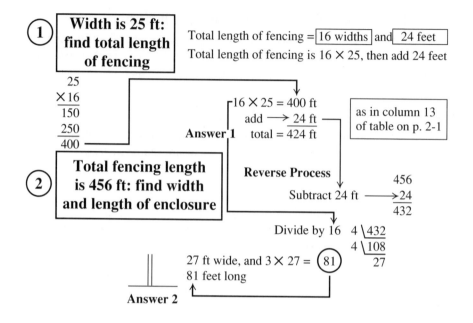

Writing it as algebra

Writing problems of this kind can be shortened by using the first, or some convenient letter, to stand for each original measurement or quantity in the problem. You could write *w* for width. Length is specified as 3 times the width, so write length as *3w*, meaning 3 times *w*. Finally, write *f* for the total length of fencing.

When using algebra, as in arithmetic, always be careful that units are consistent. Here they are all in feet. Do not use some in inches or yards, and others in feet.

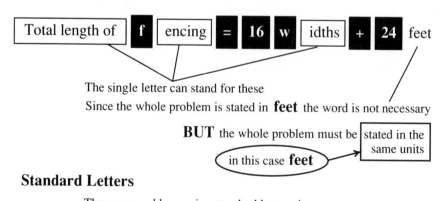

Standard Letters

The same problem, using standard letters, is

$$y = 16x + 24$$

where: x = width of enclosure **in feet**

y = total length of fencing **in feet**

When you start in algebra, a lot of questions arise that many teachers don't answer, so most people eventually give up. Through this book, I try to answer questions as they logically arise.

Different ways of writing in arithmetic and algebra

In arithmetic, we write numbers using figures in a row. Thus, 23 is 2 tens plus 3 ones. In algebra, *ab* isn't *a* tens and *b* ones, but *a* multiplied by *b*. Get used to this, because it's the way everybody does it.

As you look at such new things, ask yourself, "What's this new way of writing mean?" That's one difference between arithmetic and algebra, of which you'll see more as you go on.

Different kinds of algebra exist. The kind that's been taught in schools for years is "school algebra." In it, *ab* is *a* times *b* and *xy* is *x* times *y*. Sometimes, a period, often above the line, is used for multiplication. Sometimes, even a times sign is used, but a times sign is too easily confused with the letter *x*.

Computer algebra is somewhat different from "regular" algebra. One difference is that multiplication is shown by an asterisk between the letters: *a∗b*."

ARITHMETIC   ALGEBRA

163	means	100		16w	means	10
	plus	60			plus	6
	plus	3			all multiplied by	w

To multiply 16 by 3 :			To multiply 16 by x :		
multiply 6 by 3	18		multiply 6 by x	6x	
multiply 10 by 3	30		multiply 10 by x	10x	
Add together	48		A total of 16 times x	16x	

16w + 24 means:
16 is multiplied by w,
but 24 is not

What does 16w24 mean?   It has no definite meaning

Brackets or parentheses

When many older folks learned algebra, the use of brackets or parentheses, was standardized. Some uses that you will see today were considered unnecessary.

Computers use algebra that has changed that somewhat. But the principles haven't changed.

Parentheses are used today because we often want to know which to do first, add or multiply, for instance.

Some remember the rule, "do the inside brackets first." Others think better about why to do that. What is inside the parentheses is regarded as a single quantity. Thus, "*w* + 6" is enclosed in parentheses, meaning that all of it is multiplied by the "2" outside the parentheses.

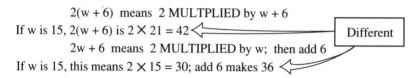

2(w + 6) means 2 MULTPLIED by w + 6

If w is 15, 2(w + 6) is 2 × 21 = 42 ⟵   Different

2w + 6 means 2 MULTIPLIED by w; then add 6

If w is 15, this means 2 × 15 = 30; add 6 makes 36 ⟵

That's why brackets are used

In arithmetic

23 means 20
plus   3

In algebra

ab means a times b
If a is 2 and b is 3;
ab = 2 × 3 = 6

Rewriting the Problem

Inner fence: 2 sections w and 2 sections 3w
= 2w + 6w = $\boxed{8w}$

Outer fence: 2 sections w + 6 and 2 sections 3w + 6
= $\boxed{2(w + 6)}$ + $\boxed{2(3w + 6)}$

Total fence = 8w + 2(w + 6) + 2(3w + 6)

= 8w + 2w + 12 + 6w + 12

= (8 + 2 + 6)w + 12 + 12

= 16w + 24 ⟵ (as before)

Using more than one set

In the previous section, two expressions in parentheses used the same *w* to derive lengths in the outer fence for width and length. The inner fence didn't involve parentheses. If you want to calculate how much lumber it requires, assume that you need 12 feet of lumber for each foot run of fence, whether inner or outer. Multiply the total length of fence by 12, so put a different pair of parentheses around this section of equation.

Four kinds of parentheses are commonly used. Some call them all parentheses, some call them all brackets. Call them what you like, so long as you know how to use them.

HOW MUCH LUMBER FOR STAKES?

12 feet needed for every foot run of fence
Total footage of lumber = 12[8w + 2(w+ 6) + 2(3w + 6)] ft

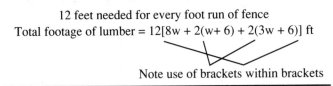

Note use of brackets within brackets

Four kinds of brackets can be used, when needed.

Parenthesis, or round $\longrightarrow$ (w + 6)

Square $\longrightarrow$ [w + 6]

Brace $\longrightarrow$ { w + 6 }

Vinculum $\longrightarrow$ $\overline{w + 6}$

A problem expressed by algebra

For any kind of algebra to be useful, you must be able to write a problem into it. Don't make *x* mean anything for the moment so that it can mean anything about which you might have a problem. *x* comes into the problem three times. First, as the number itself. Then, as a number that is 4 times another number that is 5 more than *x*. Finally, as a number that is 2 times a number that is 3 less than *x*. These three numbers added together are 210. How do we solve it to find out what *x* was to begin with?

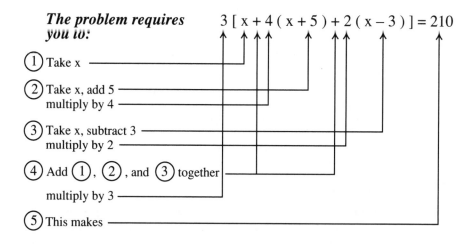

The problem requires you to:

$3 [x + 4 (x + 5) + 2 (x - 3)] = 210$

1. Take x
2. Take x, add 5
   multiply by 4
3. Take x, subtract 3
   multiply by 2
4. Add ①, ②, and ③ together
   multiply by 3
5. This makes

Removing the parentheses to solve it

We solve the previous equation by removing the parentheses, following the rule as explained, starting from the inside ones.

4 times x + 5 is $4x$ + 20.   2 times x − 3 is $2x$ − 6.

Multiply each part of what's inside the parenthesis by what's outside it. Then, collect the pieces inside the big parentheses. Add the three x terms: 1 (don't write the 1 in algebra), 4, and 2. That makes $7x$. Add two plain numbers, 20 and − 6. $20 − 6 = 14$. Now, you can multiply the whole thing by 3 more easily: $21x$ + 42.

An easier, more direct way exists in this case. If 3 times what's in the big parentheses is 210, then each must be 1/3 of 210 (70). $7x$ + 14 = 70. If you subtract 14 from both, the new numbers will still be equal. $14 − 14$ is 0. So, $7x$ + 14 − 14 = 70 − 14. $70 − 14 = 56$. So $7x = 56$. Divide both sides by 7: $x = 8$. That's the number you wanted to know.

Check it. First, you have 8. Then, multiply 5 more than 8 (which is 13) by 4. That makes 52. Finally, multiply 3 less than 8 (which is 5) by 2. That makes 10. Adding together: $8 + 52 + 10 = 70$. $3 \times 70 = 210$, which is what the problem said to start with.

Solving the Problem

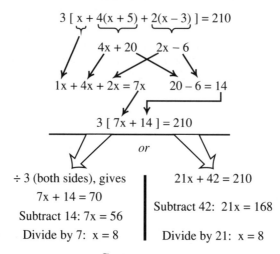

Same answer

Putting a problem into algebra

Here's a problem where algebra could help. Eleven young people (some boys, some girls) went to eat together. Each boy ordered something that cost $1.25. Each girl ordered something that cost $1.60. The total check came to $16.20.

One way of writing the problem is shown. Each price was written as a number of nickels. You could also write it in dollars, but that would involve decimals. Whatever units you use, stick to them throughout the problem.

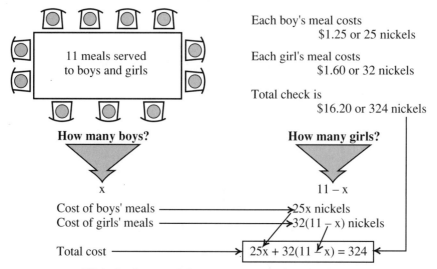

Each boy's meal costs
$1.25 or 25 nickels

Each girl's meal costs
$1.60 or 32 nickels

Total check is
$16.20 or 324 nickels

How many boys?

x

How many girls?

11 − x

Cost of boys' meals ⟶ 25x nickels
Cost of girls' meals ⟶ 32(11 − x) nickels

Total cost ⟶ 25x + 32(11 − x) = 324

This is the problem expressed in algebra

Solving it by removing the parentheses

After removing the parentheses, the −x on the left is bigger than the + x, so the x term is minus. You can change all the signs. If two minus quantities are equal, the same two plus quantities will be equal too. You can look at the same thing as taking away the minus quantity from both sides. If you subtract a minus quantity, it is the same as adding a plus quantity. See example at the top of page 132.

Think about it carefully. Understand it, then make the rule you can most easily use yourself. However you do it, it comes down to 28 = 7x or 7x = 28. Dividing both sides by 7, x = 4 (meaning 4 boys), which leaves 7 girls.

Checking your answer and your work

Always check your answer against the original problem statement. 4 boys at $1.25 is $5. 7 girls at $1.60 is $11.20. Added together, this makes $16.20, which the problem gave as the total bill (as shown on page 132).

Now, look at the algebraic statement to see what it means. In that statement, 352 − 7x = 324 (or the equivalent, if you did it in dollars), 352 nickels is the price it would have been if they were all girls (i.e., no boys, the x figure was zero). That's $17.60. The 7x means that every boy makes the bill 7 nickels (or 35 cents) less. The 324 nickels are the actual bill ($16.20).

Solving the Problem

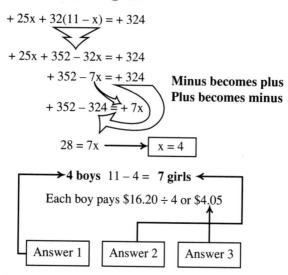

$$+ 25x + 32(11 - x) = + 324$$

$$+ 25x + 352 - 32x = + 324$$

$$+ 352 - 7x = + 324$$

Minus becomes plus
Plus becomes minus

$$+ 352 - 324 = + 7x$$

$$28 = 7x \longrightarrow \boxed{x = 4}$$

4 boys $11 - 4 =$ **7 girls**

Each boy pays $16.20 \div 4$ or $4.05

| Answer 1 | Answer 2 | Answer 3 |

CHECKING YOUR ANSWER

4 boys at $1.25 = $5.00
7 girls at $1.60 = $11.20

Total check $16.20

Checking each statement

$$25x + 352 - 32x = 324$$

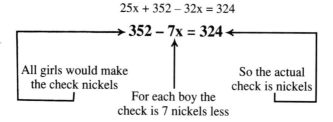

352 − 7x = 324

All girls would make
the check nickels

So the actual
check is nickels

For each boy the
check is 7 nickels less

Magic by algebra

This trick is good to use at a party. Ask each person to think of a number (not say what it is) and write it on a piece of paper. Then, tell them to: Add 5. Multiply by 2. Subtract 4. Multiply by 3. Add 24. Divide by 6. Subtract the number you first thought of. Give them time to do each, before you give the next instruction. Finally you announce that their answer is 7.

The fact that they all thought of different numbers, but all have the same answer (if they didn't make a mistake) seems like magic. However, you can prove that it works by using x to stand for any number. The fact that x disappears at the end shows that the answer will work for any number.

Think of a Number TRICKS

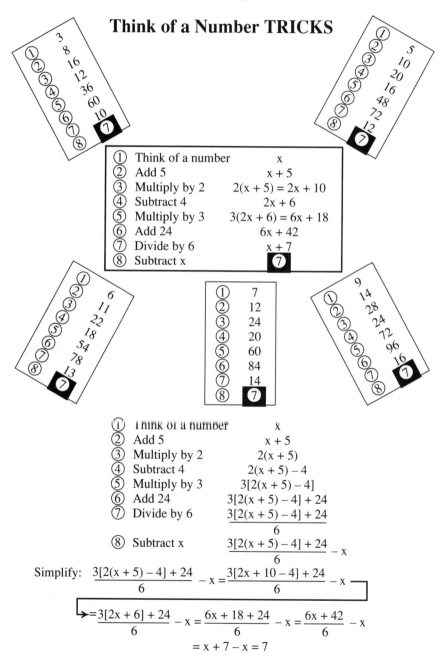

①	Think of a number	x
②	Add 5	x + 5
③	Multiply by 2	2(x + 5) = 2x + 10
④	Subtract 4	2x + 6
⑤	Multiply by 3	3(2x + 6) = 6x + 18
⑥	Add 24	6x + 42
⑦	Divide by 6	x + 7
⑧	Subtract x	**7**

①	Think of a number	x
②	Add 5	x + 5
③	Multiply by 2	2(x + 5)
④	Subtract 4	2(x + 5) – 4
⑤	Multiply by 3	3[2(x + 5) – 4]
⑥	Add 24	3[2(x + 5) – 4] + 24
⑦	Divide by 6	$\dfrac{3[2(x + 5) - 4] + 24}{6}$
⑧	Subtract x	$\dfrac{3[2(x + 5) - 4] + 24}{6} - x$

Simplify: $\dfrac{3[2(x + 5) - 4] + 24}{6} - x = \dfrac{3[2x + 10 - 4] + 24}{6} - x$

$= \dfrac{3[2x + 6] + 24}{6} - x = \dfrac{6x + 18 + 24}{6} - x = \dfrac{6x + 42}{6} - x$

$= x + 7 - x = 7$

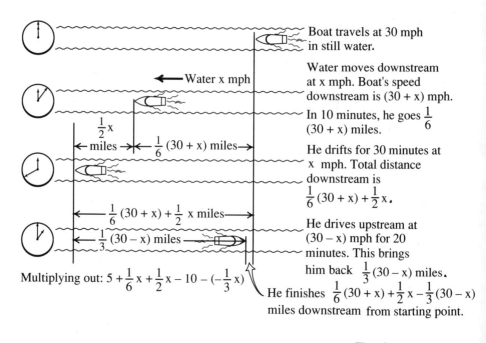

Boat travels at 30 mph in still water.

Water moves downstream at x mph. Boat's speed downstream is (30 + x) mph.

In 10 minutes, he goes $\frac{1}{6}$ (30 + x) miles.

He drifts for 30 minutes at x mph. Total distance downstream is $\frac{1}{6}$ (30 + x) $+ \frac{1}{2}$ x.

He drives upstream at (30 − x) mph for 20 minutes. This brings him back $\frac{1}{3}$ (30 − x) miles.

Multiplying out: $5 + \frac{1}{6} x + \frac{1}{2} x - 10 - (-\frac{1}{3} x)$

He finishes $\frac{1}{6}$ (30 + x) $+ \frac{1}{2}$ x $- \frac{1}{3}$ (30 − x) miles downstream from starting point.

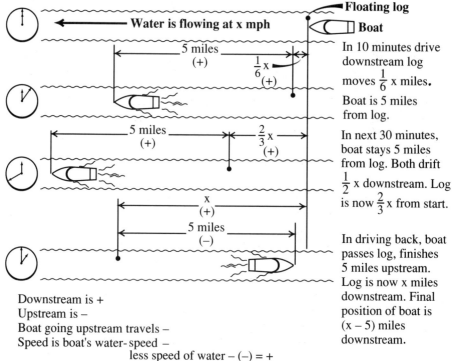

Floating log

Boat

In 10 minutes drive downstream log moves $\frac{1}{6}$ x miles.
Boat is 5 miles from log.

In next 30 minutes, boat stays 5 miles from log. Both drift $\frac{1}{2}$ x downstream. Log is now $\frac{2}{3}$ x from start.

In driving back, boat passes log, finishes 5 miles upstream. Log is now x miles downstream. Final position of boat is (x − 5) miles downstream.

Downstream is +
Upstream is −
Boat going upstream travels −
Speed is boat's water- speed −
        less speed of water − (−) = +
        because water is still going downstream

$- \times - = +$

Minus times a minus makes a plus

This problem is difficult to understand without something you can visualize. Suppose a man rents a boat with a motor, which will drive it at 30 miles per hour in still water. First, he drives it downstream for 10 minutes. Then, he drifts with the current for 30 minutes. Finally, he heads upstream for the remaining 20 minutes. Where does he finish, relative to his starting point?

You don't know how fast the river is flowing, so write an x for the river's rate of flow in miles per hour. This variable gives us an expression for his final position, in miles downstream: $5 + 1/6x + 1/2x - 10 - (-1/3x)$. The minus is times a minus because heading upstream is the opposite direction. If the river was not flowing, he'd go 10 miles upstream (the -10). Since the river is flowing at x miles/hour, he floats less than 10 miles upstream by $1/3x$ miles, which turns out to be downstream, because that's the way the river is flowing. Since he's heading upstream, it slows him down: minus times minus.

Solving the problem

Now, use an imaginary floating log to solve a problem of that kind. The boat's position, relative to the log, will be given by the numbers without x. In one hour, the log floats x miles downstream. You can substitute values for x to get a variety of "answers."

Collecting the number terms, $+5$ and -10, he finishes 5 miles upstream from wherever the log finishes. If the stream is flowing at 5 miles/hour, the log

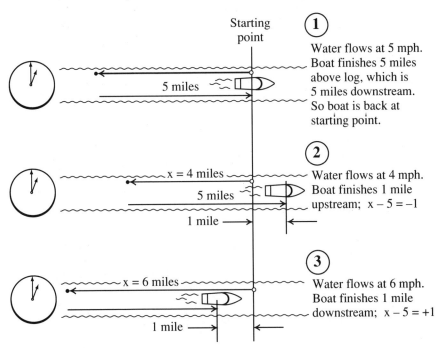

Starting point

1 Water flows at 5 mph. Boat finishes 5 miles above log, which is 5 miles downstream. So boat is back at starting point.

5 miles

2 Water flows at 4 mph. Boat finishes 1 mile upstream; $x - 5 = -1$

x = 4 miles
5 miles
1 mile

3 Water flows at 6 mph. Boat finishes 1 mile downstream; $x - 5 = +1$

x = 6 miles
1 mile

travels 5 miles downstream, so he finishes where he started. If the stream flows at 4 mph, he finishes one mile upstream from his starting point. If the stream flows at 6 miles per hour, he finishes one mile downstream from his starting point.

Arithmetic numbers in algebra

Earlier in this chapter, the difference in writing arithmetic and algebraic numbers was shown: numbers in a row represent, for example, thousands, hundreds, tens, and ones. Algebraic letters in a row represent numbers multiplied together—in school algebra, not always, as you shall see later. To understand this distinction better, use algebra to solve number problems.

If you don't know what the digits of a number are, you would write them, for example $100a + 10b + c$; a is the hundreds' digit, b the tens' digit, and c the ones' digit.

ab means a times b

56 means 5 tens plus 6 ones

If a is 5 and b is 6

10a + b is 56

10a + b could stand for ANY two-figure number in arithmetic

Any three-figure number could be 100a + 10b + c
where a, b, and c, are the figures

Number problems

Suppose someone notices that a certain number's ones' digit is twice the tens' digit, but adding 18 to the first number reverses its digits. What number have you got? You could try a few numbers until you find the one that "works." Algebra gives you a more direct route.

Assume the ten's digit is x, then the tens' digit means $10x$. The ones' digit is twice x $(2x)$. So, the whole number is $12x$. Now, add 18. That's $12x+18$. What was the units' digit is now the tens' digit. The tens' digit in the new number is $20x$ instead of $10x$. The unit's digit is just x instead of $2x$, so the new number is $21x$. Write an equation putting these two descriptions together:

$$12x + 18 = 21x.$$

Subtract $12x$ from both sides or move the $12x$ to the other side by changing its sign (however you prefer to think of that), and get $9x = 18$ (or $18 = 9x$). x is 2 and the original ones' digit is twice that (4). The numbers are 24 and 42. Check: $42 - 24 = 18$.

That problem was easy. They're not all that easy, but the same method works. Working with them helps you to understand the differences.

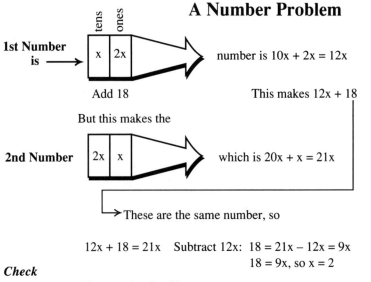

A Number Problem

1st Number is → | x | 2x | number is 10x + 2x = 12x

Add 18   This makes 12x + 18

But this makes the

2nd Number | 2x | x | which is 20x + x = 21x

→ These are the same number, so

12x + 18 = 21x   Subtract 12x: 18 = 21x − 12x = 9x

18 = 9x, so x = 2

Check

First number is   24
Add   18
Total is   42, which is the original number reversed

Questions and problems

1. Form a table to show the length of fencing for a double enclosure, in which the length of the inner enclosure is twice the width, and the spacing is 5 feet.

2. Check your tabulation (question 1) by writing a formula for it and making a graph of the results. From either (or both), find what size of enclosure 1120 feet of fencing will make.

3. Suppose the outer enclosure has a length twice the width. Construct a new table, make a new graph, write a new formula, and find the dimensions of the enclosure built with 1120 feet of fencing.

4. If $a = 5$ and $b = 7$, write the following quantities in the opposite form:

$a + b$; $a \times b$; 57; 75; $b - a$.

5. Simplify the following expressions:

$5[3x - 2(5x + 7)] - 9$

$14 + 2[x + 5(2x + 3)]$

$3/2\{3/2[3/2(3x + 1) + 1] + 1\} + 1$

6. To check that last expression, simplify it to $81/8x + 65/8$ or $10\ 1/8x + 8\ 1/8$. What is the smallest numerical value that x can have so that it will be a whole number with no fractional part? (HINT: 8 1/8 does not change, so to make an exact whole number, 10 1/8x will have to be a whole number and 7/8.)

7. Using the value of x from question 6, evaluate the original expression.

8. A puzzle story goes like this:
   "On evening, 3 men and a monkey found a bag of pecans, but they decided to count them the next day. In the night, one man awoke, decided to count the pecans into 3 piles, and found 1 leftover. He gave the leftover pecan to the monkey, put 2 piles back in the bag, and took the third pile for himself. Later, each not knowing what the others did, did the same thing, always found 1 leftover (which the monkey got), and put the other 2 piles back in the bag. Finally in the morning, they all count the pecans together, again find 1 leftover, and take the final third each. What is the smallest number of pecans that could have been in the bag at the beginning?"
   Working back from the end of the story, satisfy yourself that x in the last expression in question 5 represents the answer.

9. Write down and simplify an expression for the following: a number has 5 added, then is multiplied by 3; the same number has 6 added, then is multiplied by 4; finally, it has 7 added, then is multiplied by 50. Add these three results together, then show that, whatever the first number was, the result is always divisible by 2, but never by 4.

10. A number has 3 added, then is multiplied by 4; the same number is multiplied by 4, then 3 is added; these two results are added, multiplied by 5, and 6 is added to the total. Write down this expression and simplify it. If the total is 361, what was the original number?

11. A professional society's membership is $20 per year for full members and $8 per year for student members. Membership totals 2000 with annual dues of $35,200. How many members and how many student members do they have?

12. Copying down a "think of a number" sequence, I have:
   "Think of a number. Double it. Add 4. Multiply it by 3. Add 12. Multiply it by 5. Add 300. Divide it by 10. Subtract 15." The next instruction I didn't get, then: "subtract the number you first thought of. And the answer is . . . ." and I forget the answer. By using x as the number, find the omitted instruction and what the answer will be.

13. The ones' digit of a number is 2 more than its tens' digit. Multiplied by 3, the tens' digit is what the ones' digit was, and the ones' digit is 3 times what it was. What was the number? (HINT: use x for the tens' digit.)

14. A two-figure number has a one's digit that is 1 more than the tens' digit. Multiplied by 4, the ones' digit is what the tens' digit was, and the tens' digit is 3 times the first ones' digit. What was the original number? (HINT: use x for the original tens' digit.)

15. Use algebra to show that in any number where the ones' digit is 1 more than the tens' digit, adding 9 will reverse the digits.

16. Use algebra to show that in any number where the ones' digit is greater than the tens' digit, adding 9 times the difference between the digits reverses them. (HINT: use a for the tens' digit, $a + x$ for the units' digit.)

17. In adding a string of money figures, an error occurs because two digits were transposed. The total is incorrect by $270. Which two digits were transposed, and by how much did the two digits differ?

18. By substituting various values of x into the following two expressions, say what is different about them. Show why the second is unique:

(a) $(3x + 7)/(x + 1)$
(b) $(3x + 9)/(x + 3)$

19. In each of the following expressions, y is on one side and an expression containing x is on the other. In each case, make a transposition that will put x by itself on one side, with the correct expression containing y on the other.

$$y = x + 5 \qquad y = \frac{1 - x}{x} \qquad y = \frac{1}{x - 1}$$

$$y = \frac{x}{3} - 2 \qquad y = 3x - 7 \qquad y = \frac{x}{x + 1}$$

$$y = 6(x + 2) \qquad y = \frac{5x + 4}{3}$$

In each case, check your results (a) by substituting two different values of x into the original expressions to find y; then, (b) by substituting these values of y into your answers, see if you arrive at your original values of x.

20. Explain the rule for transposition, when equivalent to:

(a) adding to both sides; (b) subtracting from both sides; (c) multiplying both sides by something; (d) dividing both sides by something.

Developing "school" algebra

Orderly writing in algebra

Although parentheses can show multiplication, removing the parentheses to perform the multiplication needs understanding, so you can see how algebra is like arithmetic. Your pocket calculator does it in arithmetic. It's not easy to get your calculator to do it in algebra.

Here are examples, in order of increasing complexity. First, multiplying by a number (outside the parentheses). Next, multiplying by x times a number ($3x$). Then, multiplying two parentheses together, each of which contains a "term" in x (a number times x), with a plain number.

Finally, two parentheses, each containing a different "variable," one called x and the other y. When both parentheses contained the same variable (x), multiplying x by x produced x^2.

LONG MULTIPLICATION in ALGEBRA

(1) $4(x + 3)$

$$
\begin{array}{r}
x + 3 \\
4 \\
\hline
4x + 12
\end{array}
$$

(2) $3x(5x + 7)$

$$
\begin{array}{r}
5x + 7 \\
3x \\
\hline
15x^2 + 21x
\end{array}
$$

(3) $(3x + 4) (5x + 6)$

$$5x + 6$$
$$\underline{3x + 4}$$
$$20x + 24$$
$$\underline{15x^2 + 18x}$$
$$15x^2 + 38x + 24$$

(4) $(7x + 6) (5y + 4)$

$$5y + 4$$
$$\underline{7x + 6}$$
$$30y + 24$$
$$\underline{35xy + 28x}$$
$$35xy + 28x + 30y + 24$$

Indices show "place" in algebra

Compare how generations of students once wrote long multiplication in arithmetic (before pocket calculators did it for them), with something similar in algebra. Successive places in arithmetic, moving left in columns, stand for that many of successively bigger "powers" of 10. Furthest to the right are the ones (not multiplied by ten at all). Next, a number of tens; then, a number of hundreds (10 times 10); then, a number of thousands (10 times 10 times 10).

In algebra, the "places" are separated by plus (or minus) signs. Successive places, moving from right to left, contain plain numbers furthest to the right; next, a number times x; then, a number times x squared; and so on.

In arithmetic, you know the relationship between figures in successive places. They always are in steps of 10:1. In algebra, no fixed relationship between successive quantities exists. But it is consistent; x always has the same value in the same problem—even if you don't know what that value is. If x is 3, then x squared is 3 times 3, or 9, and successively higher powers are 27, 81, 324, and so on. If x is 5, then powers move up through 25, 125, 625, and so on.

Arithmetic	**Algebra**

$$
\begin{array}{r}
2\,3\,5\,4\,7 \\
6\,4\,3 \\
\hline
7\,0\,6\,4\,1 \\
9\,4\,1\,8\,8 \\
1\,4\,1\,2\,8\,2 \\
\hline
1\,5\,1\,4\,0\,7\,2\,1
\end{array}
$$

$$
\begin{array}{r}
x^2 + 3x + 2 \\
5x + 4 \\
\hline
4x^2 + 12x + 8 \\
5x^3 + 15x^2 + 10x \\
\hline
5x^3 + 19x^2 + 22x + 8
\end{array}
$$

Dimension in algebra

Different places, according to the power of x involved, also correspond to successive dimensions. When you multiply a length by a length, the result is an area. Multiply the area by another length and the result is a volume. That is why x times x is called x *squared*, and x times x times x is called x *cubed*. A cube is the simplest form of volume.

On this page, you have multiplied mixed numbers (something times x times a simple number), one of which stands for a square (this one has an x^2 in it) and the other represents a simple dimension, to get a cube. First, use numbers times x, x squared, and x cubed. Then, use letters instead of numbers, a, b, c, d, e. Here, a, b, c, d, e represent numbers, which you can fill in, if you know them.

If you substitute $a = 3$, $b = 5$, $c = 4$, $d = 7$, and $e = 6$, this is the same as the numbers you used first. The letters allow you to fill in any other numbers, and the *general form* as it is called, gives you the answer in terms of powers of x. When such letters are used, x, y, and z are *variables*, but a, b, c, etc., are called *constants*.

Constants can have different values, but these values remain constant in a particular problem.

$$
\begin{array}{l}
3x^2 + 5x + 4 \\
\underline{7x + 6} \\
18x^2 + 30x + 24 \\
21x^3 + 35x^2 + 28x \\
\hline
21x^3 + 53x^2 + 58x + 24
\end{array}
$$

$$
\begin{array}{l}
ax^2 + bx + c \\
\underline{dx + e} \\
aex^2 + bex + ce \\
adx^3 + bdx^2 + cdx \\
\hline
adx^3 + (ae + bd)x^2 + (be + cd)\,x + ce
\end{array}
$$

GENERAL FORM

Putting. $a = 3$ $b = 5$ $c = 4$ $d = 7$ $e = 6$

$ad = 21$ $ae + bd = 18 + 35 = 53$ $be + cd = 30 + 28 = 58$ $ce = 24$

Both methods agree

Expressions, equations, etc.

In old-time school algebra, students learned equations. In that use, an equation is a type of statement with an expression on either side and an equals sign ($=$) in the

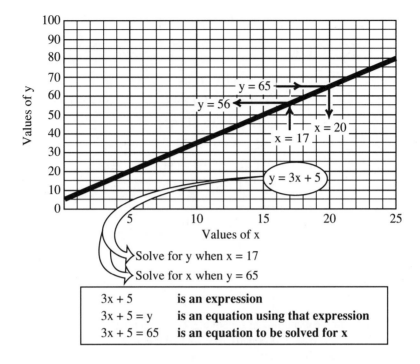

3x + 5	**is an expression**
3x + 5 = y	**is an equation using that expression**
3x + 5 = 65	**is an equation to be solved for x**

middle. For the problem being worked, the quantities on either side of the equals sign are equal.

That sounds obvious? Well, later came inequations and more. An *inequation* is like an equation, except that two expressions are not equal, which school algebra shows with an $\neq$. Some statements use the signs > meaning "is greater than," and < meaning "is less than," to be more specific.

Nest came "truth" statements, which simply tell whether such equations or inequations are true or false. Computers use such statements with similar signs, except that they use > < or < > (take your pick, they both mean the same) instead of $\neq$.

You might not understand these pages yet, but you should begin to think about them and try to figure out what they mean.

An equation as an action statement

A school algebraic equation is a simple statement. It might be true or not true. It never represents a state of change. In school algebra, states of change come later in studies called *calculus*.

Computer programs use an equals sign quite differently—as an action statement. It looks like an equation, but it isn't. Failure to understand this difference causes many problems for would-be computer programmers.

STATEMENTS

SCHOOL		COMPUTER
=	EQUATION	=
≠	INEQUATIONS	>< or <>
>	GREATER THAN	>
<	LESS THAN	<
≯	NOT GREATER THAN	=<
≮	NOT LESS THAN	=>

TRUTH TABLE			

Here is a simple example. In school algebra, the equation $x + 2 = 9$ can be true only if x is 7. If true, it is equally true written the other way around: $9 = x + 2$. Complicate it a little with an equation, such as, $x + 2 = 2x - 3$. You can still find a value of x for which the equation is true. Try 5. No other value "works" in that equation. It's a good school equation.

Now look at this: $x = x + 2$. In school algebra it's impossible. It cannot be "true." No value of x can make it 2 more than itself. But to a computer that, written exactly like an equation in school algebra, is an "action statement." After the computer "reads" the statement, it means x is 2 more than it was before.

It would go like this. Maybe x has a value of 7. The computer reads $x = x + 2$. Now, x has a value of 9—2 more than it was before.

In school algebra, $x = 7$ and $7 = x$ both mean the same thing. A computer could read $x = x + 2$, but not $x + 2 = x$. However, it could read $x = x - 2$. Confusing? Start thinking about it now, you will understand it better when you get to use it.

Using an equation to solve a problem

Suppose a problem reduces to the following facts: three consecutive numbers, the first is divided by 2, added to the next and divided by 3, added to the third and divided by 4, which yields the 4th consecutive number. You need to find these numbers, which would be difficult by arithmetic.

In algebra, write the first three numbers as x, $x + 1$, and $x + 2$. Then, doing what the problem says yields the 4th number, which will be $x + 3$. Algebra derives the right numbers directly. In arithmetic, you can only guess at various numbers until you hit the right ones.

Using an Equation to Solve a Problem

A number, x, is divided by 2 $\dfrac{x}{2}$

The next number, x + 1, is divided by 3 $\dfrac{x+1}{3}$

The next one, x + 2, is divided by 4 $\dfrac{x+2}{4}$

These results are added: $\dfrac{x}{2} + \dfrac{x+1}{3} + \dfrac{x+2}{4}$

$$= \frac{1}{12} \{\, 6x + 4(x + 1) + 3(x + 2)\,\}$$

$$= \frac{1}{12} \{\, 6x + 4x + 4 + 3x + 6\,\}$$

$$= \frac{1}{12} \{\, 13x + 10\,\}$$

This gives the result of doing (THIS) to any three consecutive numbers.

But the three we want give a fourth consecutive number, x + 3; so:

$$\frac{1}{12}\{\, 13x + 10\,\} = x + 3$$

Multiply by 12: 13x + 10 = 12(x + 3) = 12x + 36

Subtract 12x + 10: x = 26

CHECK $\dfrac{26}{2} + \dfrac{27}{3} + \dfrac{28}{4} = 13 + 9 + 7 = 29$ *As required*

Simultaneous equations

Often a problem can be solved with only one variable. Sometimes it is easier to use two or more variables. If you know that 4 times one number plus 5 times another number add up to 47, and 5 times the first number plus 4 times the second number add up to 43: how would you find the answers?

The numbers can be found in several ways. Don't think that one way is the only right way. Sometimes you can even spot a way that's easier than the "textbook" way.

Here the textbook way "eliminates" one variable. To do so, multiply one equation, both sides, by 4, and the other one by 5, then subtract one product from the other. That gets rid of one variable: 4 times the 1st is $16x + 20y = 188$. 5 times the 2nd is $25x + 20y = 215$. Subtracting the 1st from the 2nd is $9x = 27$. So $x = 3$. From either equation you can then get $y = 7$. Check it for yourself.

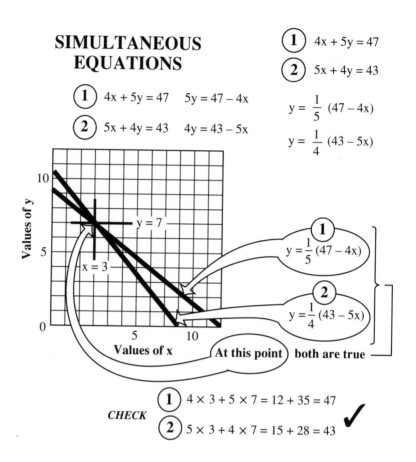

SIMULTANEOUS EQUATIONS

(1) $4x + 5y = 47$

(2) $5x + 4y = 43$

(1) $4x + 5y = 47$ $5y = 47 - 4x$

(2) $5x + 4y = 43$ $4y = 43 - 5x$

$y = \frac{1}{5}(47 - 4x)$

$y = \frac{1}{4}(43 - 5x)$

10

Values of y

$y = 7$

5

$x = 3$

0

5   10

Values of x

(1) $y = \frac{1}{5}(47 - 4x)$

(2) $y = \frac{1}{4}(43 - 5x)$

At this point both are true

CHECK

(1) $4 \times 3 + 5 \times 7 = 12 + 35 = 47$

(2) $5 \times 3 + 4 \times 7 = 15 + 28 = 43$ ✓

The previous method is graphical. However, one student who couldn't do algebra saw 9 "miscellaneous unknowns" in both equations, but one had 5 x's and 4 y's, and the other had the other combination. He concluded that y must be 4 more than x. Putting $y = x + 4$ into either one produced the equation $9x = 27$. From that point, the method was the same.

Simultaneous equations solve a fraction problem

Suppose you have a fraction. You don't know what its numerator and denominator are, but someone says that subtracting 1 from both creates the fraction 1/2, but adding 1 to both makes the fraction 3/5.

Write x/y for the original fraction. Subtracting 1 from each gives the first fact (equation). Adding 1 to each gives the second fact (equation). See how these answers are converted into a pair of simultaneous equations.

Fraction is $\dfrac{x}{y}$

First fact: $\dfrac{x-1}{y-1} = \dfrac{1}{2}$

Second fact: $\dfrac{x+1}{y+1} = \dfrac{3}{5}$

(1) Multiply by $2(y-1)$:
$$2(x-1) = y-1$$
$$2x-2 = y-1$$
$$2x - y = 1$$

(2) Multiply by $5(y+1)$:
$$5(x+1) = 3(y+1)$$
$$5x+5 = 3y+3$$
$$5x - 3y = -2$$

Solving the problem

Here are a few examples, the first of which is the fraction problem from the previous section that shows how to eliminate one variable. First, multiply one or both equations by numbers that will make the variable to be eliminated have the same *coefficient* (a fancy word for the number in front of the variable). Then, either add or subtract the two equations in that form. If the signs are the same, subtract. If they are opposite, add.

Here is a third problem, reduced to two equations so that you can concentrate on the algebraic method. Study each of the three examples on these pages, so you understand how to use this method, called *eliminating one variable*. Always check your results.

SOLVING SIMULTANEOUS EQUATIONS - Method 1

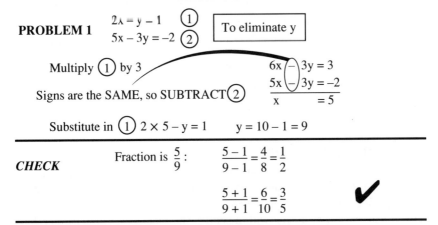

PROBLEM 1 $2x = y - 1$ **(1)** $5x - 3y = -2$ **(2)**   To eliminate y

Multiply **(1)** by 3

$$6x - 3y = 3$$
$$5x - 3y = -2$$
$$x = 5$$

Signs are the SAME, so SUBTRACT **(2)**

Substitute in **(1)** $2 \times 5 - y = 1$ $y = 10 - 1 = 9$

CHECK   Fraction is $\dfrac{5}{9}$: $\dfrac{5-1}{9-1} = \dfrac{4}{8} = \dfrac{1}{2}$

$\dfrac{5+1}{9+1} = \dfrac{6}{10} = \dfrac{3}{5}$ ✔

continued

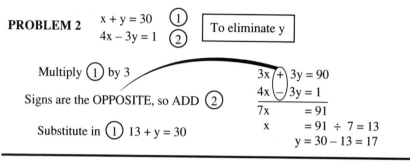

PROBLEM 2 $x + y = 30$ ①
$4x - 3y = 1$ ② | To eliminate y |

Multiply ① by 3

Signs are the OPPOSITE, so ADD ②

Substitute in ① $13 + y = 30$

$3x + 3y = 90$
$4x - 3y = 1$
$7x \quad = 91$
$x \quad = 91 \div 7 = 13$
$y = 30 - 13 = 17$

CHECK

$13 + 17 = 30$
$4 \times 13 - 3 \times 17 = 52 - 51 = 1$ ✔

SOLVING SIMULTANEOUS EQUATIONS
Method 1

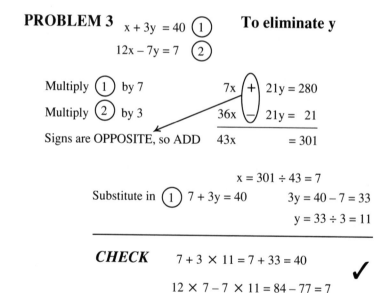

PROBLEM 3 $x + 3y = 40$ ①   **To eliminate y**
$12x - 7y = 7$ ②

Multiply ① by 7

Multiply ② by 3

Signs are OPPOSITE, so ADD

$7x + 21y = 280$
$36x - 21y = 21$
$43x \quad = 301$

$x = 301 \div 43 = 7$

Substitute in ① $7 + 3y = 40$ $3y = 40 - 7 = 33$
$y = 33 \div 3 = 11$

CHECK $7 + 3 \times 11 = 7 + 33 = 40$ ✔

$12 \times 7 - 7 \times 11 = 84 - 77 = 7$

Solving by substitution

In the three sets of equations so far, you used the first method, eliminating one variable. The following method can often be a short cut. The section of simultaneous equations introduced it at the end. Interestingly, the student who "discovered" it was having problems with algebra! The method is demonstrated here more formally. Again, don't forget to check your result.

Most teachers came to rely on published answers to check their results. This led to students doing the same thing—if they could get the answer book. Checking your own result not only avoids such "cheating": In life, when you use mathematics, you don't have an "answer book."

In recent years, many engineering projects have failed because the designers don't know how to check their work. The Golden Gate bridge has celebrated 50 years of operations and it's still going strong. Many more recent bridges have plunged people into the drink below because they failed. The habit of checking your work could save lives when you get into the work world!

SOLVING SIMULTANEOUS EQUATIONS
Method 2

PROBLEM 4 $7x + 2y = 90$ ①   **Rearrange ② to give**

$8x - y = 1$ ②

Value for y: $y = 8x - 1$

Substitute in ① $7x + 2(8x - 1) = 90$

$7x + 16x - 2 = 90$

$23x = 92$ $x = 92 \div 23 = 4$

Substitute in ② $y = 8 \times 4 - 1 = 31$

CHECK $7 \times 4 + 2 \times 31 = 28 + 62 = 90$ ✓

$8 \times 4 - 31 = 32 - 31 = 1$

Solving for reciprocals

The example at the top of page 150 shows a different kind of simultaneous equation: one in which variables appear as reciprocals, or in which it is easier to solve for reciprocals. Study this example in which we solve for $1/x$ and $1/y$, instead of for x and y. Notice particularly why solving for the reciprocals is easier for some kinds of problems.

One thing you should learn from studying simultaneous equations is that you can often save time by using common sense to find the easiest way, rather than following a set routine for all such equations.

SOLVING for RECIPROCALS

PROBLEM 5 $\dfrac{2}{x} + \dfrac{3}{y} = \dfrac{14}{15}$ ①

$\dfrac{5}{x} - \dfrac{4}{y} = \dfrac{1}{30}$ ②   **To eliminate** $\dfrac{1}{y}$

Multiply ① by 4 $\dfrac{8}{x} + \dfrac{12}{y} = \dfrac{56}{15}$

Multiply ② by 3 $\dfrac{15}{x} - \dfrac{12}{y} = \dfrac{3}{30} = \dfrac{1}{10}$

Signs are OPPOSITE, so ADD $\dfrac{23}{x} = \dfrac{56}{15} + \dfrac{1}{10} = \dfrac{112 + 3}{30} = \dfrac{115}{30} = \dfrac{23}{6}$

So $\dfrac{1}{x} = \dfrac{1}{6}$ $x = 6$

Substitute in ② $\dfrac{5}{6} - \dfrac{4}{y} = \dfrac{1}{30}$ $\dfrac{4}{y} = \dfrac{5}{6} - \dfrac{1}{30} = \dfrac{25 - 1}{30}$

$\dfrac{4}{y} = \dfrac{24}{30} = \dfrac{4}{5}$ So $\dfrac{1}{y} = \dfrac{1}{5}$ $y = 5$

Alternative

Multiply ① by 15xy $30y + 45x = 14xy$

Multiply ② by 30xy $150y - 120x = xy$

The other way is much easier

CHECK $\dfrac{2}{6} + \dfrac{3}{5} = \dfrac{1}{3} + \dfrac{3}{5} = \dfrac{5 + 9}{15} = \dfrac{14}{15}$

$\dfrac{5}{6} - \dfrac{4}{5} = \dfrac{25 - 24}{30} = \dfrac{1}{30}$ ✓

Long division clarifies how algebra works

These days long division by algebra has little practical use. It does help to understand how algebra works, and it is particularly helpful in developing an understanding of dimension. Compare the two ways of doing long division shown at the top of the facing page, in arithmetic and algebra. You should be able to "figure it out," and do the examples in the Question and Problem section at the end of this chapter. Look at this exercise to see how people used to solve these problems.

Arithmetic   Algebra

$$
\begin{array}{r}
279 \\
153\overline{)42687} \\
306 \\
\hline
1208 \\
1071 \\
\hline
1377 \\
1377 \\
\end{array}
$$

$$
\begin{array}{r}
4x^2+ \; 7x - 3 \\
3x+5\,\overline{)12x^3 + 41x^2 + 26x - 15} \\
12x^3 + 20x^2 \\
\hline
21x^2 + 26x \\
21x^2 + 35x \\
\hline
- \; 9x - 15 \\
- \; 9x - 15 \; . \\
\end{array}
$$

Long division finds factors in algebra

This chapter covered finding factors in arithmetic—you still need that. Students of mathematics will find it important in algebra, but not in everyday use, as it was once. Here is a parallel between finding factors, in arithmetic and in algebra for you to study. Try the simple exercises that follow.

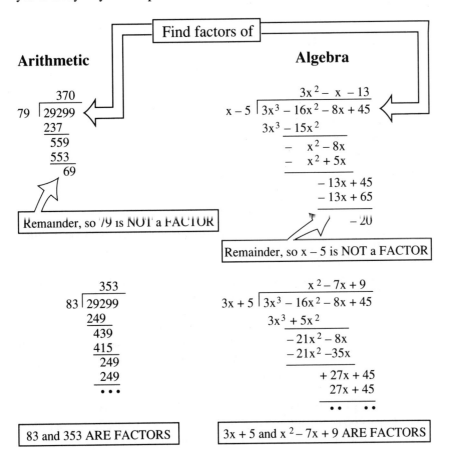

Find factors of

Arithmetic

$$
\begin{array}{r}
370 \\
79\,\overline{)29299} \\
237 \\
\hline
559 \\
553 \\
\hline
69 \\
\end{array}
$$

Remainder, so 79 is NOT a FACTOR

Algebra

$$
\begin{array}{r}
3x^2 - x - 13 \\
x-5\,\overline{)3x^3 - 16x^2 - 8x + 45} \\
3x^3 - 15x^2 \\
\hline
- \quad x^2 - 8x \\
- \quad x^2 + 5x \\
\hline
- 13x + 45 \\
- 13x + 65 \\
\hline
- 20 \\
\end{array}
$$

Remainder, so x − 5 is NOT a FACTOR

$$
\begin{array}{r}
353 \\
83\,\overline{)29299} \\
249 \\
\hline
439 \\
415 \\
\hline
249 \\
249 \\
\hline
\bullet \bullet \bullet \\
\end{array}
$$

83 and 353 ARE FACTORS

$$
\begin{array}{r}
x^2 - 7x + 9 \\
3x+5\,\overline{)3x^3 - 16x^2 - 8x + 45} \\
3x^3 + 5x^2 \\
\hline
-21x^2 - 8x \\
-21x^2 - 35x \\
\hline
+ 27x + 45 \\
27x + 45 \\
\hline
\bullet \bullet \quad \bullet \bullet \\
\end{array}
$$

3x + 5 and $x^2 - 7x + 9$ ARE FACTORS

Questions and problems

1. Perform the following multiplications:

$(x + 1)(x - 3)$ $(x + 3)(x - 1)$

$(x - 3)(x - 5)$ $(3x + 5)(5x - 3)$

$(x + 1)(x - 1)$ $(x + a)(x - a)$

$(x + 1)(x^2 - x + 1)$ $(x - a)(x^2 + ax + a^2)$

$(x^2 + \sqrt{2x} + 1)(x^2 - \sqrt{2x} + 1)$ $(x^2 - 1.5x + 1)(x^2 + 2x + 1)$

2. Perform the following divisions:

$$\frac{x^2 - 2x - 3}{x + 1} \qquad \frac{x^2 + 2x - 3}{x - 1}$$

$$\frac{x^2 - 8x + 15}{x - 3} \qquad \frac{15x^2 + 16x - 15}{3x + 5}$$

In each case, check your result by multiplication.

$$\frac{x^2 - y^2}{x + y} \qquad \frac{x^2 - y^2}{x - y}$$

$$\frac{x^3 - a^3}{x + a} \qquad \frac{x^3 - c^3}{x - c}$$

$$\frac{x^4 + 2x^3 - x^2 + 4}{x + 2} \qquad \frac{x^5 - 8x^2 - x + 2}{x - 2}$$

3. Use long division to find the factors of:

$x^2 + 2x - 35$ $21x^2 + 8xy - 45y^2$

$x^3 + x^2 - 5x - 5$ $x^3 - 2x^2 + 5x - 10$

$x^3 + x^2 - 7x - 3$ $x^4 - x^3 - 5x^2 - 76x + 5$

NOTE: each of the above has two factors.

4. When working separately, two men, A and B, complete a certain job in different times. If A works 6 days, B can finish the job in 9 days, but if A works 8 days, B can finish it in 6 days. How long would each take to do the job completely on his own? HINT: use x and y for the number of days each would take, then solve for $1/x$ and $1/y$.

5. A rectangle has certain dimensions. Making it 2 feet wider and 5 feet longer increases its area by 133 square feet. Making it 3 feet wider and 8 feet longer increases its area by 217 square feet. What were its original dimensions? A good question for simultaneous equations.

6. Two men working together can finish a job in 12 days. If A starts the job and spends 15 days on it, B can finish it in 10 days. How long would each take to do it completely alone? Use the same method as for question 4.

7. Divide the number *c* into two parts, such that *a* times one part is equal to *b* times the other part. If *c* is 28, *a* is 3 and *b* is 4, what are the parts? It might be easiest to find the second part first.

8. A man has 45 coins (quarters, dimes, and nickels), with a total value of $6.30. In counting his change, he makes a mistake, transposes the numbers of quarters and dimes, and figures the amount to be $5.70. How many of each coin does he really have?

9. In a 3-digit number, the ones' digit is the sum of the hundreds' and tens' digits. If the hundreds' digit is moved to the ones' place, and the other two digits are moved one place to the left, the new number is less than the original number by 432. The sum of the digits in the number is 18. Find the number.

10. A fraction is somewhere between 3/4 and 4/5. Adding 3 to both the numerator and the denominator makes the fraction equal to 4/5, but 4 subtracted from each makes the fraction equal to 3/4. What is the fraction?

11. In another fraction, adding 1 to both the numerator and the denominator makes the fraction equal to 4/7, but 1 subtracted from each makes the fraction equal to 5/9. What is the fraction?

12. Adding 1 to numerator and denominator of another fraction makes it equal 7/12, but 1 subtracted from each makes it equal 9/16. What is the fraction?

13. The highest two of four consecutive numbers, multiplied together, produce a product that is 90 more than the lowest two multiplied together. What are the numbers?

14. Work with five consecutive numbers yields the fact that if the middle three numbers are multiplied together, they are 15 more than the first, middle, and last numbers multiplied together. What are the numbers?

15. A man has an option on a piece of land. He was told that the measurements were 50 feet longer than it is wide. The survey shows that it is 10 feet less in width than he was told. The seller offers him an extra 10 feet in length. Does he get the same total area? If not, how much does he lose on the deal? Does it depend on the actual dimensions?

16. The man in question 15 spots that an extra 20 feet in length would give him the same area. What were the dimensions that he was originally told and what did he finally get?

10
CHAPTER

Quadratics

Problems with two or more answers

Simultaneous equations come from problems that have two or more answers at the same time. *Quadratics* begin a kind of problem where different answers are possible, not at the same time, but as alternatives: either of two answers or sets of answers could be correct.

To see what these definitions mean, suppose you have 80 feet of fencing to enclose a four-sided area. What area does it enclose?

Assuming it is rectangular in shape, it has two sides of each of two dimensions. So, if the sides are W feet wide by L feet long, $2W + 2L = 80$, or $W + L = 40$. Its area is W times L. By substituting $L = 40 - W$, the area can be written $W(40 - W)$ square feet, which will multiply out to $40W - W^2$. Or using the conventional x, for the first dimension (which was width) and y for area, the equation is: $y = 40x - x^2$.

A more specific comparison calls the equations in the last chapter *linear*, referring to a form that could be written: $y = ax + b$ (where a and b are constants). Then, quadratics can take the form: $y = ax^2 + bx + c$. Writing in letters for constants makes such equations a "standard form" into which any problem can be expressed.

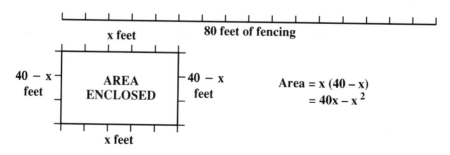

x	0	5	10	15	20	25	30	35	40	
40 − x	40	35	30	25	20	15	10	5	0	**1st Method**
Area x(40 − x)	0	175	300	375	400	375	300	175	0	

Same Answers

x	0	5	10	15	20	25	30	35	40	
40x	0	200	400	600	800	1000	1200	1400	1600	
x²	0	25	100	225	400	625	900	1225	1600	**2nd Method**
Area 40x − x²	0	175	300	375	400	375	300	175	0	

From LINEAR to QUADRATIC

$$y = ax + b \quad y = ax^2 + bx + c$$

Quadratic graph is a symmetrical curve

If you plot the values of $40x - x^2$, against x, you get a curve. If x is either 0 or 40, $40x - x^2$ has a value of zero. When x is 20, $40 - x^2$ reaches a maximum of 400.

In this particular problem, x represented one side of a rectangle whose two sides added up to 40. The two pairs of values were the same, just reversed. Later, you will come to pairs of answers that are not simple reversals.

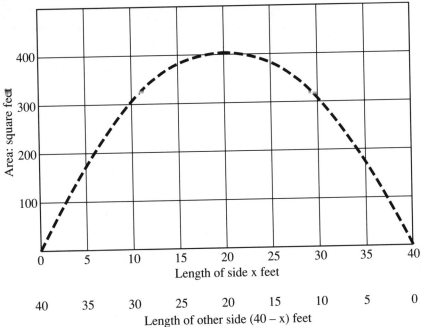

Solving a quadratic equation

Suppose you know that, as well as using 80 feet of fencing, it encloses an area of 300 square feet. One way to solve this problem would be to write: $40 - x^2 = 300$. Then, rearrange this equation to make something on one side equal to zero on the other side. The equation reduces to: $x^2 - 40x + 300 = 0$.

Why would we do that? This procedure will be clearer later. For now, notice what happens if you plot values of the left-hand side, $x^2 - 40x + 300$. Here, the curve is inverted and the position of the values are changed to straddle the horizontal zero line. The zero line becomes the locator of the solutions because the solution was written: $x^2 - 40x + 300 = 0$.

This changing around is called *transposition*. It reproduces the same curve in a different position.

$$40x - x^2 = 300$$

Transpose: $-$

$$40x - x^2 - 300 = 0$$

or $\qquad x^2 - 40x + 300 = 0$

x	0	5	10	15	20	25	30	35	40
x^2	0	25	100	225	400	625	900	1225	1600
40x	0	200	400	600	800	1000	1200	1400	1600
$x^2 - 40x$	0	−175	−300	−375	−400	−375	−300	−175	0
$x^2 - 40x + 300$	300	125	0	−75	−100	−75	0	125	300

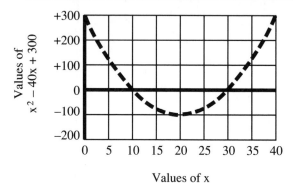

Values of $x^2 - 40x + 300$

Values of x

TRANSPOSITION: Same Curve – Different Position

Using factors to solve equations

The expression: $x^2 - 40x + 300$ can be factorized into two linear factors: $(x - 10)(x - 30)$. Multiplying those expressions, $x - 10$ and $x - 30$, yields our first expression: $x^2 - 40x + 300$. Now, see something else about the curve on the previous page.

Plotting lines that represent $x - 10$ and $x - 30$, these become zero when $x = 10$ or when $x = 30$. Notice that the quadratic curve passes through the zero line at the same values of x as $x - 10 = 0$ or $x - 30 = 0$.

So, finding the factors of an expression, formed by transposing an equation to something equal to zero, gives the solutions of that equation. If the factors have a minus sign (as here), the solutions are the corresponding plus quantities. Later, you'll see that if the factors have (or one of them has) a + sign, the corresponding solution has a − sign.

FACTORS

$$x^2 - 40x + 300 = (x - 10)(x - 30)$$

x	0	5	10	15	20	25	30	35	40
x − 10	−10	−5	0	+5	+10	+15	+20	+25	+30
x − 30	−30	−25	−20	−15	−10	−5	0	+5	+10
(x − 10)(x − 30)	+300	+125	0	−75	−100	−75	0	+125	+300

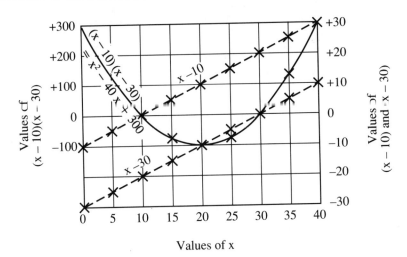

**Combined value is zero
when EITHER factor is zero**

Finding factors to solve quadratics

To find factors, you need rules to help. You find these rules by examining the coefficients of each term to see how they are constructed. Each factor, for the kind of expression you are using (of the form: $ax^2 + bx + c = 0$, where b and c might be either $+$ or $-$) will be "so many" x, plus or minus a number.

In each expression, numbered (1) through (4), the coefficient of x^2 is 3. If it has simple factors, one of them must be x plus or minus a number, and the other must be $3x$ plus or minus a number. In all of these expressions, the numeric term is 15. So, in the two factors, one must have plus or minus 3 and the other must have plus or minus 5. Or else one must have plus or minus 1 and the other must have plus or minus 15.

If the sign in front of the number (in this case 15) is plus, then the signs in the factors must be either both plus or else both minus, and the coefficient of x is the sum of the two cross products.   If the sign in front of the number is minus, the signs in the factors must be opposite; one minus and one plus, and the coefficient of x is the difference between the two cross products.

Study this system carefully, until you can find the factors fairly easily.

Finding FACTORS to SOLVE QUADRATICS

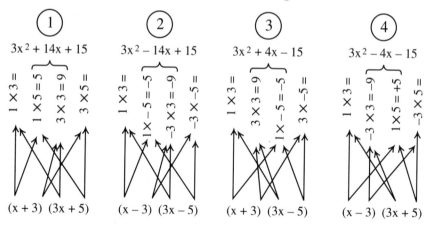

$$3x^2 + 14x + 15$$

$$3x^2 - 14x + 15$$

$$3x^2 + 4x - 15$$

$$3x^2 - 4x - 15$$

$(x + 3)$ $(3x + 5)$ $(x - 3)$ $(3x - 5)$ $(x + 3)$ $(3x - 5)$ $(x - 3)$ $(3x + 5)$

How factors solve quadratics

Those first factors in the section "Using factors to solve equations" were simple. Those in the last section get a little more difficult. The lines for the factors on the graph are not parallel lines. Taking example 2, the thing that is the same as in the previous section is that the solutions come at the two places where the straight lines cross the zero line.

Make sure you understand the points these diagrams show:

1. How the solutions are derived from the points where the graphs cross the zero line.

2. How the root (answer) that corresponds to the factor has the opposite sign.

3. Its value is the numerical part of the factor, divided by the coefficient of x (or whatever variable you use).

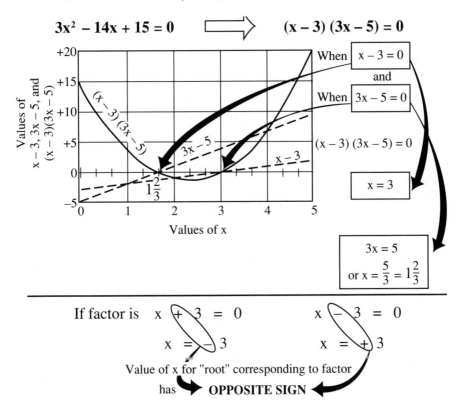

When factors are even more difficult to find

Factorizing is a simple method of solving quadratics when factors can be found easily. Then, show some that are not so easy. For the expression: $x^2 - 6x + 6$, you need two numbers of the same sign (which will be minus) that, multiplied together, make 6 and are added together to also make 6.

You can try fractions, but they never work out exactly. The other example is also difficult. You need a way to find such factors more directly than by trial and error.

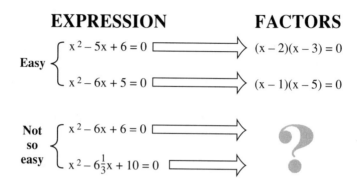

PAGE 2-42

Completing the square

The whole purpose of adopting algebra was to find more direct ways of working. Remember that. By completing the square, you can stop searching for factors. It can be understood by using a geometrical figure.

Your expression starts in the form: $ax^2 + bx + c = 0$. Notice that if $N = (x + n)$, then $N^2 = (x + n)^2$ which, multiplied out, is: $N^2 = x^2 + 2nx + n^2$. Here, you show the correspondence between the algebraic expression and a geometrical figure. Study it carefully.

Go back to those equations that did not factor easily. Take them in the reverse order. Because $2n$ from our formula is 6 1/3, n must be half of that, 3 1/6. To find n^2 square 3 1/6. 3 1/6 is 19/6. Squaring that (both numerator and denominator) is 361/36 (10 1/36).

COMPLETING THE SQUARE
A direct method for solving quadratic equations

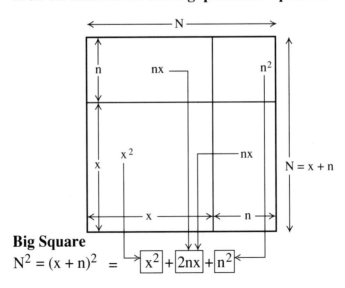

Big Square
$$N^2 = (x + n)^2 = \boxed{x^2} + \boxed{2nx} + \boxed{n^2}$$

Completing the Square

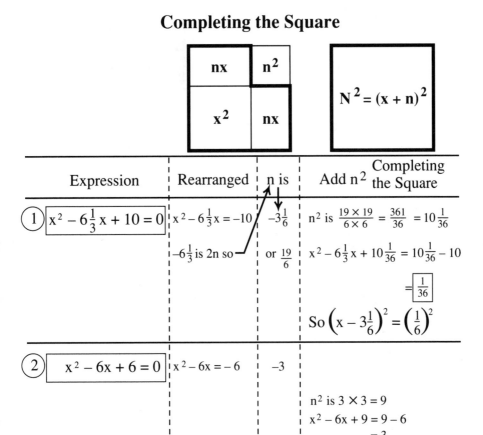

Expression	Rearranged	n is	Add n^2	Completing the Square
① $x^2 - 6\frac{1}{3}x + 10 = 0$	$x^2 - 6\frac{1}{3}x = -10$	$-3\frac{1}{6}$	n^2 is $\frac{19 \times 19}{6 \times 6} = \frac{361}{36} = 10\frac{1}{36}$	
	$-6\frac{1}{3}$ is $2n$ so	or $\frac{19}{6}$	$x^2 - 6\frac{1}{3}x + 10\frac{1}{36} = 10\frac{1}{36} - 10$	
			$= \boxed{\frac{1}{36}}$	
			So $\left(x - 3\frac{1}{6}\right)^2 = \left(\frac{1}{6}\right)^2$	
② $x^2 - 6x + 6 = 0$	$x^2 - 6x = -6$	-3		
			n^2 is $3 \times 3 = 9$	
			$x^2 - 6x + 9 = 9 - 6$	
			$= 3$	
			So $(x - 3)^2 = \boxed{3}$	

Both sides need to be changed so that the left side has the number 10 1/36. Then, 1/36 is on the right side, which is the square of 1/6. The other side isn't so easy, although it looked easier at first. The square is 3, so the root must be the square root of 3.

Completing the solution by completing the square

The previous section had two equations in which both sides of each equation was a square, although the second one didn't look like it. A square can have a root (the number that makes that square) that is either positive or negative. So, the first one can have $x - 3$ 1/6 equal to either $+1/3$ or $-1/3$. The possible answers are 3 1/3 or 3. The second one leads to 3 plus root 3 or 3 minus root 3 as the possible answers.

(1) $(x - 3\frac{1}{6})^2 = (\frac{1}{6})^2$

(FIRST EQUATION)

| − X − = + |
| + X + = + |

So $(x - 3\frac{1}{6}) = +\frac{1}{6}$ or $-\frac{1}{6}$

If $x - 3\frac{1}{6} = +\frac{1}{6}$ $x = 3\frac{1}{6} + \frac{1}{6} = \boxed{3\frac{1}{3}}$

If $x - 3\frac{1}{6} = -\frac{1}{6}$ $x = 3\frac{1}{6} - \frac{1}{6} = \boxed{3}$

$\left.\right\}$ **Two Answers**

(2) $(x - 3)^2 = 3$

(SECOND EQUATION)

So $(x - 3) = +\sqrt{3}$ or $-\sqrt{3}$

If $x - 3 = +\sqrt{3}$ $x = \boxed{3 + \sqrt{3}}$

If $x - 3 = -\sqrt{3}$ $x = \boxed{3 - \sqrt{3}}$

$\left.\right\}$ **Two Answers**

Checking the answers

Don't forget to check your answers. Go through each of the original equations, substitute in each solution you found to see that the equations are true.

Checking the Answers

(1) $x = 3\frac{1}{3}$ $x^2 = \frac{10 \times 10}{3 \times 3} = \frac{100}{9} = 11\frac{1}{9}$

$6\frac{1}{3}x = \frac{10}{3} \times \frac{19}{3} = \frac{190}{9} = 21\frac{1}{9}$

$\boxed{x^2 - 6\frac{1}{3}x = 11\frac{1}{9} - 21\frac{1}{9} = -10}$

Both answers check

or $x = 3$ $x^2 = 9$

$6\frac{1}{3}x = 3 \times \frac{19}{3} = 19$

$\boxed{x^2 - 6\frac{1}{3}x = 9 - 19 = -10}$

(2) $x = 3 + \sqrt{3}$ $x^2 = 3^2 + 2 \times 3 \times \sqrt{3} + \sqrt{3}^2 = 9 + 6\sqrt{3} + 3 = 12 + 6\sqrt{3}$

$6x = 6(3 + \sqrt{3}) = 18 + 6\sqrt{3}$

$$x^2 - 6x = 12 + 6\sqrt{3} - (18 + 6\sqrt{3})$$
$$= 12 + 6\sqrt{3} - 18 - 6\sqrt{3} = -6$$

$x = 3 - \sqrt{3}$ $x^2 = 3^2 - 2 \times 3 \times \sqrt{3} + \sqrt{3}^2 = 9 - 6\sqrt{3} + 3 = 12 - 6\sqrt{3}$

$6x = 6(3 - \sqrt{3}) = 18 - 6\sqrt{3}$

Both answers check

$$x^2 - 6x = 12 - 6\sqrt{3} - (18 - 6\sqrt{3})$$
$$= 12 - 6\sqrt{3} - 18 + 6\sqrt{3} = -6$$

What the answers mean

You might wonder how completing the square can have two answers. The construction in "Completing the bill" explains. Only one solution seems obvious. After all, one area will fill the space.

Look at the significance of $N^2 = (x - n)^2$. Here, x is the big square and the final square that represents N^2 is smaller, so x^2 is diminished by two quantities nx. Notice how drawing the two rectangles, which represent n times x overlap, by another square that is x^2.

So, as you found in algebra, the geometrical construction supports it: $N^2 = x^2 - 2nx + n^2$.

Go over this formula carefully to be sure you understand it.

Geometric Interpretation for the Two Solutions

x could be either 3 or $3\frac{1}{3}$

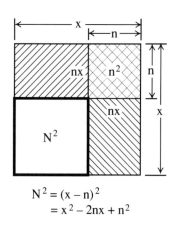

$N^2 = (x - n)^2$
    $= x^2 - 2nx + n^2$

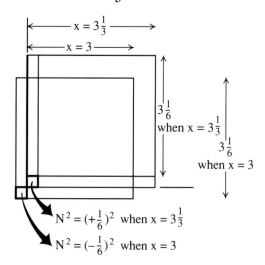

$N^2 = (+\frac{1}{6})^2$ when $x = 3\frac{1}{3}$

$N^2 = (-\frac{1}{6})^2$ when $x = 3$

Comparing methods

If a problem can be solved in two or more ways, making a comparison can help in several ways: you can pick the best method for a particular problem and, you can gain a better overall understanding of the methods.

What the factor method means, when looked at as geometry, is shown below. Writing an equation with zero on one side, if it factors, can represent a rectangle where each factor represents one of its dimensions. When either of its sides is zero, its area is zero. Factors find what those values are.

Look at the completing the square method in steps. You start with a similar general form: $x^2 + bx + c = 0$. The coefficient of x^2 is optional, but eliminating it makes the equation easier to solve. First, rearrange it so that the number is on the other side with the sign changed. Now, the left side has an incomplete square. The third step is to complete the square, and the fourth is to add the same to both sides. Now, you have a complete square on the left, so the fifth step is to take the square root of both sides. Finally, transpose the roots so that you have a statement that lists the two values of x (or whatever variable your equation uses).

METHODS of SOLVING QUADRATICS

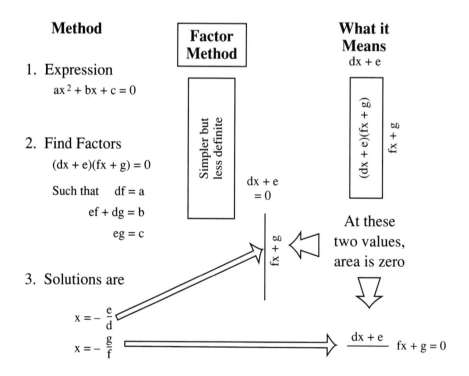

The geometry shows what it means. This method is more involved than just finding the factors, but it always gets an answer, which finding the factors cannot do so easily.,

"Completing the Square" Method

Method

What it means

1. Expression

$$x^2 + bx + c = 0$$

2. Rearranged

$$x^2 + bx = -c$$

3. n is $\frac{1}{2}$ b, so $n^2 = \frac{1}{4} b^2$

4. Complete Square

$$x^2 + bx + \frac{1}{4} b^2 = \frac{1}{4} b^2 - c$$

5. Take square root

$$\left(x + \frac{1}{2} b \right) = +\sqrt{\frac{1}{4} b^2 - c}$$

$$\text{or} \quad -\sqrt{\frac{1}{4} b^2 - c}$$

6. Transpose

$$x = -\frac{1}{2} b \pm \sqrt{\frac{1}{4} b^2 - c}$$

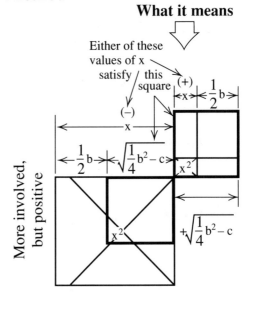

Either of these values of x satisfy this square

Formula method

The *formula method* really applies the completing the square method to derive a formula, into which you substitute the constants if the problem can be expressed in that general form. The form of the answer shown on this page might look different from the one that is generally given in textbooks. This example is to relate it better to the geometric way of visualizing it. The more common form is:

$$x = \frac{-b \pm \sqrt{b^2 - 4ac}}{2a}$$

Formula Method
Using Algebra on Algebra

(1) General Form of Quadratic Equation:

$$ax^2 + bx + c = 0$$

(2) Reduce so Coefficient of x^2 is 1: divide by a

$$x^2 + \frac{b}{a}x + \frac{c}{a} = 0$$

(3) Transpose for Completing the Square:

$$x^2 + \frac{b}{a}x = -\frac{c}{a}$$

(4) Complete the Square: n is $\frac{b}{2a}$ $n^2 = \frac{b^2}{4a^2}$

$$x^2 + \frac{b}{a}x + \frac{b^2}{4a^2} = \frac{b^2}{4a^2} - \frac{c}{a}$$

(5) Take the Square Root:

$$x + \frac{b}{2a} = \pm\sqrt{\frac{b^2}{4a^2} - \frac{c}{a}}$$

(6) Transpose to get Final Formula:

$$x = -\frac{b}{2a} \pm \sqrt{\frac{b^2}{4a^2} - \frac{c}{a}}$$

Solving by formula

Once you reduce the problem to the standard form, the formula method is just a matter of substituting the numbers. Work through the same equations solved by factors. Pay particular attention to the signs, since you should have already learned to do them by factors.

Here are some more quadratics that are solved by formula. No. 5 (on page 168) follows the same method of examples 1 through 4. You might not immediately spot the square root of 169, but you can check it by multiplying 13 by 13. No. 6 (on page 168) needs to be rearranged before the formula can be used, by multiplying through by x and then transposing.

No. 7 (on page 168) is the first of a kind where the final result on the right doesn't have a simple root. You need the square root of 8, which is something less than 3. So, one answer is a small fraction (an unending decimal) and the other is slightly less than 6. That sign (which you'll find on a calculator button) means the square root of, and it is called a *surd*.

SOLVING BY FORMULA

Important!

$$+ \boxed{\begin{array}{c} - \\ \text{Watch} \\ \text{the} \\ \text{Signs} \\ + \end{array}} +$$
$$- \quad\quad -$$

For $ax^2 + bx + c = 0$

$$x = -\frac{b}{2a} \pm \sqrt{\frac{b^2}{4a^2} - \frac{c}{a}}$$

---

① $3x^2 + 14x + 15 = 0$

$$x = -\frac{14}{6} \pm \sqrt{\frac{196}{36} - \frac{15}{3}}$$

$$= -\frac{7}{3} \pm \sqrt{\frac{49}{9} - \frac{45}{9}} = -\frac{7}{3} \pm \sqrt{\frac{4}{9}} = -\frac{7}{3} \pm \frac{2}{3} = \boxed{-\frac{5}{3}} \text{ or } \boxed{-3}$$

     a = 3
     b = 14
     c = 15

② $3x^2 - 14x + 15 = 0$

$$x = \frac{14}{6} \pm \sqrt{\frac{196}{36} - \frac{15}{3}}$$

$$= \frac{7}{3} \pm \sqrt{\frac{4}{9}} = \frac{7}{3} \pm \frac{2}{3} = \boxed{\frac{5}{3}} \text{ or } \boxed{3}$$

     a = 3
     b = − 14
     c = 15

③ $3x^2 + 4x - 15 = 0$

$$x = -\frac{2}{3} \pm \sqrt{\frac{4}{9} + \frac{15}{3}}$$

$$= -\frac{2}{3} \pm \sqrt{\frac{4}{9} + \frac{45}{9}} = -\frac{2}{3} \pm \sqrt{\frac{49}{9}} = -\frac{2}{3} \pm \frac{7}{3} = \boxed{-3} \text{ or } \boxed{+\frac{5}{3}}$$

     a = 3
     b = 4
     c = − 15

④ $3x^2 - 4x - 15 = 0$

$$x = \frac{2}{3} \pm \sqrt{\frac{4}{9} + \frac{15}{3}}$$

$$= \frac{2}{3} \pm \sqrt{\frac{49}{9}} = \frac{2}{3} \pm \frac{7}{3} = \boxed{-\frac{5}{3}} \text{ or } \boxed{+3}$$

     a = 3
     b = − 4
     c = − 15

continued

(5) $6x^2 - 5x = 6$

Rearrange: $6x^2 - 5x - 6 = 0$

$a = 6$
$b = -5$
$c = -6$

$x = \dfrac{5}{12} \pm \sqrt{\dfrac{25}{144} + \dfrac{6}{6}}$

$x = \dfrac{5}{12} \pm \sqrt{\dfrac{25 + 144}{144}}$

$\qquad\qquad\qquad\begin{array}{r} 13 \\ 13 \\ \hline 39 \\ 13 \\ \hline 169 \end{array}$

$= \dfrac{5}{12} \pm \sqrt{\dfrac{169}{144}}$

$= \dfrac{5}{12} \pm \dfrac{13}{12} = -\dfrac{8}{12} \text{ or } +\dfrac{18}{12} = \boxed{-\dfrac{2}{3}} \text{ or } \boxed{+\dfrac{3}{2}}$

(6) $x + \dfrac{1}{x} = 5\dfrac{1}{5}$

Multiply by x: $x^2 + 1 = \dfrac{26}{5}x$

Rearrange: $x^2 - \dfrac{26}{5}x + 1 = 0$

$a = 1$

$b = -\dfrac{26}{5}$

$c = 1$

$x = \dfrac{13}{5} \pm \sqrt{\dfrac{169}{25} - 1} = \dfrac{13}{5} \pm \sqrt{\dfrac{169 - 25}{25}}$

$= \dfrac{13}{5} \pm \sqrt{\dfrac{144}{25}} = \dfrac{13}{5} \pm \dfrac{12}{5} = \boxed{\dfrac{1}{5}} \text{ OR } \boxed{\dfrac{25}{5}} = 5$

(7) $x + \dfrac{1}{x} = 6$

Multiply by x: $x^2 + 1 = 6x$

Rearrange: $x^2 - 6x + 1 = 0$

$a = 1$
$b = -6$
$c = 1$

$x = 3 \pm \sqrt{9 - 1}$

$= 3 \pm \sqrt{8} = \boxed{3 - \sqrt{8}} \text{ or } \boxed{3 + \sqrt{8}}$

Checking results

Always check your results, by substituting back into the original equation or the original problem. In quadratics, you have two solutions to check.

(1) $\boxed{x = -\dfrac{5}{3}}$ $\quad 3x^2 = 3 \times \dfrac{25}{9} = \dfrac{25}{3}\quad 14x = -\dfrac{14 \times 5}{3} = -\dfrac{70}{3}$

$$3x^2 + 14x + 15 = \dfrac{25}{3} - \dfrac{70}{3} + 15 = -\dfrac{45}{3} + 15 = -15 + 15 = 0$$

$\boxed{x = -3}$ $\quad 3x^2 = 3 \times 9 = 27 \quad 14x = -14 \times 3 = -42$

$$3x^2 + 14x + 15 = 27 - 42 + 15 = 42 - 42 = 0 \quad \text{Both Check } ✔$$

(5) $\boxed{x = -\dfrac{2}{3}}$ $\quad 6x^2 = 6 \times \dfrac{4}{9} = 2 \times \dfrac{4}{3} = \dfrac{8}{3} \quad 5x = -\dfrac{10}{3}$

$$6x^2 - 5x = \dfrac{8}{3} - \left(-\dfrac{10}{3}\right) = \dfrac{8 + 10}{3} = 6$$

$\boxed{x = \dfrac{3}{2}}$ $\quad 6x^2 = 6 \times \dfrac{9}{4} = 3 \times \dfrac{9}{2} = \dfrac{27}{2} \quad 5x = \dfrac{15}{2}$

$$6x^2 - 5x = \dfrac{27}{2} - \dfrac{15}{2} = \dfrac{12}{2} = 6 \qquad \text{Both Check } ✔$$

(6) $\boxed{x = \dfrac{1}{5}}$ $\quad \dfrac{1}{x} = 5 \quad x + \dfrac{1}{x} = 5\dfrac{1}{5}$

$\boxed{x = 5}$ $\quad \dfrac{1}{x} = \dfrac{1}{5} \quad x + \dfrac{1}{x} = 5\dfrac{1}{5}$

$\left. \right\}$ Both Check ✔

(7) $\boxed{x = 3 - \sqrt{8}}$ $\quad \dfrac{1}{x} = \dfrac{1}{3 - \sqrt{8}} = \dfrac{3 + \sqrt{8}}{(3 - \sqrt{8})(3 + \sqrt{8})} = 3 + \sqrt{8}$

$$x + \dfrac{1}{x} = 3 - \sqrt{8} + 3 + \sqrt{8} = 6$$

$$\begin{array}{r} 3 + \sqrt{8} \\ 3 - \sqrt{8} \\ \hline -3\sqrt{8} - 8 \\ 9 + 3\sqrt{8} \\ \hline 9 \qquad -8 = 1 \end{array}$$

$\boxed{x = 3 + \sqrt{8}}$ $\quad \dfrac{1}{x} = \dfrac{1}{3 + \sqrt{8}} = \dfrac{3 - \sqrt{8}}{(3 + \sqrt{8})(3 - \sqrt{8})} = 3 - \sqrt{8}$

$$x + \dfrac{1}{x} = 3 + \sqrt{8} + 3 - \sqrt{8} = 6 \qquad \text{Both Check } ✔$$

A quadratic problem

Sometimes a problem leading to a quadratic equation seems to have only one real answer. Being realistic people, we naturally ask what the other answer means. If we don't, we should.

For example, a picture to be framed is twice as long as it is high. The frame provides a 3-inch margin around all sides. The total area, frame and picture, is 260 square inches. What are the dimensions of the picture?

Making x the height and $2x$ the length of the picture, the dimensions of the frame will be $x + 6$ and $2x + 6$, which multiply to $(x + 6)(2x + 6) = 2x^2 + 18x + 36$ as the area. The question lists the area as 260 square inches. So, the equation can be reduced to $x^2 + 9x - 112 = 0$. Solving it, you find the two answers that are characteristic of quadratics, 7 or -16.

The positive answer is easy. The picture is $7'' \times 14''$. The frame is $13'' \times 20''$, which multiplies to 260 square inches.

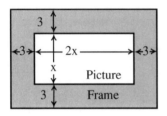

A PROBLEM What are the picture dimensions?

Total area = 260 square inches

$(x + 6)(2x + 6) = 260$
$2x^2 + 18x + 36 = 260$
$2x^2 + 18x - 224 = 0$
$x^2 + 9x - 112 = 0$

By formula $a = 1$
$b = 9$
$c = -112$

$$x = -\frac{9}{2} \pm \sqrt{\frac{81}{4} + 112}$$

$$= -\frac{9}{2} \pm \sqrt{\frac{81 + 448}{4}}$$

$$= -\frac{9}{2} \pm \sqrt{\frac{529}{4}} = -\frac{9}{2} \pm \frac{23}{2}$$

$$= 7 \text{ or} - 16$$

$$
\begin{array}{r}
23 \\
23 \\
\hline
69 \\
46 \\
\hline
529
\end{array}
$$

CHECK

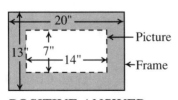

POSITIVE ANSWER

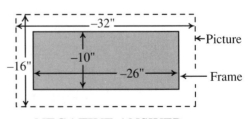

NEGATIVE ANSWER

Obviously a picture with negative dimensions has no practical meaning. However, it could have a mathematical meaning. The picture dimensions are $-16'' \times -32''$. By adding twice $3''$ each way, the frame dimensions are $-16 + 6'' \times -32 + 6''$, or $-10'' \times -26,''$ which also multiplies to $+260$ square inches.

Questions and problems

1. Find the factors and thus say what values of x make the following expressions equal to zero:

$x^2 + 7x - 8$ $x^2 - 8x + 7$

$3x^2 - 16x + 13$ $7x^2 + 50x + 7$

$7x^2 - 48x - 7$ $15x^2 - 34x + 15$

$30x^2 - 73x + 40$ $\dfrac{5}{7x^2} + \dfrac{24}{35x} - \dfrac{7}{5}$

2. Solve the following by completing the square:

$x^2 - 4x = 45$ $x^2 + 5x = 150$

$x^2 - 6 = x$ $x^2 + 39 = 8x$

$x^2 - 7x + 7 = 0$ $x^2 + 8x = 8$

$x^2 - 12x = 4$ $x^2 - 10x + 16 = 0$

3. Solve the following quadratics by formula:

$5x^2 - 2x - 7 = 0$ $2x^2 - 5x - 3 = 0$

$7x^2 - 4x - 3 = 0$ $x + \dfrac{1}{x} = \dfrac{82}{9}$

$x - \dfrac{1}{x} = \dfrac{24}{5}$ $5x^2 + 2x - 3 = 0$

$x + \dfrac{1}{x} = 10$ $x - \dfrac{1}{x} = 10$

4. Solve the following quadratics and explain anything unusual you observe about the solutions:

a. $5x^2 - 3x - 2 = 0$ b. $5x^2 - 3x = 0$ c. $5x^2 - 3x + \dfrac{9}{20} = 0$

5. A quantity is required, such that adding twice its reciprocal will produce a sum of 4. What is the quantity? Leave surds in your answer and check both results.

6. In a given number, the ones' digit is 2 more than the tens' digit; by reversing the digits and multiplying the two numbers together, you have 4725. What is the

original number? This answer can be simplified by dividing the resulting equation through by 11, twice. Explain why only one of the two solutions would normally be accepted.

7. An enclosure's length is 10 feet less than twice its width. Its area is 2800 square feet. Find its dimensions. Explain the negative answer as well as checking the positive answer.

8. Extending each side of a square area by 6 feet makes its area 4 times as big. Find the original side length and explain the negative answer.

9. Find three successive numbers whose sum is 3/8 the product of the lower two numbers. (HINT: take *x* as the middle number.) Explain the less obvious solution.

10. In mowing a lawn 60 × 80 feet, how wide of a strip around the edge must be mowed for half the grass to be cut? Explain the second answer.

11. A motorboat goes upstream 8 miles against a current of 15 miles per hour and returns with the current in a total time of 1 hour. What speed can the boat do in still water?

12. At a party someone tried to run a "think of a number" game and gave the instructions: think of a number, double it, subtract 22, multiply it by the number you first thought of, divide it by 2, add 70, and subtract the number you first thought of. The answer (he said) was 35. Only two people, who had used different numbers, had that answer. What two numbers did those two use?

13. A man has to buy $18 worth of a certain part. Since the purchase was authorized, the cost per part has risen by 4 cents. As a result, he gets 5 fewer parts for the money. What price per part did he pay?

14. The height of a small box is 1 inch less than its width and the length is 2 inches more than its width. If the total area of its sides is 108 square inches, what are its dimensions?

15. The negative solution to question 14 leads to another set of dimensions whose total surface area is also 108 square inches. What are these dimensions?

11

CHAPTER

Finding short cuts

Difference of squares is always sum time difference

You probably already noticed that when two factors have the same terms with only the sign between them changed, the resulting product is the difference between the terms, each squared. This statement, like any other in school algebra, can be reversed. You can factorize $a^2 - b^2$ into $(a + b)(a - b)$. Or you can multiply the same terms, $(a + b)(a - b)$, to yield $a^2 - b^2$.

This statement might seem unimportant, but it begins some surprising short cuts that you can make in calculating. Using a calculator doesn't tell you about them. If you ask people who "do it in their heads" quicker than you can hit your calculator's buttons, you will find they use such short cuts.

$$\boxed{a^2 - b^2} \; = \; \boxed{(a + b)(a - b)}$$

Example

$a = 9 \quad b = 6$

$a^2 = 81$
$b^2 = 36$

$a + b = 15$
$a - b = 3$

So, $81 - 36 = 3 \times 15$

Proof

$$
\begin{array}{r}
a - b \\
a + b \\
\hline
ab - b^2 \\
a^2 - ab \phantom{{}- b^2} \\
\hline
a^2 - b^2
\end{array}
$$

$$
\begin{array}{r}
81 \\
- 36 \\
\hline
45
\end{array}
\qquad
\begin{array}{r}
15 \\
\times 3 \\
\hline
45
\end{array}
$$

DIFFERENCE OF SQUARES
= SUM × DIFFERENCE

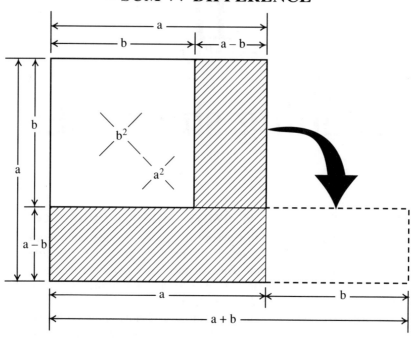

MULTIPLICATION
Short Cuts

Multiply 37×43:
$$37 = 40 - 3$$
$$43 = 40 + 3$$
$$\text{So } 37 \times 43 = (40 - 3)(40 + 3) = 40^2 - 3^2$$
$$= 1600 - 9$$
$$= 1591$$

Check by Long Multiplication

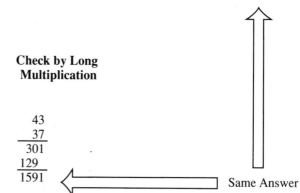

Same Answer

Sum and difference in geometry

The *sum and difference* principle can be seen in geometry, as well as in algebra. If you take the smaller square from the bigger square here, you are left with the shaded area. Cutting the upright piece off and laying it end-on to the other piece, the resulting rectangle has the dimensions: $a + b$ and $a - b$.

It's just a matter of spotting the easier way once you know the principle. Suppose you have to multiply 37 by 43. If you spot that 37 is $40 - 3$ and 43 is $40 + 3$, it's quicker to think, 40 squared is 1600, 3 squared is 9, so 37 times 43 is $1600 - 9$, which is 1591.

If you aren't sure, check it by long multiplication or with your calculator. It's always good to check, anyway. See examples on the facing page.

Difference of squares finds factors

It also provides short cuts in algebra. $x^4 + x^2 + 1$ has no odd powers of x, so you might think of x^2 as the variable and thus think of it as $(x^2)^2 + (x^2) + 1$. $x^2 + x + 1$ doesn't have factors (not simple ones, anyway), so you think that making x a square doesn't make it have factors, either.

However, if you notice that $(x + 1)^2$ is $x^2 + 2x + 1$, then $[(x^2) + 1]^2$ is $x^4 + 2x^2 + 1$. That is x^2 more than the expression you want factors for, and x^2 is also a square. So, the expression: $x^4 + x^2 + 1$ is the difference between two squares.

Written that way, the factors are: $[(x^2 + 1) + x]$ and $[(x^2 + 1) - x]$, which you can turn into a more conventional order: $(x^2 + x + 1)(x^2 - x + 1)$. Check the equation by multiplying out.

Factorize: $x^4 + x^2 + 1$

$x^4 + 2x^2 + 1 = (x^2 + 1)^2$

So:

$x^4 + x^2 + 1 = (x^2 + 1)^2 - x^2$

Factors are: $(x^2 + 1 + x)(x^2 + 1 - x)$

More usually written:

$$(x^2 + x + 1)(x^2 - x + 1)$$

$$\text{CHECK} \quad \begin{array}{r} x^2 - x + 1 \\ x^2 + x + 1 \\ \hline x^2 - x + 1 \\ + x^3 - x^2 + x \\ x^4 - x^3 + x^2 \\ \hline x^4 \qquad + x^2 \qquad + 1 \end{array}$$

One way to find a square root

Years ago, students learned a routine to find square roots that was not explained to them. Understanding it paves the way for understanding not only square roots, but a lot of other things that computers do.

Earlier in this book, I asked for the square root of 8. The square root of 9 is easy. First, look at squares and the square roots of easier numbers in arithmetic, algebra, and geometry, using numbers that "work out."

See how this works for a square root that is a 2-digit number (the square has 3 or 4 digits). Here the problem 37 squared is 1369 is worked through forward and backward.

You can use the same method to find square roots that have more than 2 digits. Here, a third digit is added to the root, and the square has 6 digits. Now, look at how math students used to set it—before calculators would do it for them.

The first thing they did was to mark off the digits in pairs from the right. Why do that? For example, the square root of 9, which is 1 digit, is 3; but the square root of 90 (2 digits) is more than 9^2 (81) and less than 10^2 (100). So, a 1-digit root can become a square that has 1 or 2 digits. That is why you mark the digits in pairs; to know whether the first digit of the root is taken from a number between 1 and 10 or one between 10 and 100.

Next, enter the first digit of the root, on top, and try for the second digit. If the first digit of the root is a, we subtract a^2 from the square line. What is left must be $2ab + b^2$. So, double a, leave space for b, try various values for b and multiply both a and b by b, which gives $2ab + b^2$ for that place. Subtract again.

Now, the first two digits of the root are a and the remainder is a new $2ab + b^2$. Go on like that through however many digits you need.

SQUARES

Arithmetic	**Algebra**	**Geometry**

Square of 32 **Square of (a + b)**

$$\begin{array}{r} 32 \\ 32 \\ \hline 64 \\ 96 \\ \hline \end{array}$$

3 × 3 in 100's
Twice 3 × 2 in 10's
2 × 2 in 1's

$$\begin{array}{r} a + b \\ a + b \\ \hline ab + b^2 \\ a^2 + \ ab \\ \hline a^2 + 2ab + b^2 \end{array}$$

Area = $a^2 + 2ab + b^2$

Root is 10a + b
Square is 100a² + 20ab + b²

FIND the SQUARE ROOT of 1369

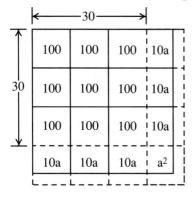

Tens: 30^2 is 900; 40^2 is 1600

Square root is between 30 and 40

2 times $30a + a^2$ is $1369 - 900 = 469$

2 times 30 is 60

$$
\begin{array}{r}
7 \\
\hline
60 \,|\, 469 \\
42 \\
\hline
49
\end{array}
$$

So $a = 7$
$a^2 = 49$

$$30^2 = 900$$
$$2 \times 30 \times 7 = 420$$
$$7^2 = 49$$
$$(30 + 7)^2 = 1369$$

The square root of 1369 is 37

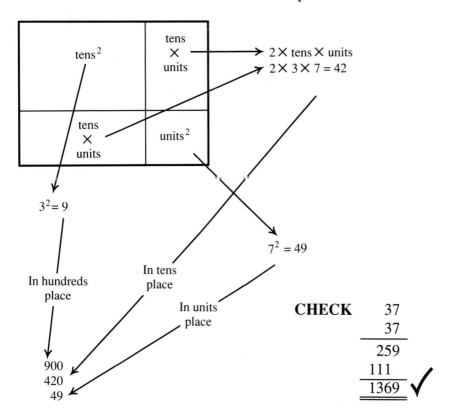

$3^2 = 9$

In hundreds place

$7^2 = 49$

In tens place

In units place

$2 \times$ tens $\times$ units
$2 \times 3 \times 7 = 42$

900
420
49

CHECK

$$
\begin{array}{r}
37 \\
37 \\
\hline
259 \\
111 \\
\hline
1369
\end{array} \checkmark
$$

Find the Square Root of 139,876

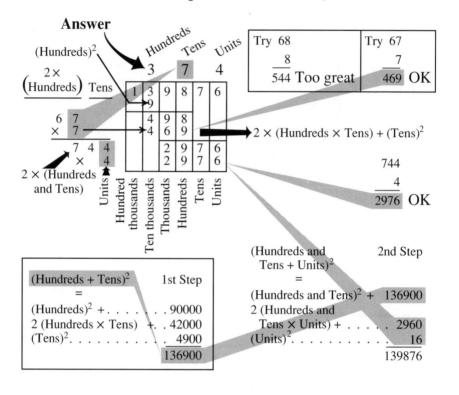

"CONTINUED" SQUARE ROOT

Find the Square Root of 2

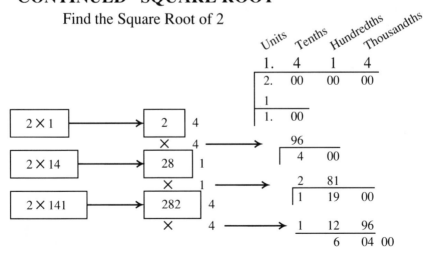

"Continued" square root

In the procedure where the square is a number of digits with no decimal fraction, you marked digits in pairs from the right. Where the square has a decimal fraction part or where the root might do so, mark places in the square from the decimal point each way. Where you have no decimal point, the right end of the number is where the decimal point would be. Where the root continues, perhaps indefinitely, you have a decimal part to the root.

The square root of 2 (a well-known number in mathematics) is found by this method. Later, other methods of finding, not only square roots, but other roots are shown. See bottom of page 178.

Importance of place in square root

It is always important to watch place. This has occurred before—in division, for example. But with square roots, using the wrong pairing can produce a wrong set of digits and a wrong decimal place. Remember, mark off the digits in pairs—each direction from the decimal point.

A similar, but much more complicated method, can find cube roots. Years ago (before my time) some schools taught that method. Now, when a calculator will probably do it anyway (a lot faster) knowing how to do it "by hand" is rather pointless. Being more complicated, it has more opportunities for making mistakes!

In SQUARE ROOT, watch "PLACE"!

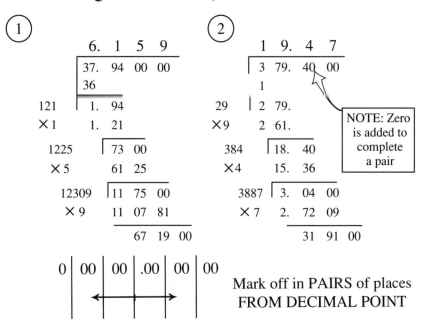

Mark off in PAIRS of places
FROM DECIMAL POINT

Importance of signs in successive roots

Looking at this pattern leads to discovering a whole new branch of mathematics. The square root of a positive number can be positive or negative. If you pursue a cube root, you'll find that the cube root of a positive number must be positive and the cube root of a negative number must be negative.

Now take the fourth root, which is the square root of the square root. Obviously the root can be either positive or negative, but something seems to be missing. The first square root can be either positive or negative. We can take the root of the positive root again, but is the negative root just left hanging sort of uselessly?

It begins to look as though you should have 2 square roots, 3 cube roots, 4 4th roots, and so on.

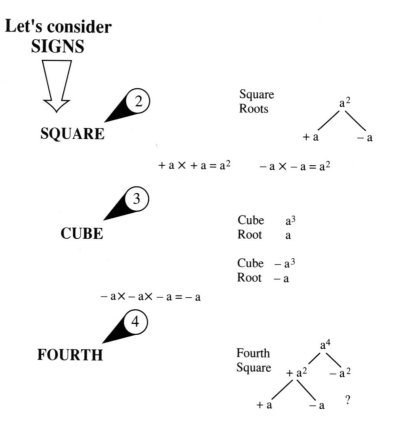

Imaginary numbers

Studying this problem led to a concept that is now called imaginary numbers. Once negative numbers weren't "allowed": they didn't represent real things. Then, it was found that they could be used in calculation to get valid answers.

The roots on the previous page are "real" roots—roots that mean something physical (even the negative ones).

So you can handle these imaginary numbers, but you need a sign to separate them from real numbers, just as the minus sign separates negative numbers from positive numbers. The symbol mathematicians picked to do this was the letter *i*. They write an *i* in front of the number. Thus, just as " − " "times" " − " is a positive, *i* times *i* is a negative.

Just as a negative times a negative is a positive, and 3 negatives multiplied together is another negative, *i* times *i* is a negative. 3 *i*'s multiplied together is −*i*; 4 *i*'s multiplied together is another positive again, and so on.

IMAGINARY NUMBERS

Real roots $\sqrt{+1} = +1$ or -1

Imaginary roots $\sqrt{-1} = +i$ or $-i$

$$i^2 = -1$$

Fourth Roots

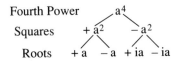

Fourth Power a^4

Squares $+a^2$ $-a^2$

Roots $+a$ $-a$ $+ia$ $-ia$

Cube Roots

Suppose $(a + ib)^3 = 1$

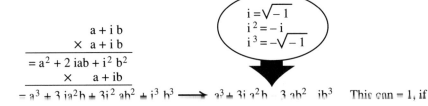

$$a + i b$$
$$\times \ a + i b$$
$$= a^2 + 2\,iab + i^2\,b^2$$
$$\times \quad a + ib$$
$$= a^3 + 3\,ia^2b + 3i^2\,ab^2 + i^3\,b^3 \longrightarrow a^3 + 3i\,a^2b - 3\,ab^2 - ib^3 \quad \text{This can} = 1, \text{ if}$$

$$i = \sqrt{-1}$$
$$i^2 = -i$$
$$i^3 = -\sqrt{-1}$$

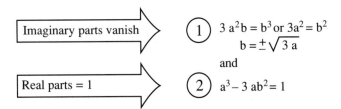

Imaginary parts vanish	⟶	① $3\,a^2b = b^3$ or $3a^2 = b^2$
		$b = \pm\sqrt{3}\,a$
		and
Real parts = 1	⟶	② $a^3 - 3\,ab^2 = 1$

Imaginary numbers find the other two cube roots

The pattern suggested that we should have 3 cube roots, but the old arithmetic only found one. We can approach finding the other two in several ways. First, try algebra. Let's look for the cube root of $+1$.

Write $(a + ib)^3 = 1$ and multiply it. Notice the sign changes (including from imaginary to real, or vice versa). This gives $a^3 + 3ia^{2b} - 3\,ab^2 - ib^3$. This must be 1, which has no imaginary part.

The real part is $a^3 - 3ab^2 = 1$ because the equation can only be true if the imaginary parts "disappear." So, the imaginary part must satisfy another equation: $3a^{2b} - b^3 = 0$. That can be true if $b = 0$, which makes the real part what we are familiar with: $a^3 = 1$ ($a = 1$). Taking ib out as a factor, the equation can also be zero if: $b^2 = 3a^2$. Now substitute this piece into the real equation. This equation is: $a^3 - 9a^3 = 1$, which turns to $8a^3 = -1$. So $a = -1/2$. So, b^2 is 3 times a^2 ($-3/4$). Then, b is the root of $-\,-3/4$:

$$\frac{\pm i \sqrt{3}}{2}$$

You can take the square root of the 4, but write root 3 with a surd.

A "real" number, as mathematicians call them, does not have an i, but an imaginary number does. A number that has both parts is called a *complex number*. Imaginary numbers and complex numbers appear in a later chapter. In this chapter, it's enough to realize they exist.

IMAGINARY NUMBERS Three cube roots

Substitute for

(1) $b^2 = 3a^2$ (2) $a^3 - 3ab^2 = 1$

$a^3 - 9a^3 = 1$

$-8a^3 = 1$

$a^3 = -\dfrac{1}{8}$

$a = -\dfrac{1}{2}$ $b = \pm\dfrac{\sqrt{3}}{2}$

$ib = \pm\dfrac{\sqrt{-3}}{2}$

Cube

Roots $-\dfrac{1}{2} + \dfrac{\sqrt{-3}}{2}$ $-\dfrac{1}{2} - \dfrac{\sqrt{-3}}{2}$

Check $\left(-\dfrac{1}{2} + \dfrac{\sqrt{-3}}{2}\right)^3 = -\dfrac{1}{8} + \dfrac{3}{8}\sqrt{-3} + \dfrac{9}{8} - \dfrac{3}{8}\sqrt{-3}$

$= \dfrac{9}{8} - \dfrac{1}{8} = 1$

$\left(-\dfrac{1}{2} - \dfrac{\sqrt{-3}}{2}\right)^3 = -\dfrac{1}{8} - \dfrac{3}{8}\sqrt{-3} + \dfrac{9}{8} + \dfrac{3}{8}\sqrt{-3}$

$= \dfrac{9}{8} - \dfrac{1}{8} = 1$

Simultaneous quadratics

Sometimes a problem with two variables has two possible solutions that are given by quadratic equations. Suppose a rectangular area is enclosed on only 3 sides. Those 3 sides require 20 feet of fencing to enclose an area of 48 square feet. What are the dimensions of the fence?

From this problem, you get the equations shown, using dimensions a and b: a for the two sides and b for the one side.

Equation 1: $2a + b = 20$ Equation 2: $ab = 48$

This problem looks difficult, compared with what you have done before. One way to eliminate work here is to square equation 1. $(2a + b)^2 = 400$, multiplies to:

$$4a^2 + 4ab + b^2 = 400$$

By subtracting 8 times equation 2: $8ab = 384$, you have:

$$4a^2 - 4ab + b^2 = 16.$$

Now, taking the square root, you get $2a - b = 4$, but that answer can be ± 4. This equation leads to 2 sets of answers. The square root of $4a^2 + 4ab + b^2 = 400$ can only be $+20$, because that was in our original equation. Now, you have two sets of answers: 8 by 6 or 12 by 4.

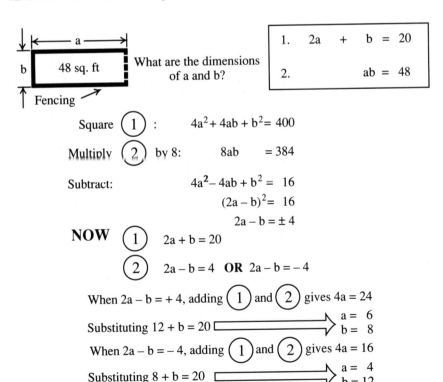

What are the dimensions of a and b?

1. $2a + b = 20$

2. $ab = 48$

Square ①: $4a^2 + 4ab + b^2 = 400$

Multiply ② by 8: $8ab = 384$

Subtract: $4a^2 - 4ab + b^2 = 16$

$$(2a - b)^2 = 16$$

$$2a - b = \pm 4$$

NOW ① $2a + b = 20$

② $2a - b = 4$ **OR** $2a - b = -4$

When $2a - b = +4$, adding ① and ② gives $4a = 24$
a = 6
Substituting $12 + b = 20$ b = 8

When $2a - b = -4$, adding ① and ② gives $4a = 16$
a = 4
Substituting $8 + b = 20$ b = 12

Always check!

I can never stress checking enough. Again, don't use the answer book; it doesn't check your work. Go back over your work yourself. Verify it by using the two sets of answers in the original problem.

Two sides 6 feet and one side 8 feet add up to 20 feet of fencing, which enclosed 48 square feet. Two sides 4 feet and one side 12 feet also add up to 20 feet of fencing and it also encloses 48 square feet. This time, because the problem is not symmetrical, you find 2 answers that are not reversals of one another.

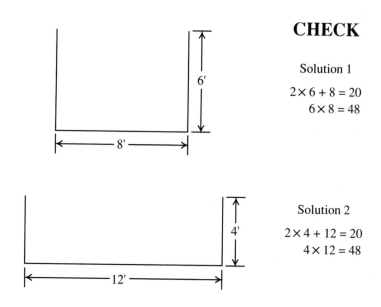

CHECK

Solution 1

$$2 \times 6 + 8 = 20$$
$$6 \times 8 = 48$$

Solution 2

$$2 \times 4 + 12 = 20$$
$$4 \times 12 = 48$$

Questions and problems

1. Using the difference-of-squares method, find the following products:

98 × 102	63 × 77
395 × 405	85 × 95
76 × 84	117 × 123
190 × 210	193 × 207
28 × 32	49 × 51

Use long multiplication (or a calculator) to check your results.

2. Use the difference of squares method to find the factors of:

$x^4 + 3x^2 + 4$	$x^4 - 5x^2 + 4$
$x^4 + 4$	$4x^4 + 4^2 + 25$
$9x^4 - x^2 + 16$	$x^4 + x^2 + 169$

3. Find the square root of the following numbers:

17,956	179,776
200,704	20,164
45,796	456,976
8,010.25	9,920.16
127,449	12,769

4. Find the square root of the following numbers (correct to three decimal places:

3	30
5	50
6	60
7	70
8	80

5. Find three cube roots of (*a*) + 8, (*b*) − 8. Check your results by multiplying each root to find its cube.

6. What are the four possible fourth roots of 16? How many are real, imaginary, or complex?

7. Solve the following simultaneous equations:

$x + y = 20$	$x - y = 5$
$xy = 96$	$x^2 + y^2 = 53$
$3x + y = 34$	$x - y = 6$
$xy = 63$	$x^2 + y^2 = 26$
$x + y = 40$	$7x - 4y = 9$
$x^2 + y^2 = 808$	$xy = 70$
$3x - y = 26$	$x^2 + 3y^2 = 40$
$x^2 + y^2 = 170$	$\sqrt{3}x - y = 0$

Each of the pairs have a pair of solutions; if the method is not obvious, refer back to the end of chapter 9 for hints. Check each pair of solutions in the original equations.

8. A cubic container has a volume of 480 cubic inches and a surface area (all six rectangular sides added together) of 376 square inches. It is 6 inches high. Find the other two dimensions.

9. A rectangular area has to be enclosed with fencing. It is known to have an area of 10 acres (435,600 square feet). Assuming that the area is square, with 660 feet on each side, 2640 feet of fencing are bought. Since the area is rectangular, the

fencing covers only 3 sides and exactly half the fourth side. What are the dimensions of the area?

10. Suppose, in the previous question that the fencing was 110 feet short of completing the enclosure. What are the dimensions? Why do the alternative answers in question 9 differ, but for this question, the dimensions are the same in opposite order?

11. As a variant on the "think-of-a-number" game, try this one: ask your audience to think of a number between zero and 10. First, multiply it by 8 and remember the answer. Now, add 3 to the original number and multiply by 2; add 10 and multiply it by 2 again. Add 8 and subtract the number that you remembered previously. Finally, divide the answer by 4. Ask each person what answer he or she has. Subtract the number from 10, and tell him that the remainder is the number he first thought of. Show, by algebra, why this system works.

12. Two numbers multiplied together are 432; one divided by the other leaves a quotient of 3. What are the numbers?

12
CHAPTER

Mechanical mathematics

Relationship between force and work

Force is a measure of "push." A truck with a heavy load begins to move steadily when it is pushed. If the truck has a lighter load, the same "push" will make it go much faster. See the illustration on page 188.

Alternatively, more push can move the heavier load more quickly, too.

Measure of force

Force is needed to start and stop movement. However, little force is needed to maintain movement if the "bearings are well oiled" to offset friction. With no friction, movement would continue unchanged indefinitely, until some force would change it. For simplicity, assume that no friction exists; force is needed only to start, stop, or change motion. See the example at the top of page 189.

From observation, force is proportional to: the weight to be moved, and to the acceleration (rate at which motion increases) or the deceleration (rate at which motion decreases).

One unit for measuring force is the *poundal*. It is the force needed to accelerate (or decelerate) a weight (usually called *mass*) of 1 pound so that it changes its velocity by 1 foot per second every second.

The metric unit of force is a *dyne*. 1 dyne accelerates a mass of 1 gram 1 centimeter per second per second.

Whatever units you measure force with, they are mass times acceleration—mass multiplied by distance, divided by time squared. That is what "per second per second" means (it isn't a misprint!). You will better understand this concept later.

FORCE and WORK

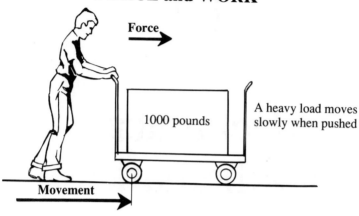

A heavy load moves
slowly when pushed

1000 pounds

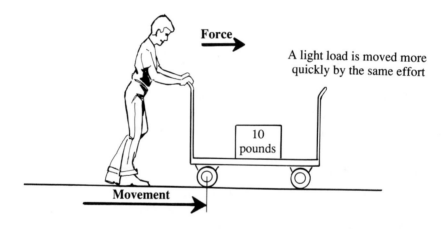

A light load is moved more
quickly by the same effort

10 pounds

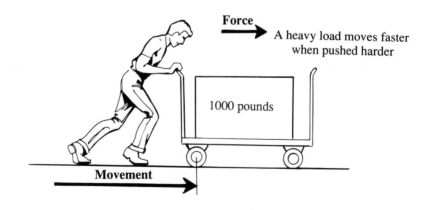

A heavy load moves faster
when pushed harder

1000 pounds

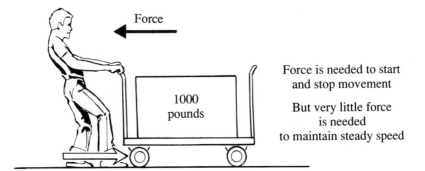

Force is needed to start
and stop movement

But very little force
is needed
to maintain steady speed

Force is proportional to: ① weight moved

② acceleration
(or deceleration)

Poundal is force that will accelerate

1 pound by a velocity of
1 foot per second, in every
1 second

Dimension of Force

• In poundal units: $\dfrac{\text{pounds} \times \text{feet}}{\text{seconds}^2}$

$\dfrac{\text{Mass} \times \text{length (or distance)}}{\text{time}^2}$

• In metric units: Dyne is $\dfrac{\text{grams} \times \text{centimeters}}{\text{seconds}^2}$

Speed and distance

That "per second per second" sounds confusing at first—as if I repeated myself by mistake. Look at it in a way that avoids this repetition.

Assume that you travel by car at 40 miles per hour. For the next minute, you accelerate steadily so that 1 minute later, you are moving at 60 miles per hour. This acceleration is 20 miles per hour per minute. Although exact repetition is avoided, "per hour, per minute" is still used. These measures are two different units of time. "Per hour per minute" is not a standard unit of acceleration, but it helps to understand the principle by not using confusing repetition.

If acceleration is steady, the average speed during that minute will be midway between the start and finish speeds of 50 miles per hour. The distance travelled during that minute will be the same as if this average speed had been used for the whole minute: 5/6 mile.

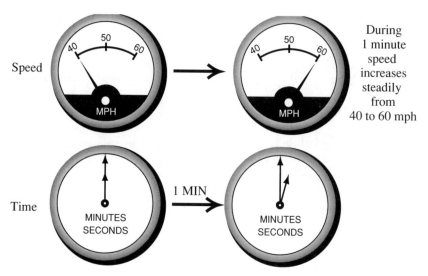

During 1 minute speed increases steadily from 40 to 60 mph

SPEED and DISTANCE

Average speed for 1 minute is $\dfrac{40 + 60}{2} + 50$ mph

1 minute is $\dfrac{1}{60}$ hour

So distance is $50 \times \dfrac{1}{60} = \dfrac{5}{6}$ mile

Acceleration was 20 miles per hour per minute

How Far do You Travel

Acceleration and distance

A standard unit of acceleration is feet per second per second (ft/sec^2). Assume that a steady acceleration of 10 ft/sec^2 from a standstill is used.

At the start, you are not moving. After 1 second, you will have accelerated by 10 ft/sec^2 to a velocity of 10 ft/sec. The average velocity for the first second will be 5 ft/sec, half way between 0 and 10. So, you will travel 5 feet during this first second.

During the next second, the velocity will change from 10 ft/sec to 20 ft/sec, an average of 15 ft/sec. Thus, you will travel 15 feet in the 2nd second, a total of 20 feet from our start. The average over 2 seconds is 10 ft/sec, which will take you 20 feet in 2 seconds—the same result.

During the 3rd second, the speed increases from 20 ft/sec to 30 ft/sec, an average of 25 ft/sec, to travel 25 feet, a total of 45 feet. Over the 3 seconds, the average is 15 ft/sec (45 feet).

You can tabulate distances travelled for any number of seconds from the start. If we plot the result as a graph, the resulting curve is a quadratic.

Acceleration in feet per second per second (ft/sec^2)

Assume acceleration is 10 ft/sec^2

Number of second from start	Speed			Total distance from start
	At beginning of second	At end of second	Average for second	
	Feet per second			
			Also distance in feet	
1st	0	10	5	5
2nd	10	20	15	20
3rd	20	30	25	45
4th	30	40	35	80
5th	40	50	45	125

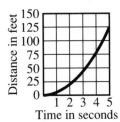

Curve is Quadratic

ACCELERATION and DISTANCE

	General Formula	Specific Example
Total time from start	t	5 seconds
Speed at start	0	0
Speed at end of time	at	50 feet/second
Average speed	$\frac{1}{2}at$	25 feet/second
Total distance from start	$\frac{1}{2}at^2$	125 feet

Note on Dimension	Algebraic Symbol		Quantities	Units
Acceleration	a	is	$\dfrac{\text{length (distance)}}{\text{time}^2}$	$\dfrac{\text{feet}}{\text{second}^2}$
Speed (Velocity)	at or v	is	$\dfrac{\text{length (distance)}}{\text{time}}$	$\dfrac{\text{feet}}{\text{second}}$
Distance	$\frac{1}{2}at^2$ or d	is	length (distance)	feet

Mathematics develops formulas to give the answer directly, from any particular set of facts. If t is the total time from the start (when speed is 0, a standstill), the speed at the end of this time is at (a times t) feet per second (ft/sec). So, the average speed between a start of 0 and a finish of at ft/sec is 1/2at ft/sec. Since the time is t seconds, the distance travelled from the start is 1/2at^2 feet.

The results will agree with those worked out step by step on the previous page if you use $a = 10$. Notice that acceleration uses units of length (distance) per time per time, such as feet per second per second. Thus, acceleration has the dimensions of length/time^2, or in units: feet/second^2.

Velocity results from multiplying acceleration by time, and thus it is length/time (length divided by time once), in units, such as feet/second or miles/hour. Distance covered, measured in feet or miles, is obtained by multiplying velocity by time.

Force and work

Force is a measure of push or pull, as stated earlier. *Work* (as a mathematical term) is a measure of what is done by that push or pull. If nothing moves, as when you lean against something, the force is there, but no work is performed. Work results when an applied force causes movement.

Force is mass times acceleration. Work is proportional to applied force and distance moved. Work is force times distance, so it must be mass times acceleration times distance. Distance moved, using constant acceleration from a standstill, is 1/2at^2. So, work is mass times acceleration times 1/2at^2. Since velocity is acceleration multiplied by time, this can be simplified to 1/2mv^2.

FORCE is mass × acceleration

Symbols	Quantities	Units
$f = ma$	$\dfrac{\text{mass} \times \text{length}}{\text{time}^2}$	$\dfrac{\text{pounds} \times \text{feet}}{\text{seconds}^2}$ or **poundals**
$= \dfrac{mv}{t}$		$\dfrac{\text{grams} \times \text{centimeters}}{\text{seconds}^2}$ or **dynes**

WORK is the result of FORCE applied for DISTANCE

$w = mad$ $\dfrac{\text{mass} \times \text{length}^2}{\text{time}^2}$ $\dfrac{\text{pounds} \times \text{feet}^2}{\text{seconds}^2}$ or **foot-poundals**

$d = \dfrac{1}{2}\,at^2$, so $w = \dfrac{1}{2}\,ma^2\,t^2$

$\dfrac{\text{grams} \times \text{centimeters}^2}{\text{seconds}^2}$ or **ergs**

$a = \dfrac{v}{t}$, or $v = at$,

so $w = \dfrac{1}{2}\,mv^2$

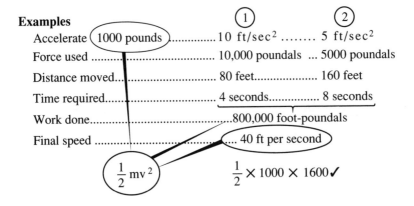

Examples

	①	②
Accelerate (1000 pounds)...............	10 ft/sec²	5 ft/sec²
Force used	10,000 poundals ...	5000 poundals
Distance moved............................	80 feet.................	160 feet
Time required.............................	4 seconds...............	8 seconds
Work done.................................	...800,000 foot-poundals	
Final speed	.... 40 ft per second	

$$\frac{1}{2}\,mv^2 \qquad \frac{1}{2} \times 1000 \times 1600 \checkmark$$

Units of work are the *foot-poundal*, (the work done by a force of 1 poundal moving through 1 foot), and the *erg*, the work done by 1 dyne moving through 1 centimeter.

The formula $1/2mv^2$ represents the work of bringing a specific weight to a certain velocity, regardless of the acceleration. If the acceleration is 10 ft/sec², 80 feet and 4 seconds are required to reach a velocity of 40 ft/sec. If the acceleration is 5 ft/sec², it requires 8 seconds and 160 feet. Either way, 800,000 foot-poundals will move 1000 pounds from standstill to 40 ft/sec, although time and distance differ. Work is given directly by the formula $1/2mv^2$. You do not have to know time or distance—only the mass and final speed.

Work and energy

Work and energy use the same units, because they are the same at different times. For example, an archer pulls back the string of his bow. The string is pulled back by a force that is equal to (or a little greater than) the tension of the string. The archer's energy transfers to the bow string as work. The amount of work that is needed to pull the bow string back is stored in the bow as energy, which ultimately sends the arrow on its flight.

When the archer releases the arrow, the string's thrust accelerates it to flight velocity. This work transfers the energy of the drawn bow to the energy of the arrow in flight.

Energy is a capacity for doing work and, conversely, work is the transfer of energy, from one form or place to another. So, both use the same units: foot-poundals or ergs, according to the system of units employed.

When the energy is in the form of a mass in motion, the appropriate formula is $1/2mv^2$. This formula can be used for the work needed to attain this motion or for the energy stored by it.

WORK and ENERGY

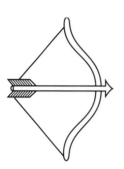

ENERGY	**WORK**
is a capacity for doing work	is the transfer of ENERGY from one form to another

Units are the same for both

foot-poundals ———————— OR ———————— ergs

$$\frac{1}{2}\,mv^2$$

Energy and power

Power is a rate of doing work, or of transferring energy from one form to another. As work is force applied over a distance, power is force applied over a distance within a specified time.

You already know the units used, but an example will illustrate the relation between power and the other quantities. Assume the question relates to horsepower vs. the weight of a car.

For example, one motor unit develops a power of 290,000 foot-poundals per second, and another has twice the power, 580,000 foot-poundals per second. Coupled with these powers, different weights must be moved. One is 1500 pounds, the other is 3000 pounds.

Energy is measured in the form $1/2\,mv^2$. So, power must be in the form: $1/2\,mv^{2/t}$ or $mv^{2/2t}$. Transposing this, using the symbol p for power, the time for a mass to reach a given velocity is: $t = mv^{2/2p}$. From this formula you can find the time taken by:

1. The smaller power unit with the smaller weight
2. The smaller power unit with the greater weight
3. The greater power with the greater weight

POWER is a rate of doing WORK
  or of transferring ENERGY

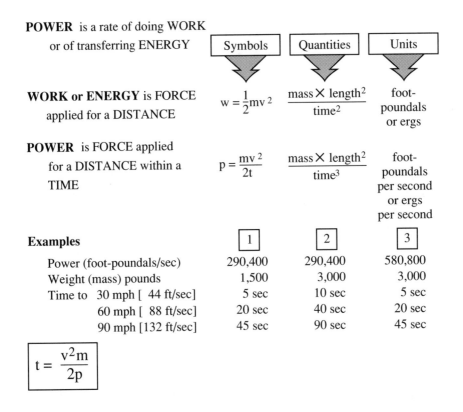

	Symbols	Quantities	Units
WORK or ENERGY is FORCE applied for a DISTANCE	$w = \frac{1}{2}mv^2$	$\dfrac{mass \times length^2}{time^2}$	foot-poundals or ergs
POWER is FORCE applied for a DISTANCE within a TIME	$p = \dfrac{mv^2}{2t}$	$\dfrac{mass \times length^2}{time^3}$	foot-poundals per second or ergs per second

Examples

	1	2	3
Power (foot-poundals/sec)	290,400	290,400	580,800
Weight (mass) pounds	1,500	3,000	3,000
Time to 30 mph [44 ft/sec]	5 sec	10 sec	5 sec
60 mph [88 ft/sec]	20 sec	40 sec	20 sec
90 mph [132 ft/sec]	45 sec	90 sec	45 sec

$$t = \frac{v^2 m}{2p}$$

If you want to, you can complete the set by taking the greater power with the smaller weight!

Tabulate the time needed in each case to reach 44, 88, and 132 ft/sec, which are the speeds that correspond to 30, 60, and 90 miles/hour. Notice that the time needed is related to the square of the speed to be reached. At constant acceleration, speed is proportional to time. At constant power, acceleration must diminish as speed increases.

Gravity as a source of energy because of position

To keep the units basic (1 pound, 1 ft/sec² etc.), gravity has been left out of force, work, and power. This force can be realized only by working along a level road or surface horizontally. The constant vertical force of gravity that acts around us, however, provides a convenient means of storing and concentrating energy.

A pile driver illustrates this principle. First, work is done by lifting a weight against the force of gravity. The weight is not accelerated upwards, but it is lifted steadily against a constant force—gravity pulling downwards. Just as energy or work is force times distance, this energy takes the form of distance lifted times weight.

GRAVITY PROVIDES SOURCE
of ENERGY
due to POSITION

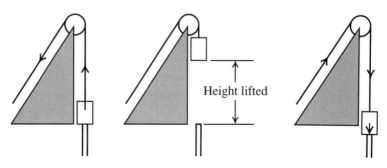

Height lifted

Work lifting weight stores energy

Loosing weight,
gravity speeds weight down.
Energy drives the pile

Moving twice the distance requires twice the work, stores twice the energy. Velocity is not involved . . . yet. Less power requires more time to do the same amount of work.

When the weight reaches the top, it is released to drop on the pile. Gravity is a mutual pull between earth and any mass. Doubling the mass doubles the pull (weight). So, in freefall, any object will drop at the same acceleration—approximately 32 ft/sec^2.

As the object accelerates downwards, it stores energy at the rate of $1/2\,mv^2$, as a result of its motion. When it hits the pile, this stored energy is concentrated for a very short time, thrusting the pile downward. The momentary force is many times the weight caused by the steady force of gravity on that mass. In metric units, the force of gravity is equivalent to an acceleration by a force of about 981 dynes per gram.

Weight as force

From that illustration, you can see that weight provides a steady force acting downwards as a result of the gravitational pull between the earth and any mass. The force is found from the mass on which gravity acts, multiplied by the acceleration of gravity, which produces 32 feet/sec^2 or 981 centimeters/sec^2. The force that is needed to prevent the weight from falling is equal to the pull of gravity on the weight.

As the pull of gravity accelerates the weight downwards at 32 ft/sec^2, the force of gravity on a mass of 1 pound must be 32 poundals. On a 2-pound weight, the force will be 64 poundals, and so on.

WEIGHT is a STEADY FORCE
acting DOWNWARDS

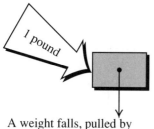 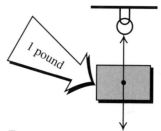

A weight falls, pulled by
gravity, with an acceleration
of 32 feet per second per second

Force needed to prevent weight
falling is equal to gravity's pull
on the weight

On a 1-pound weight, the force of gravity is 32 poundals

On a 2-pound weight, the force of gravity is 64 poundals

A pound, the gravitational unit of force,

is equal to 32 poundals

Thus, a mass of 1 pound provides a gravitational force of 32 poundals. Otherwise stated, when you use gravity on a mass of 1 pound, it becomes a 1-pound weight, exerting a force of 1 pound (32 poundals in absolute nongravitational units).

For this reason, in basic or absolute force units, the *mass* is 1 pound. However, in gravitational units, the *weight* is 1 pound.

Gravitational measure of work

In gravitational measure of work, force does not have to accelerate a mass. Gravity exerts a force continuously on everything, pulling it downwards. If something doesn't fall downwards, it's because an equal force supports it, pushing it up.

If a 10-pound weight (gravity acting on a 10-pound mass) rests on the floor, it presses on the floor with a weight (force) of 10 pounds. Correspondingly, the floor pushes upward against the weight with a force of 10 pounds to prevent it from falling.

Does the floor change its upward force according to what is on it? Yes. If you hold the 10-pound weight, your feet press on the floor, and the floor presses on your feet with a force that is 10 pounds more than just you standing on the floor.

All the time that these forces balance, they are forces in equilibrium. If the floor cannot provide that much upward push, it collapses, and work (although probably destructive) is done. See illustration at top of next page.

Gravitational Measure of Work
FOOT-POUNDS

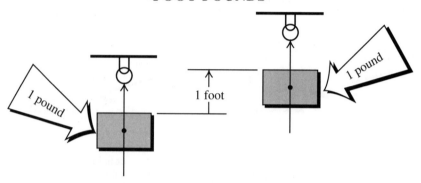

To lift a weight against the pull of gravity requires work.
One pound requires a force of one pound.
To lift it 1 foot requires 1 foot-pound of work.

A foot-pound is = **the gravitational unit of work**

1 foot-pound = 32 foot-poundals

HOW MUCH ENERGY is NEEDED
to MAINTAIN CONSTANT ACCELERATION?

Time

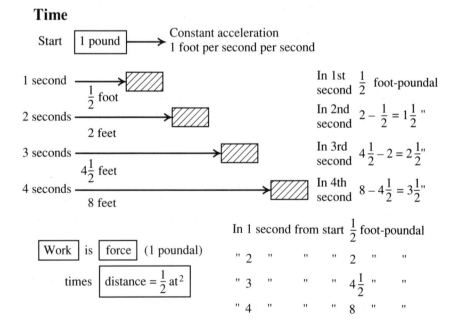

Energy for constant acceleration

It is simplest to assume that acceleration is constant, which means that velocity increases at a uniform rate, such as 1 ft/sec^2. This strategy is convenient because as well as being a steady growth in velocity, acceleration represents a steady force. However, it does not correspond with a constant rate of work, transfer of energy, or power. The faster an object goes, the more power is needed to maintain the given force.

Work is force times distance. So, maintaining the same force at higher speeds requires (or produces, depending on viewpoint or situation) more work, energy, or power. As shown in the example at the bottom of the facing page.

Kinetic energy and velocity

At constant acceleration, such as when a weight falls by the pull of gravity, energy builds in proportion to time squared. This rule occurs because energy is proportional to velocity squared.

ENERGY is PROPORTIONAL
to VELOCITY SQUARED

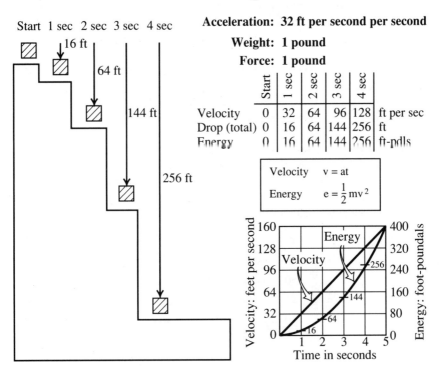

	Start	1 sec	2 sec	3 sec	4 sec	
Velocity	0	32	64	96	128	ft per sec
Drop (total)	0	16	64	144	256	ft
Energy	0	16	64	144	256	ft-pdls

Velocity $v = at$

Energy $e = \frac{1}{2}mv^2$

Viewed another way, energy is proportional to force times distance. But as distance, at constant acceleration, increases in proportion to time squared, energy and distance both increase as time squared. Constant acceleration means velocity grows in direct proportion to time.

Thus, the two ways of referring to energy conform: kinetic energy is energy caused by motion; it is proportional to velocity squared. Potential energy is energy caused by position; it is proportional to force and distance moved against that force.

Potential energy must be built from movement. As long as start and finish are both the same or at some constant velocity, movement is not involved in the calculation, as it is with kinetic energy.

In the pile driver, for instance, a little more force is needed to start its upward movement. While it ascends at a constant rate, force and movement are both constant. A little less force is used to reach the top, if it stops before being released. The overall work needed to lift it is weight times height lifted, in foot-pounds.

Acceleration at constant power

The rate of work (power) at constant acceleration, increases with velocity, requiring progressively more power during acceleration. For many purposes, acceleration at constant power is closer to what happens. By rearranging the formula that relates kinetic energy and power:

$$1/2\,mv^2 = pt$$

So if power p and mass m are both constant, velocity must increase with the square root of time.

Assume that constant power enables the accelerated mass to reach 100 ft/sec in 20 seconds. You can calculate the velocity at any time during the 20 seconds. It has been calculated here for 4-second intervals. You can do it without knowing the mass or power involved.

For the full 20 seconds, velocity reaches 100 ft/sec. 100 squared is 10,000. Divide this number in proportion to time: 2000 for 4 seconds, 4000 for 8 seconds, 6000 for 12 seconds, and 8000 for 16 seconds. Then, take the square root to find the velocity at each of these times in ft/sec.

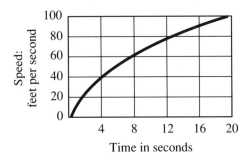

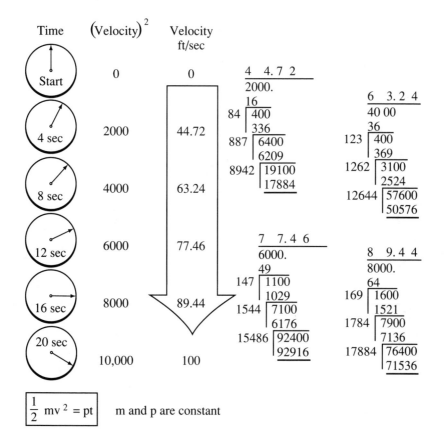

Time	$(\text{Velocity})^2$	Velocity ft/sec
Start	0	0
4 sec	2000	44.72
8 sec	4000	63.24
12 sec	6000	77.46
16 sec	8000	89.44
20 sec	10,000	100

$$\boxed{\dfrac{1}{2}\,mv^2 = pt}$$ m and p are constant

Notice that acceleration is much faster at the beginning. Then, as velocity builds, acceleration drops. That particular figure was not included, but half the final speed is reached in only a quarter of the time. The faster an object goes, the slower its speed increases.

A stressed spring stores energy

Another way to store potential energy is with a spring, rather similar to the archer's bow from earlier in this chapter. First, assume that the spring supports only the weight that is attached to it. This weight will figure in the energy interchanges discussed in the following pages.

Now, progressively apply more force (than the 1 pound) to compress the spring. To compress it 3 inches requires an additional 2 pounds of force. A 6-inch compression requires 4 pounds, 9 inches, an 8-pound extra force.

The force that is applied to compress this spring by 1 foot, uniformly grows from 0 at the start to 8 pounds at the finish. So, the average force over the 1 foot of compression must be 4 pounds. Thus, the energy stored in the spring, when it is compressed, is 4 foot-pounds (4 × 32 = 128 foot-poundals). The energy will remain stored as long as the 8-pound force holds the spring compressed.

A Stressed Spring Stores Energy

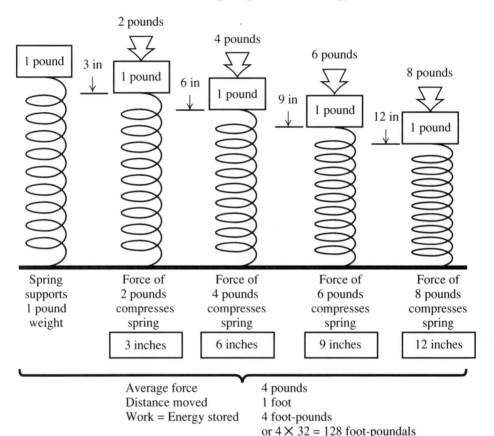

Spring supports 1 pound weight	Force of 2 pounds compresses spring	Force of 4 pounds compresses spring	Force of 6 pounds compresses spring	Force of 8 pounds compresses spring
	3 inches	6 inches	9 inches	12 inches

Average force   4 pounds
Distance moved   1 foot
Work = Energy stored   4 foot-pounds
                                    or 4 × 32 = 128 foot-poundals

Spring transfers energy

Assume that the force holding the spring compressed is suddenly released. The spring starts to accelerate the 1-pound weight upward with a force of 8 pounds (which was just removed). As it goes upward, the accelerating force will diminish, but the velocity will continue to grow because the accelerating force disappears only when the extra 1-foot compression has all been decompressed. To find what happens to velocity, remember that both methods of considering energy must always agree. The total energy doesn't change.

When the spring is half decompressed (to 6 inches), the force has dropped to 4 pounds (128 poundals). The average force represented in this compression is 2 pounds, (64 poundals) and the distance over which this average force is compressed is 6 inches (1/2 foot). So, the remaining potential energy is 64 × 1/2 = 32 foot-poundals. Of the original 128 foot-poundals, 96 must have been turned into kinetic energy. This equation must be $1/2\,mv^2$. m is 1 pound, so v^2 must be 2 × 96 = 192. So, v is the square root of 192 (13.856 ft/sec). When the spring is fully decompressed, all the energy is kinetic, so now $v^2 = 256$ ($v = 16$ ft/sec).

TRANSFER of ENERGY

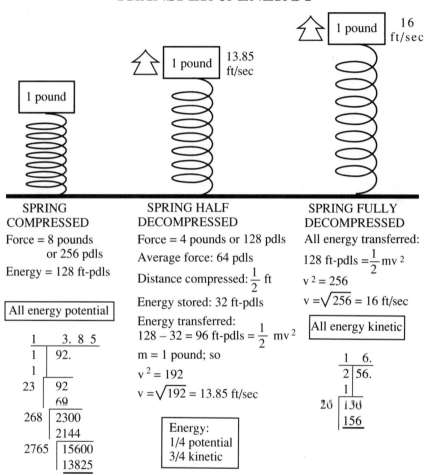

SPRING COMPRESSED	SPRING HALF DECOMPRESSED	SPRING FULLY DECOMPRESSED
Force = 8 pounds or 256 pdls	Force = 4 pounds or 128 pdls	All energy transferred:
Energy = 128 ft-pdls	Average force: 64 pdls	$128 \text{ ft-pdls} = \frac{1}{2}mv^2$
	Distance compressed: $\frac{1}{2}$ ft	$v^2 = 256$
	Energy stored: 32 ft-pdls	$v = \sqrt{256} = 16$ ft/sec

SPRING COMPRESSED

All energy potential

```
   1    3. 8 5
   1  | 92.
   1  |
  23  | 92
      | 69
 268  | 2300
      | 2144
2765  | 15600
      | 13825
```

SPRING HALF DECOMPRESSED

Energy transferred:
$128 - 32 = 96 \text{ ft-pdls} = \frac{1}{2}mv^2$
m = 1 pound; so
$v^2 = 192$
$v = \sqrt{192} = 13.85$ ft/sec

Energy:
1/4 potential
3/4 kinetic

SPRING FULLY DECOMPRESSED

All energy kinetic

```
       1   6.
       2 | 56.
       1 |
  26  | 130
      | 156
```

Resonance cycle

That transfer of energy from potential to kinetic in the spring and weight arrangement forms the first part of a resonance cycle. The weight moves upwards at 16 ft/sec. The spring now starts to decelerate the weight because the spring is going into tension (pulling down, instead of pushing up). For each 3 inches upward above the neutral position, it will apply a tension of 2 pounds until it reaches a foot, where the tension becomes 8 pounds.

ENERGY TRANSFER
Resonance Cycle

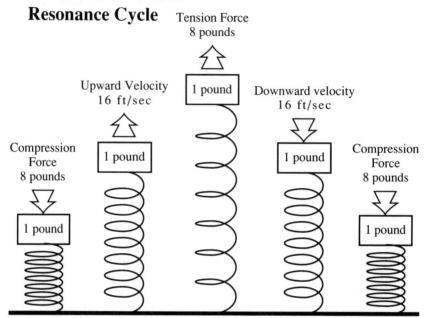

As with the compression, the average force of tension over the foot of move-ment is 4 pounds, so the potential energy will again be 4 pounds (128 foot-poundals). All of the energy is again potential and the weight is momentarily stationary.

Having reached this extreme, an equal acceleration downward starts the sec-ond half of the cycle. A similar interchange of energy continues until the neutral position is again reached. At this point, all the energy is again kinetic and the velocity will be 16 ft/sec downward, steady (for the moment)—neither accelerat-ing nor decelerating. Then, as it continues downward, compression starts again, until the weight comes to rest fully compressed, 1 foot down, with an 8-pound force pushing it back up.

This process would go on forever, but the energy gradually transfers to other forms. Friction will absorb some of the energy. The excursion and velocity slowly diminish and the weight eventually stops.

Travel and velocity in resonance system

You started with an assumed compression of 1 foot, which led to a maximum velocity of 16 ft/sec. Suppose the initial compression is only 6 inches, or that friction has decreased the excursion to this magnitude. The maximum force is now 4 pounds, instead of 8 (128 poundals, instead of 256). The distance over which the average force was 2 pounds (64 poundals) is now compressed to 6 inches, instead of 1 foot. So, the maximum potential energy is 32 foot-poundals.

When the weight passes through the neutral position, all this energy will become kinetic. v^2 will not be 64, so v is 8 ft/sec.

Notice that halving the travel also halves the maximum velocity reached. The object travels half the distance at half the speed, so it performs the entire cycle in the same time. Interestingly, regardless of the magnitude of the oscillation, resonance still requires the same time.

This principle is used in the balance wheel of clocks or watches, the pendulum of grandfather clocks, and many similar devices—not just mechanical, but also electrical, electronic, and atomic.

Maximum Travel and Maximum Velocity

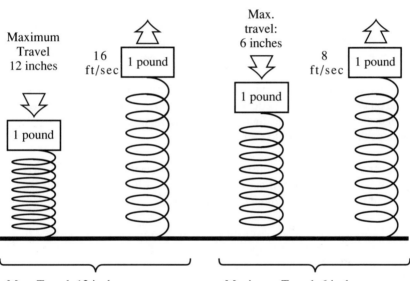

Max. Travel: 12 inches
Max. force: 8 pounds
Energy stored: 4 ft-pounds, 128 ft-pdls

Maximum Travel: 6 inches
Max. force: 4 pounds
Energy stored: 1 ft-pound, 32 ft-pdls

Max. velocity:
$$v^2 = 256$$
$$v = 16 \text{ ft/sec}$$

Max. velocity:
$$v^2 = 64$$
$$v = 8 \text{ ft/sec}$$

Aver. force = 4 pounds
Distance = 1 foot

Aver. force = 2 pounds
Distance = $\frac{1}{2}$ foot

Half the travel, half the speed / Same
Twice the travel, twice the speed / Time

Questions and problems

1. Suppose a car accelerates uniformly from standstill to 40 miles/hour in 3 minutes. How far will it travel in that 3 minutes?

2. In the next 6 minutes, the car increases its speed at a steady rate, from 40 miles/hour to 60 miles/hour. How far will it travel in these 6 minutes?

3. The same car brakes to a stop in 30 seconds. If the deceleration is uniform during this 30 seconds, how far will the car travel before it stops?

4. From the fact that 1 mile = 5280 feet and 1 hour = 3600 seconds, find the speed in miles/hour that corresponds to 88 ft/sec.

5. During takeoff, an aircraft builds up a thrust that accelerates it at 16 feet/sec^2. Its take-off speed is 240 miles/hour. Find the time from releasing the brakes until the plane lifts into the air. How much runway is required?

6. A gun can use cartridges with pellets of two sizes, one that is twice the weight of the other. If the heavier pellet leaves with a muzzle velocity of 150 ft/sec, find the muzzle velocity of the lighter pellet, assuming that the explosive charge develops the same energy in each case.

7. A car's motor and transmission develops constant power during maximum acceleration. This particular car can reach 60 miles/hour in 20 seconds. In how long will it reach 30 miles/hour? 45 miles/hour?

8. If the weight of car and driver (in question 7) were 3000 pounds, what time is necessary to reach the three speeds, 30, 45, and 60 miles/hour, when an additional load of 1000 pounds is carried?

9. Find the power developed by the motor and transmission of the same car in foot-poundals per second and in foot-pounds per second.

10. From the second answer in question 9, find the maximum speed that the car can attain up a 10% grade, without the extra 1000-pound load. (A 10% grade is one in which it rises 1 foot for every 10 feet forward).

11. Another car, which also develops constant power at all speeds, can maintain 30 miles/hour up a 10% grade. In how long does it accelerate from 0 to 60 miles/hour on a level road?

12. If a car accelerates from 0 to 60 miles/hour in 15 seconds at constant power, what is its acceleration at the end of the 15 seconds, when it is travelling 60 miles/hour?

13. If the same car, instead of using constant power, used only the acceleration it reached at 60 miles/hour from the start, in how long would it reach 60 miles/hour?

14. A spring and weight resonance system can be changed, either by altering the weight or the spring. By figuring the effect of such change on maximum velocity reached from a given starting deflection, deduce the effect of (1) doubling the weight, (2) halving it, (3) making the spring twice as strong, (4) halving its strength on the period that is taken for each complete oscillation.

15. A spring produces a pressure of 1 ounce for each inch that it is stretched or compressed from neutral. If it is compressed 8 inches and has a 4-ounce weight attached to it, what will be the weight's maximum velocity?

16. A car travelling 60 miles/hour and weighing 3000 pounds is braked to a standstill on a level road. The energy of the car's motion must be dissipated as heat in the brake drums (or discs). How much heat is developed if 1 BTU (British thermal unit, the recognized unit of heat) is equivalent to 24,200 foot-poundals?

17. A car with passengers weighs 4000 pounds. Its motor and transmission develop a maximum forward thrust that is equivalent to 36,300 foot-pounds per second. What is the maximum speed that this car will reach up a 10% grade?

18. With the car of question 17, how long will it take to reach 40 miles/hour on level road? 80 miles/hour?

19. When the car reaches a speed of 40 miles/hour up a 10% grade, how much of its maximum power will be needed to maintain that steady speed in foot-pounds per second? As a percentage?

20. If maximum power is used to maintain 40 miles/hour on 10% upgrade, what will be the car's acceleration at 40 miles/hour on level road?

21. Using the details of question 17, compare acceleration on the following gradients:
> level road;
> 5% grade;
> 10% grade;

at each of the following speeds;
> the start;
> 20 miles/hour;
> 40 miles/hour;
> 60 miles/hour.

Tabulate the 12 answers.

22. By equating the energy gained in descent to that which was caused by the velocity increase, find the acceleration down a 10% grade, with no brakes (assuming no friction or wind resistance).

23. A roller-coaster ride uses a starting grade that drops a vertical height of 64 feet, then levels off before going into turns, banks, upgrades, etc. As the cars reach the level at the bottom of the grade (assuming no friction), what speed will they be doing in ft/sec? Miles/hour?

24. A spring is made in two parts: for the first 6 inches of its compression, each inch requires a force of 5 pounds; at 6 inches of compression, the force stiffens, and requires a force of 10 pounds for every additional inch. What will be: (a) the force exerted by the spring at a compression of 1 foot? (b) the initial acceleration when the spring is released with a 2-pound weight attached to it? (c) the velocity of the 2-pound weight as it passes the 6-inch point? (d) the velocity when the spring is fully decompressed?

25. Assuming that the spring (of question 24) extends beyond the starting point without any change in the rate at which force increases with extension in tension (5 pounds for every inch, indefinitely), how far will the weight travel in the opposite direction before the spring starts to bring it back?

13
CHAPTER

Ratio in mathematics

Proportion or ratio

Thus far, proportion and ratio have been covered with respect to the simple idea of similarity. However, isolating ratio or proportion from the quantities that proportion relates will be covered in this chapter.

A fraction is one way to express a ratio: it is the relationship between the numerator and denominator. Different numbers can be used to write the same value as a fraction—the same relationship between the numerator and denominator. In earlier times, not used so much these days, proportion was more explicitly conveyed with colons and an equals sign.

One application that sets the scene for later concepts is the use of ratio or proportion, relative to such things as projecting images, shown here. This concept is sometimes called *aspect ratio*. The standard TV screen aspect ratio is 4:3. Movie screens have various aspect ratios, particularly since wide screens became popular.

In school arithmetic, text books used exercises of the type shown here. The student had to complete the statement or fill in the gap. Always, the relationship was between specific quantities, such as dimensions. We must realize, that a ratio (proportion) has no dimension of itself. The ratio might even be the relationship between two dimensions, in inches, feet, numbers, etc.

Manipulation of ratio

The manipulations on the previous pages were done in arithmetic textbooks. As school textbooks work into algebra, formulas replace rules when dealing with numbers. So, here is the algebraic way of thinking about what was done arithmetically on the previous page.

SIZES are DIFFERENT
but PROPORTIONS
are THE SAME

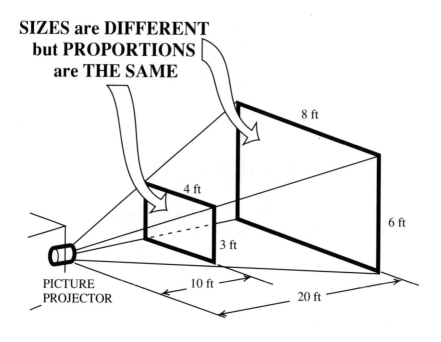

$$\frac{3}{4} = \frac{6}{8} \qquad 3 : 4 = 6 : 8$$

$$\frac{3}{10} = \frac{6}{20} \qquad 3 : 10 = 6 : 20$$

$$\frac{4}{10} = \frac{8}{20} \qquad 4 : 10 = 8 : 20$$

PROPORTION
or RATIO

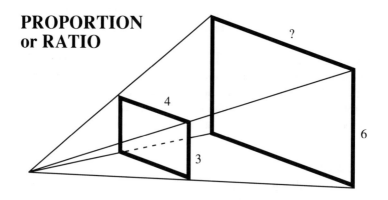

$$\frac{3}{4} = \frac{6}{?} \quad \text{or} \quad 3 : 4 = 6 : ?$$

LETTERS can be USED to REPRESENT RATIOS

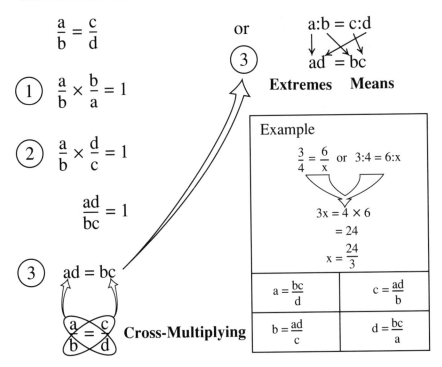

$$\frac{a}{b} = \frac{c}{d} \qquad \text{or} \qquad a{:}b = c{:}d$$

(3)

$$ad = bc$$

Extremes   Means

(1) $\frac{a}{b} \times \frac{b}{a} = 1$

(2) $\frac{a}{b} \times \frac{d}{c} = 1$

$$\frac{ad}{bc} = 1$$

(3) $ad = bc$

$\frac{a}{b} \times \frac{c}{d}$ **Cross-Multiplying**

Example

$$\frac{3}{4} = \frac{6}{x} \quad \text{or} \quad 3{:}4 = 6{:}x$$

$$3x = 4 \times 6$$
$$= 24$$
$$x = \frac{24}{3}$$

$a = \dfrac{bc}{d}$	$c = \dfrac{ad}{b}$
$b = \dfrac{ad}{c}$	$d = \dfrac{bc}{a}$

Applying the principle to bigger problems

A man's will states that his estate is to be divided between his three sons, in proportion to their ages. The amount of the estate is $78,000, and their ages are 53, 47, and 30. How much does each get?

First, assume that the basis of the proportion is $x per year of age to each person. This means the sons get 53x, $47x, and $30x, respectively. That adds up to $130x. You know that total is $78,000. So, the equation is:

$$130x = 78,000$$
$$x = 600$$

Substituting into the statements, they receive $31,800, $28,200, and $18,000. To check, add these amounts. They add to $78,000 and prove that the answer is correct.

A PROBLEM based on PROPORTION

Ages...   53   47   30

Money... $78,000.   How much each?

$x per year of age

$$53x + 47x + 30x = 130x = 78{,}000$$

$$x = \frac{78{,}000}{130} = 600$$

Son aged 53 years gets $600 × 53 = $31,800

"   "   47   "   " $600 × 47 = $28,200

"   "   30   "   " $600 × 30 = $18,000

CHECK→ add up $78,000

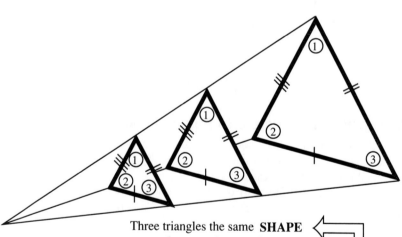

Three triangles the same **SHAPE**
use the same three angles

Size is fixed by length of sides
Each has three sides

in the same proportion

same three angles
same shape

Shape and size

Ratio and proportion form a good basis for showing the distinction between shape and size. This principle can be shown with triangles, the simplest geometric figure with straight lines for a boundary (see example at bottom of facing page).

If a triangle is expanded in proportion, its respective sides maintain the same ratio, one to another. It has the same shape, but it differs in size. Since it has the same shape, it also has the same angles. All angles marked 1 are equal, those marked 2 are also equal, as are those marked 3.

The sides with one crossmark have the same proportion to the sides with two crossmarks in each triangle. The proportion between sides with two and three crossmarks, or between sides with one and three crossmarks, are the same in each triangle—or with any others that have the same shape.

So, triangles that have the same shape have the same angles. Although their sides might be longer or shorter, they are in the same proportion.

About angles in triangles

When two straight lines cross (mathematicians use the word *intersect*), the opposite angles are equal. To prove this statement, draw square corners at the intersection, based on each of the lines. You can easily see that any pair of angles numbered 1 and 2 have a total angle of two square corners. One pair of square

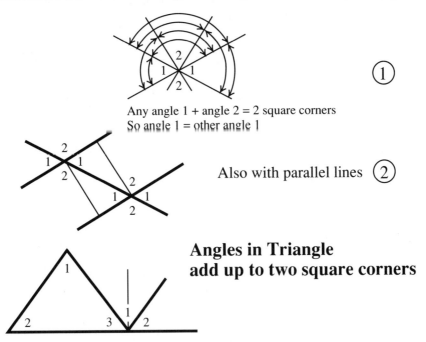

Any angle 1 + angle 2 = 2 square corners
So angle 1 = other angle 1

Also with parallel lines ②

**Angles in Triangle
add up to two square corners**

Shape of triangle fixed by TWO angles

corners consists of top angle 2 and left side angle 1. Another pair consists of top angle 2 and right side angle 1. The total is the same, two square corners, and both use the same angle (2), so the two angles (1) must be the same. Similarly, you can show that the two angles (marked 2) are equal.

Next, if two parallel lines intersect a third line, the angles at the intersections will be equal, if taken in correct pairs. To show this: complete a rectangle with square corners. As with the single intersection, you can now prove that angles numbered 1 are all equal, as are all angles numbered 2.

In any triangle, its three angles always add up to two square corners. To show this statement, extend one side at one corner and draw a line parallel to the opposite side. Because of their positions, relative to parallel lines, corresponding angles (marked 1 and 2) are each equal. Where the side is extended are three angles numbered 1, 2, and 3, that add up to two square corners. So, the corresponding angles inside the triangle (also numbered 1, 2, and 3) must also add up to two square corners.

Use of square-cornered triangles

A square-cornered triangle becomes a very important building block in other shapes and sizes, whether in triangles or in more-complicated shapes. Any triangle can be divided into two square-cornered triangles. Unless the original triangle is square cornered, the division into two square-cornered triangles can be made in three ways. In the triangles shown, one such division is done with a thin line, the other two with dashed lines.

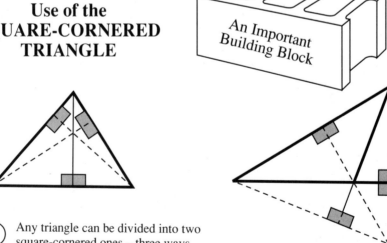

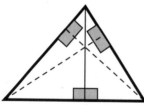

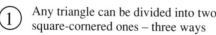

① Any triangle can be divided into two square-cornered ones – three ways

② The perpendiculars always intersect at one point

If all the angles are acute, all three ways of dividing into two square-cornered triangles are additive, so the original triangle is the two square-cornered ones added together. However, if one angle of a triangle is obtuse (wider than a square corner) two of the possible divisions require a difference, rather than a sum. The original triangle is the larger square-cornered triangle minus the smaller one.

An interesting fact about these divisions, which I will not prove here, is that the three dividing lines from the corners of the original triangle, formed by making them perpendicular to the opposite side, always intersect at a single point. *Perpendicular* means the two lines create two square-cornered angles. In an acute-angled triangle, the point of intersection is inside the triangle. In an obtuse-angled triangle, the point is found only by extending all three perpendiculars (dotted lines). This exercise begins to show the importance of square-cornered (right) triangles as building blocks.

Angles identified by ratios

Angles determine the shape of a triangle and also the ratio of its sides. In an ordinary triangle, any of the three sides can be changed to alter all three angles, so the relationship between the side ratio and the angle becomes rather involved. One angle is fixed as a square corner (right angle). Because its three angles must add up to two right angles, the other two must always add up to one right angle.

In a right triangle, the longest side opposite the right angle, is called the *hypotenuse*. A triangle that uses this angle and a right angle can have only one shape, because all three angles are fixed. Regardless of how big (or small) you draw the triangle, the ratio is the same for the particular angle of interest. This ratio identifies the angle uniquely. No other angle can have the same ratio.

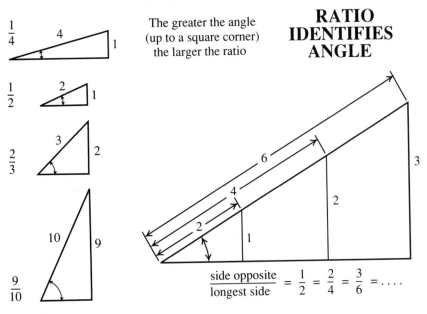

The greater the angle (up to a square corner) the larger the ratio

RATIO IDENTIFIES ANGLE

$$\frac{\text{side opposite}}{\text{longest side}} = \frac{1}{2} = \frac{2}{4} = \frac{3}{6} = \ldots.$$

If you begin with a small angle and gradually increase it toward a right angle, the fraction that represents the ratio (side opposite/hypotenuse) is always greater for greater angles.

Special fact about the right triangle

The proof shown here is the original, because it works for any square-cornered triangle, not for just one. Some proofs rely on cutting the big square into fancy-shaped pieces, then rearranging them to compose the two smaller squares. These proofs only work for that particular triangle.

The pythagorean theorem says that the area of the squares (based on the two sides alongside or adjacent to the square corner) always adds up to the area of the square on the longer side (opposite the square corner in the triangle, called the hypotenuse). This theorem is expressed by $A^2 + B^2 = C^2$.

To prove it, as shown in 1, drop a perpendicular from the square corner onto the opposite side (hypotenuse) and extend it across the big square. Now (2) you have completed two triangles, shown shaded. They must be equal. The angle between sides (1 and 2) of each triangle consist of the same angle of the original triangle, plus a square corner; so both angles must be the same. Sides (1 and 2) are the same because they are sides of the same squares. You can see that the triangles are the same, just rotated by a square corner from one another. So, they must have the same area.

In 3, the square at the right is based on (or under) the blackened line, and it is the same height (measured downwards from the line) as one of the triangles. The shaded part of the big square sits on the same base (blackened line) as the other triangle, and they are the same height, measured a different way, because you are using a different side as base. The square and the rectangle (shaded in 3 must each be double the area of the triangles (shaded in 2). As the triangles in 2

Conventional Pythagoras proof

Let ABΓ be a right-angled triangle having the angle BAΓ right; I say that the square on BΓ is equal to the squares on BA, AΓ.

For let there be described on BΓ the square BΔEΓ, and on BA, AΓ the squares HB, ΘΓ (Eucl. i. 46), and through A let AΛ be drawn parallel to either BΔ or TE, and let AΔ, ZΓ be joined. Then, since each of the angles BAΓ, BAH is right, it follows that with a straight line BA and at the point A on it, two straight lines AΓ, AH, not lying on the same side, make the adjacent angles equal to two right angles; therefore ΓA is in a straight line with AH (Eucl. i. 14). For the same reasons, BA is also in a straight line with AΘ. And since the angle ΔBΓ is equal to the angle ABA, for each is right, let the angle ABΓ be added to each; the whole angle ΔBA is therefore equal to the whole angle ZBΓ. And since ΔB is equal to BΓ, and ZA to BA, the two ΔB, BA are equal to the two BΓ, ZB respectively; and the angle ΔBA is equal to the angle ABΓ. The base AΔ is therefore equal to the base ZΓ, and the triangle ABΔ is equal to the triangle

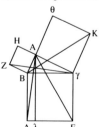

ZBΓ (Eucl. i.4). Now the parallelogram BΛ is double the triangle ABΔ, for they have the same base BΔ and are in the same parallels, BΔ, AΛ (Eucl. i. 41). And the square HB is double the triangle ZBΓ, for they have the same base AΛ and are in the same parallels ZB, HΓ. Therefore, the parallelogram BΛ is the equal to the square HB. Similarly, if AE, BK are joined, it can also be proved that the parallelogram ΓΛ is equal to the square ΘΓ. Therefore the whole square BΔEΓ is equal to the two squares HB, ΘΓ. And the square BΔEΓ is described on BΓ, while the squares HB, ΘΓ are described on BA, AΓ. Therefore the square on the side BΓ is equal to the squares on the sides BA, AΓ.

Therefore in right-angled triangles the square on the side subtending the right angle is equal to the squares on the sides containing the right angle.—Quad Erat Demonstrandum.

are equal, their doubles must be equal (i.e., the shaded square is equal to the shaded rectangular part of the big square.

Now, do the same thing in steps 4 and 5 to show that the other shaded square (5) is equal to the other part of the big square. Finally (6), since the smaller squares are equal to their respective rectangular parts of the big square, the total area of the smaller squares must equal the area of the big square. Taken slowly in steps like that, it is not difficult to see, but you might need to read it twice. Fundamentally, this principle in geometry and trigonometry has the name of the famous Greek mathematician—Pythagoras.

THE SQUARE-CORNERED PYTHAGORAS

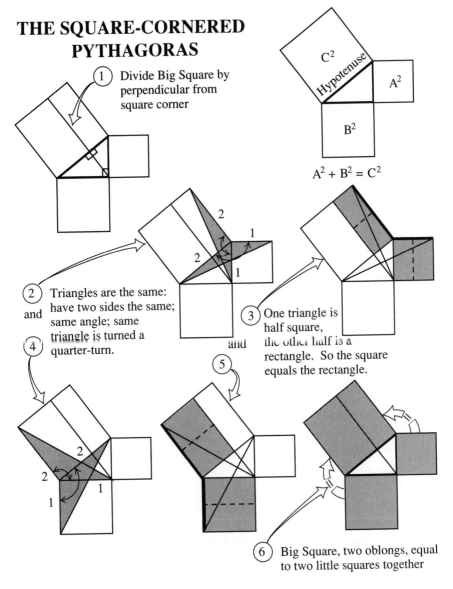

(1) Divide Big Square by perpendicular from square corner

C^2 *Hypotenuse* A^2

B^2

$A^2 + B^2 = C^2$

(2) and (4) Triangles are the same: have two sides the same; same angle; same triangle is turned a quarter-turn.

(3) One triangle is half square, and the other half is a rectangle. So the square equals the rectangle.

(5)

(6) Big Square, two oblongs, equal to two little squares together

Names for angle ratios

As you proceed in studying mathematics, the use of ratios to identify angles assumes a more important role. You need names to identify them. These names are the terminology of trigonometry.

The right triangle used to identify the ratios has three sides. The ratio of any two of its three sides identifies an angle. In picking those two sides, you have three possible choices. These three basic ratios must be defined and named.

The *sine* is opposite side/hypotenuse. Each particular angle is written with "sin," followed by a letter or symbol to identify that particular angle. A is shown by the angle. This angle ratio will be written *sin A*.

The *cosine* uses the ratio adjacent side/hypotenuse. For the same angle, write it as *cos A*.

The *tangent* uses the only remaining combination, side opposite/side adjacent. Write it as *tan A*. If you previously learned that a tangent is a line that touches the circumference of a circle, this usage is different.

The little diagrams at the left show how to remember the relationship of the ratio to the angle. The line with the arrow is the numerator. It leads to the side that forms the denominator of the ratio. Remember that these names (sine, cosine, and tangent), represent ratios, not lengths. A ratio identifies the angles, regardless of how large or how small the triangle is drawn.

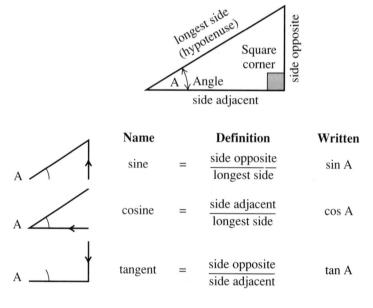

	Name		Definition	Written
	sine	=	$\dfrac{\text{side opposite}}{\text{longest side}}$	sin A
	cosine	=	$\dfrac{\text{side adjacent}}{\text{longest side}}$	cos A
	tangent	=	$\dfrac{\text{side opposite}}{\text{side adjacent}}$	tan A

These are the names used in TRIGONOMETRY

Spotting the triangle

Remembering the ratios in trigonometry, without looking back, requires as much time and practice as learning the multiplication tables in arithmetic. What can be more confusing, needing more care, is spotting the right sides for the ratio when the angle is not in the position used in the previous section. Regardless of where the angle is, you must construct a right triangle (or use one that's already there) one way or another. Then, the ratio follows the definition. A sine is always side opposite over the hypotenuse, according to how the triangle is disposed. The other ratios follow similar layouts. Here are some possible positions that you might encounter:

- The top triangle repeats the arrangement from the previous section.
- The next triangle tips the right-angled triangle the other way up.
- The third triangle is the same triangle as the first, but the angle you refer to is different.
- The fourth triangle has the angle at the bottom, with the right angle at top left.

These are only a few of the possible positions that you might encounter. You must get accustomed to identifying the sides of the appropriate right triangle so you correctly identify the ratios.

Identifying the Ratios

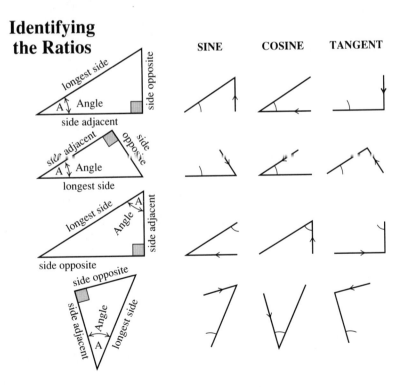

Degree measure of angles

If you already know about degree measure for angles, you might wonder why a right angle is not called a 90-degree angle. Degree measure for angles is more modern than the recognition that a certain angle is a "right angle."

Degree measure for angles requires a complete circle (rotation) in a flat surface, 360 degrees. A half rotation is half of 360 (180 degrees). A quarter rotation, which is a right angle, is thus 90 degrees. Acute angles are less than 90 degrees, and obtuse angles are more than 90 degrees (but less than 180 degrees).

Two special angles, other than 90 and 180 degrees, are based on two special triangles. The first is an equilateral triangle, which has all three equal sides. It also has all equal angles, and a unique symmetry. All three angles are 60 degrees.

The other special angle comes from a right triangle, which has two shorter equal sides. Since the three angles must add up to 180 degrees, and one of them is 90 degrees, the other two must each be 45 degrees. This triangle is called a *right-isosceles triangle*. *Isosceles* means "having two equal sides."

MEASURING ANGLES IN DEGREES

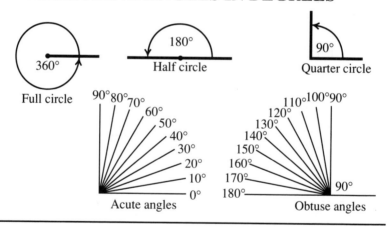

TWO SPECIAL TRIANGLES

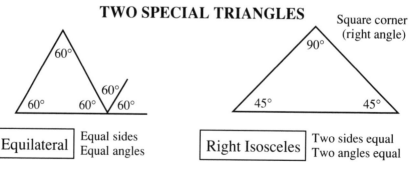

Finding trig ratios for certain angles

The special triangles from the last section help us to find ratios that correspond to special angles they contain. All angles of the equilateral triangle are 60 degrees. When you draw a perpendicular line from one corner to the opposite side, another angle forms on each side of the upright. Taking one of the right triangles, you know two of its angles, 90 and 60 degrees. So the third angle must be 30 degrees. Another way of seeing this, is that the perpendicular cuts the top angle in half.

If the equilateral triangle has sides 2 inches long, the half side will be 1 inch. From Pythagoras, calculate the perpendicular. The square on the 2-inch side is 4 square inches. The square on the 1-inch side is 1 square inch. So, the square on the perpendicular must be 3 square inches. Its length must be the square root of 3. From this fact, you can calculate all three ratios for both a 30-degree angle and a 60-degree angle.

RATIOS of 30° and 60° ANGLES

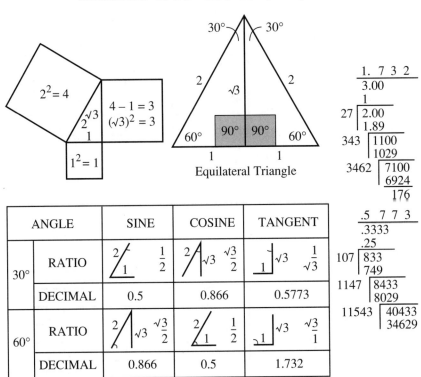

ANGLE		SINE	COSINE	TANGENT
30°	RATIO	$\frac{1}{2}$	$\sqrt{3}$ $\frac{\sqrt{3}}{2}$	$\sqrt{3}$ $\frac{1}{\sqrt{3}}$
	DECIMAL	0.5	0.866	0.5773
60°	RATIO	$\sqrt{3}$ $\frac{\sqrt{3}}{2}$	$\frac{1}{2}$	$\sqrt{3}$ $\frac{\sqrt{3}}{1}$
	DECIMAL	0.866	0.5	1.732

The right isosceles triangle

The right isosceles triangle enables you to calculate the ratios for 45 degrees. In a right isosceles triangle, the two equal angles must add up to 90 degrees, so each angle is 45 degrees. If each side by the right angle is 1 inch, the square on the hypotenuse must have an area of 2 square inches. Its length must be the square root of 2.

Two other special angles are 0 and 90 degrees. They are difficult to visualize because they have no triangle. In theory, it only has 2 sides—the third side dropped to zero length. That gives you the figures to use. Thinking of 90 degrees, the side opposite and hypotenuse are equal, and the adjacent side is zero. Now, you can make a limited table of values for angles 0, 30, 45, 60, and 90 degrees.

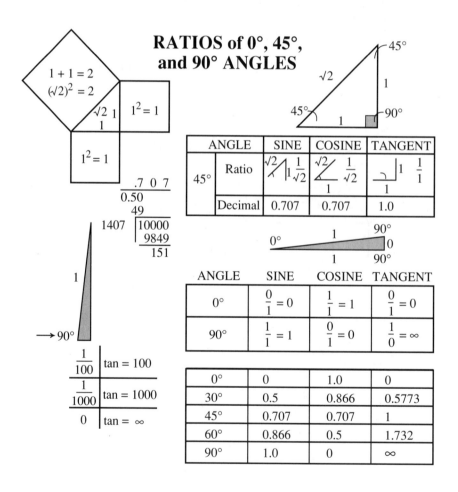

RATIOS of 0°, 45°, and 90° ANGLES

ANGLE		SINE	COSINE	TANGENT
45°	Ratio	$\frac{\sqrt{2}}{1}$ $\frac{1}{\sqrt{2}}$	$\frac{\sqrt{2}}{1}$ $\frac{1}{\sqrt{2}}$	$\frac{1}{1}$
	Decimal	0.707	0.707	1.0

ANGLE	SINE	COSINE	TANGENT
0°	$\frac{0}{1}=0$	$\frac{1}{1}=1$	$\frac{0}{1}=0$
90°	$\frac{1}{1}=1$	$\frac{0}{1}=0$	$\frac{1}{0}=\infty$

0°	0	1.0	0
30°	0.5	0.866	0.5773
45°	0.707	0.707	1
60°	0.866	0.5	1.732
90°	1.0	0	∞

Other angles

That selection is rather limited. Until recently, this book has used tables (a sample of which is on the next page). Now, it's easier to use a pocket calculator (many have these functions included). To understand mathematics, you should learn where such tables come from. A calculator works out the value for the angle you enter, but with tables, you must find the number (in the table) and perhaps to interpolate between the values (listed in the tables).

Even using a calculator, you can make mistakes with the data you enter. Knowing what to expect can help. Study the following diagram and table.

Use of SINE - COSINE - TANGENT TABLES

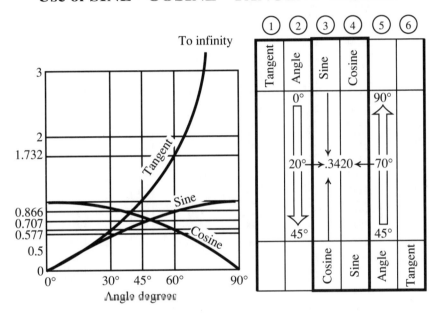

Using trigonometry in problems

Three examples shown on page 225 show how to use, either the table from the previous section or your pocket calculator to solve problems that involve trigonometry.

Example 1   An observation point at sea level is 20 miles from a mountain peak. The elevation of the peak above sea level is 5 degrees, viewed from this point. How high is the peak?

The relationship involves the adjacent side (a distance of 20 miles), the height, the opposite side, and an angle of 5 degrees. This is the tangent ratio.

Tangent	Angle	Sine	Cosine		
0.0000	0°	0.	1.0000	90°	∞
.0175	1°	.0175	.9998	89°	57.29
.0349	2°	.0349	.9994	88°	28.6363
.0524	3°	.0523	.9986	87°	19.0811
.0699	4°	.0698	.9976	86°	14.3007
.0875	5°	.0872	.9962	85°	11.4301
.1051	6°	.1045	.9945	84°	9.5144
.1228	7°	.1219	.9925	83°	8.1443
.1405	8°	.1392	.9903	82°	7.1154
.1584	9°	.1564	.9877	81°	6.3138
.1763	10°	.1736	.9848	80°	5.6713
.1944	11°	.1908	.9816	79°	5.1446
.2126	12°	.2079	.9781	78°	4.7046
.2309	13°	.2250	.9744	77°	4.3315
.2493	14°	.2419	.9703	76°	4.0108
.2679	15°	.2588	.9659	75°	3.7321
.2867	16°	.2756	.9613	74°	3.4874
.3057	17°	.2924	.8563	73°	3.2709
.3249	18°	.3090	.9511	72°	3.0777
.3443	19°	.3256	.9455	71°	2.9042
.3640	20°	.3420	.9397	70°	2.7475
.3839	21°	.3584	.9336	69°	2.6051
.4040	22°	.3746	.9272	68°	2.4751
.4245	23°	.3907	.9205	67°	2.3559
.4452	24°	.4067	.9135	66°	2.2460
.4663	25°	.4226	.9063	65°	2.1445
.4877	26°	.4384	.8988	64°	2.0503
.5095	27°	.4540	.8910	63°	1.9626
.5317	28°	.4695	.8829	62°	1.8807
.5543	29°	.4848	.8746	61°	1.8040
.5774	30°	.5000	.8660	60°	1.7321
.6009	31°	.5150	.8572	59°	1.6643
.6249	32°	.5299	.8480	58°	1.6003
.6494	33°	.5446	.8387	57°	1.5399
.6745	34°	.5592	.8290	56°	1.4826
.7002	35°	.5736	.8192	55°	1.4281
.7265	36°	.5878	.8090	54°	1.3764
.7536	37°	.6018	.7986	53°	1.3270
.7813	38°	.6157	.7880	52°	1.2799
.8098	39°	.6293	.7771	51°	1.2349
.8391	40°	.6428	.7660	50°	1.1918
.8693	41°	.6561	.7547	49°	1.1504
.9004	42°	.6691	.7431	48°	1.1106
.9325	43°	.6820	.7314	47°	1.0724
.9657	44°	.6947	.7193	46°	1.0355
1.000	45°	.7071	.7071	45°	1.0000
		Cosine	Sine	Angle	Tangent

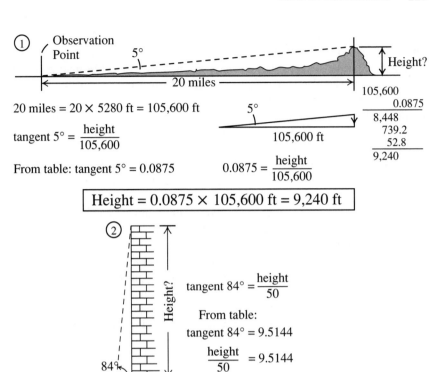

① Observation Point   5°   Height?

20 miles

20 miles = 20 × 5280 ft = 105,600 ft

5°

105,600 ft

tangent 5° = $\dfrac{\text{height}}{105,600}$

From table: tangent 5° = 0.0875

$0.0875 = \dfrac{\text{height}}{105,600}$

```
   105,600
    0.0875
    8,448
    739.2
     52.8
    9,240
```

Height = 0.0875 × 105,600 ft = 9,240 ft

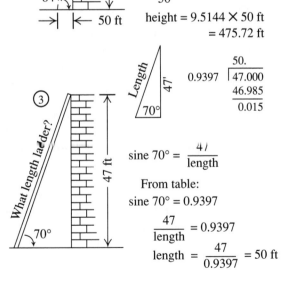

② Height?   84°   50 ft

tangent 84° = $\dfrac{\text{height}}{50}$

From table:

tangent 84° = 9.5144

$\dfrac{\text{height}}{50} = 9.5144$

height = 9.5144 × 50 ft
= 475.72 ft

Length   47'   70°

```
          50.
0.9397 | 47.000
         46.985
          0.015
```

③ What length ladder?   47 ft   70°

sine 70° = $\dfrac{47}{\text{length}}$

From table:

sine 70° = 0.9397

$\dfrac{47}{\text{length}} = 0.9397$

length = $\dfrac{47}{0.9397}$ = 50 ft

From the table or from your calculator, tangent 5 degrees is 0.0875. The calculator on my desk gives 0.0874887—a little more accurate than the table! The height is 0.0875 times the 20-mile distance. Convert the figure to feet by multiplying by 5280, and you have the peak height, 9240 feet. How accurate was that 20 miles or the 5-degree elevation do you suppose?

Example 2 A high building is viewed from 50 feet away, from its vertical wall. The angle to the horizontal line of sight is 84 degrees. How high is the building?

This problem again involves the tangent. The table gives tan 84 as 9.5144. My desk calculator says 9.514364. The table is correct to 4 decimal places. By multiplying this figure by 50 feet, the height of the building is 475.7; round it off to 476 feet.

Example 3 A ladder must reach the roof of a building 47 feet high. The slope of the ladder, when rested against the building, should be 70 degrees. What ladder length is necessary?

The solution involves ratio of opposite side to hypotenuse, which is sine, but it's the inverse. Sine 70 degrees is 0.9397 in the table, near enough. Dividing 47 by 0.9397 gives the needed ladder length as a very small fraction over 50 feet.

These three examples use two of the ratios. Any problem like this involves ascertaining which ratio you need and writing and solving the equation to find the answer.

Questions and problems

1. The aspect ratio for a television picture is 4:3. A wide-screen movie is transmitted so that the picture fills the full height of the TV screen. What proportion of the width must be lost at the sides, if the aspect ratio of the movie picture is 2:1? (Sketch this screen, to help you grasp the question).

2. Another way of transmitting the picture in question 1 would be to include the full picture width and mask off an area (top and bottom), where the picture does not fill the TV screen. What proportion of the TV screen will be masked off (top and bottom)?

3. The material for a portable movie screen comes on a roll, which is cut up to form screens of different sizes. A small screen is 24 inches high; a larger one is 54 inches wide; both are the same shape and each are to be cut from the same width of material without waste. What width is the material and what is the aspect ratio of the picture?

4. A man wills his estate to his 5 children—3 boys and 2 girls. It calls for each to get an amount proportional to his or her age at the time of his death; however, the boys get twice the rate for their ages than the girls do. When the will is made, the boys' ages are 40, 34, and 26, while the girls are aged 37 and 23. If the father dies the same year, what will each receive from an estate of $22,100?

5. If the father of question 4 lives 10 years after making the will, the estate has not changed in value, and all 5 are still living, how much will each get?

6. The two sides of a right triangle that adjoin the right angle are 5 inches and 12 inches long. What is the length of the hypotenuse (opposite the right angle)? Find the length by calculation, not by construction and measurement.

7. A highway gradient is measured as the rise in altitude divided by the distance along the pavement surface. An 8000-foot length of straight highway maintains a gradient of 1 in 8 (1/8). Find the altitude gained in this distance, and the amount by which the distance measured horizontally falls short of 8000 feet. Use the Pythagorean theorem, not tables.

8. At a distance of 8 miles, the elevation of a mountain peak, viewed from sea level, is 9 degrees. Some distance further away, still at sea level, the elevation is 5 degrees. What is the height of the peak, and the distance of the second view-point?

9. A railroad track stretches for 3 miles at a gradient of 1 in 42 up; then, 5 miles of 1 in 100 up; then, 2 miles level; followed by 6 miles of 1 in 250 down; 4 miles level, and finally 5 miles at 1 in 125 up. How much higher is the finish point than the starting point?

10. A house is to have a roof slope of 30 degrees that is gabled in the middle. The width of the house is 40 feet and the roof is to extend 2 feed beyond the wall to provide snow protection. What distance—from the ridge of the gable to the guttering—is needed for rafters?

11. By how much could the rafter length be reduced in the house (question 10) by making the roof slope 20 degrees?

12. A house wall is 50 feet high and a ladder used to scale it is 60 feet long. How far from the base of the wall must the ladder be placed for its top to just reach the top of the house wall? Finish this question two ways: first, calculate it with the tables (or a trig calculator) to the nearest degree; second, calculate it directly by the Pythagorean theorem.

13. A clever trick, intended to show that the squares of two sides adjacent to the right angle are not exactly equal to the square on the longest side, uses pieces cut out of paper. First, an 8-inch square has points marked off 2 inches from one corner along each side. These points are joined across by straight lines (see drawing) and the square is cut into four pieces. Next, with the addition of a 4-inch square, the pieces are put together a different way, to make a 9-inch square. The sum of the squares is 80 square inches, but the area of a 9-inch square is 81

square inches. The question posed is, "where did the extra square inch come from?" Show, by calculation, where the fallacy is.

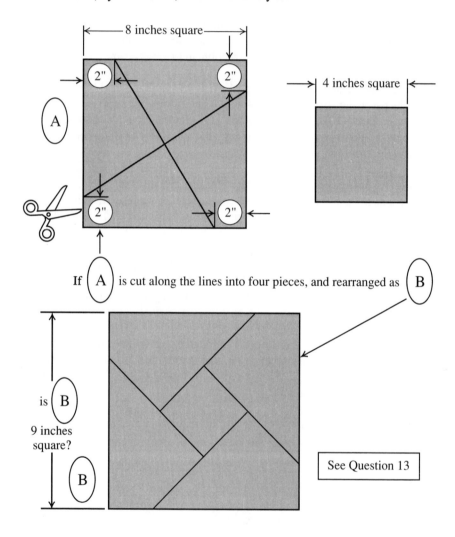

See Question 13

14. A rope extends up a 75-ft wall, over a pulley, and back to the ground, where a heavy load is attached. The load is lifted by carrying the free end of the rope horizontally away from the base of the wall. How far must the end be carried to raise the load half-way up the wall?

14
CHAPTER

Trigonometry and geometry conversions

Ratios for sum angles

As the examples showed, sometimes we need angles other than 0, 30, 45, 60, and 90 degrees. In this chapter you need to learn two things:

1. Sin($A + B$) is not equal to sin A + sin B. It doesn't work like removing the parentheses in algebra.
2. The formula for what sin($A + B$) does equal.

First to show that removing parentheses doesn't "work." Here: make A 30 degrees and B 45 degrees. Sin 30 is 0.5. Sin 45 is 0.7071. Adding the two is 1.2071.

You know that no sine (or cosine) can be more than 1. Why? The ratio has the hypotenuse as its denominator. The most that the numerator can be is equal to the denominator. A sine or cosine can never be greater than 1, so a value of 1.2071 must be wrong.

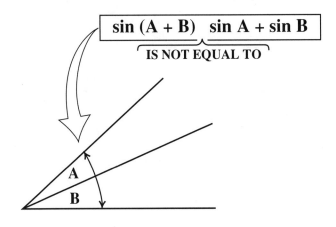

sin (A + B)   sin A + sin B

IS NOT EQUAL TO

Wanted sine, cosine, or tangent, of whole angle (A + B)

Sine 30° = 0.5

Sine 45° = 0.707

0.5 + 0.707 = 1.207

So

Sine (30° + 45°) = sine 75° = ?

Sine (30° + 45°) | does not equal | sin 30° + sin 45°

Sine of an angle is *never more than 1*

Finding sin(A + B)

The easiest way to find sin(*A* + *B*), uses the geometrical construction shown here. The big angle, (*A* + *B*), consists of two smaller ones, *A* and *B*. The construction (1) shows that the opposite side is made of two parts. The lower part, divided by the line between the angles (2), is is sin A. The line between the two angles divided by the hypotenuse (3) is cos *B*. Multiply the two together. The middle line is in both the numerator and denominator, so each cancels and leaves the lower part of the opposite over the hypotenuse (4).

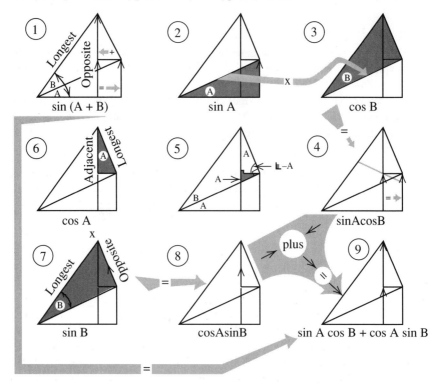

Notice the little right triangle (5). The shaded angle is *A*, because the line on its top side is parallel to the base line. Similar right triangles with an angle *A* show that the top angle, marked *A*, also equals the original *A*. The top part of the opposite (6), over the longest of that shaded triangle, is cos *A*. The opposite over the main hypotenuse (7) is sin *B*. Since the side marked "opposite" (7) is in both the numerator and denominator when cos *A* and sin *B* are multiplied together, cos *A* sin *B* is the top part of the original opposite—for (*A* + *B*)—divided by the main hypotenuse (8).

Now, put it all together (9). Sin (*A* + *B*) is the two parts of the opposite—all divided by the hypotenuse (9). Putting that into its trig form: sin(*A* + *B*) = sin *A* cos *B* + cos *A* sin *B*.

Finding cos(*A* + *B*)

A very similar construction finds the formula for the cosine of an angle made with two angles added together.

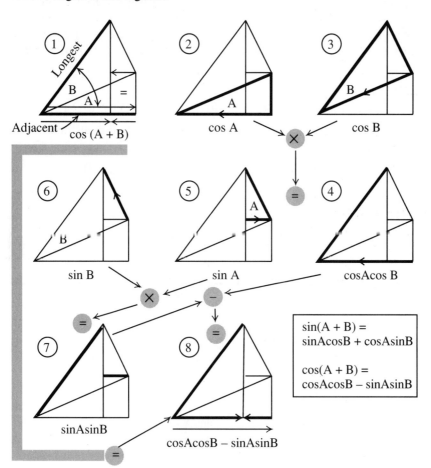

Using the same construction (1), notice that the adjacent side is the full base line (for cos *A*), with part of it subtracted at the right. Each part must use the same denominator, the hypotenuse of the (*A* + *B*) triangle.

The full base line, divided by the dividing line between angles *A* and *B*, is cos *A* (2). This dividing line, divided by the hypotenuse of (*A* + *B*) triangle, is cos *B* (3). So, the full base line divided by the hypotenuse is the product cos *A* cos *B* (4).

Now, for the little part that has to be subtracted. The shaded part (5) represents sin *A*, which multipled by the shaded part (6) is sin *B*, which produces the other piece you need (7). The subtraction produces cos(*A* + *B*) (8) so that the formula we need is:

$$\cos(A + B) = \cos A \cos B - \sin A \sin B$$

Finding tan(*A* + *B*)

A complete goemetric derivation of the formula for tan(*A* + *B*) is complicated. An easy way is to derive it from the two formulas that you have already done. In any angle, the tangent is equal to the sine divided by the cosine. Using that fact, tan(*A* + *B*) = sin(*A* + *B*)/cos(*A* + *B*). In a way that does it, but you can expand that to:

tan(*A* + *B*) =
　　　[sin *A* cos *B* + cos *A* sin *B*]/[cos *A* cos *B* − sin *A* sin *B*]

Divide through top and bottom by cos *A* cos *B*, which turns all the terms into tangents, giving:

$$\tan(A + B) = [\tan A + \tan B]/[1 - \tan A \tan B]$$

tan (A + B)

$$\frac{\text{opposite}}{\text{longest}} \div \frac{\text{adjacent}}{\text{longest}} = \frac{\text{opposite}}{\text{longest}} \times \frac{\text{longest}}{\text{adjacent}} = \frac{\text{opposite}}{\text{adjacent}}$$

$$\tan (A + B) = \frac{\sin (A + B)}{\cos (A + B)} = \frac{\sin A \cos B + \cos A \sin B}{\cos A \cos B - \sin A \sin B}$$

$$= \frac{\dfrac{\sin A \cos B}{\cos A \cos B} + \dfrac{\cos A \sin B}{\cos A \cos B}}{\dfrac{\cos A \cos B}{\cos A \cos B} - \dfrac{\sin A \sin B}{\cos A \cos B}}$$

$$= \frac{\tan A + \tan B}{1 - \tan A \tan B}$$

Ratios for 75 degrees

Show the ratios for sine, cosine, and tangent by substituting into the sum formula, then reducing the result to its simplest form, before evaluating the surds. After making the basic substitutions in each case, the rough work is in shading— to show how the result is reduced to the simplest form for evaluation.

sin 75°

$$= \sin (30° + 45°) = \sin 30° \cos 45° + \cos 30° \sin 45°$$

$$\frac{1 + \sqrt{3}}{2 \sqrt{2}} = \frac{\sqrt{2} + \sqrt{6}}{4}$$

$\sqrt{2} = 1.414$
$\sqrt{6} = 2.449$
$\ \ 3.863$

$$= \frac{1}{2} \times \frac{1}{\sqrt{2}} + \frac{3}{2} \times \frac{1}{\sqrt{2}}$$

$4\overline{)3.863}$ gives 0.966

$$= \frac{1 + \sqrt{3}}{2 \sqrt{2}} = 0.966$$

cos 75°

$$= \cos (30° + 45°) = \cos 30° \cos 45° - \sin 30° \sin 45°$$

$$\frac{\sqrt{3} - 1}{2 \sqrt{2}} = \frac{\sqrt{6} - \sqrt{2}}{4}$$

$\sqrt{6} = 2.449$
$\sqrt{2} = 1.414$
$\ \ 1.035$

$$= \frac{\sqrt{3}}{2} \times \frac{1}{\sqrt{2}} - \frac{1}{2} \times \frac{1}{\sqrt{2}}$$

$4\overline{)1.035}$ gives 0.259

$$= \frac{\sqrt{3} - 1}{2 \sqrt{2}} = 0.259$$

tan 75°

$$= \tan (30° + 45°) = \frac{\tan 30° + \tan 45°}{1 - \tan 30° \tan 45°}$$

$$\frac{\sqrt{3} + 1}{\sqrt{3} - 1} = \frac{(\sqrt{3} + 1)^2}{(\sqrt{3} - 1)(\sqrt{3} + 1)} = \frac{4 + 2\sqrt{3}}{3 - 1}$$

$$= \frac{\frac{1}{\sqrt{3}} + 1}{1 - \frac{1}{\sqrt{3}}} = \frac{\sqrt{3} + 1}{\sqrt{3} - 1}$$

$$= \frac{4 + 2\sqrt{3}}{2} = 2 + \sqrt{3}$$

$$= 3.732$$

If you use your pocket calculator for evaluation, it will probably make no difference whether you simplify the expressions first or just plow through it! Everything depends on the calculator: some do make a difference, some don't!

Ratios of angles greater than 90 degrees

So far, ratios of acute angles (between 0 and 90 degrees) have been considered. Other triangles with obtuse angles (over 90 degrees) might go over 180 degrees in later problems. To simplify classification of angles according to size, they are divided into quadrants.

A *quadrant* is a quarter of a circle. Since the circle is commonly divided into 360 degrees, the quadrants are named by 90-degree segments. 0−90 degrees is the 1st quadrant, 90−180 the 2nd, 180−270 the 3rd, and 270−360 the 4th.

Drawing in lines to represent the quadrant boundaries, with 0 or 360 horizontal to the right, 90 vertical up, 180 horizontal to the left, and 270 vertical down. Now, use this method for plotting graphs.

Progressively larger angles are defined by a rotating vector, starting from zero and rotating counterclockwise. Horizontal elements are x: positive to the right, negative to the left. Vertical elements are y: positive up, negative down. The rotating vector is r. So, the sine of an angle is y/r, the cosine x/r, and the tangent y/x. The vector r is always positive. So, the sign of the ratios can be figured for the various quadrants.

Here, the signs of the three ratios have been tabulated for the four quadrants. Also how the equivalent angle in the first quadrant "switches" as the vector passes from one quadrant to the next. In the first quadrant, the sides were defined in the ratios for sine, cosine, and tangent. As you move into bigger angles in the remaining quadrants (called the opposite side) is always the vertical (y). What was called the adjacent is always the horizontal (x). The hypotenuse is always the rotating vector (r). You will begin to see a pattern to the way these trigonometric ratios for angles vary.

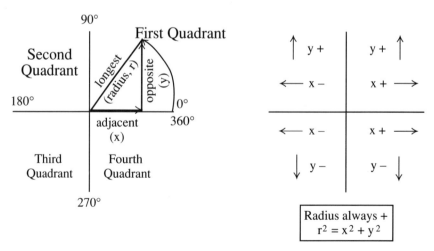

RATIOS in the FOUR QUADRANTS

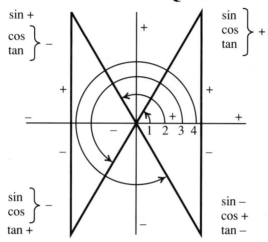

Quadrant	Angle	Sine	Cosine	Tangent
1st	A	$\frac{+}{+} = +$	$\frac{+}{+} = +$	$\frac{+}{+} = +$
2nd	$180° - A$	$\frac{+}{+} = +$	$\frac{-}{+} = -$	$\frac{+}{-} = -$
3rd	$180° + A$	$\frac{-}{+} = -$	$\frac{-}{+} = -$	$\frac{-}{-} = +$
4th	$-A$	$\frac{-}{+} = -$	$\frac{+}{+} = +$	$\frac{-}{+} = -$

Ratios for difference angles

Now, you have two ways to obtain formulas for difference angles. First, use a geometric construction, such as the one that was used for sum angles, reversing it so that $(A - B)$ is the angle B subtracted from the angle A >.

In reasoning similar to that which was used for the sum angles, presented here somewhat abbreviated, are the sine and cosine formulas:

and
$$\sin(A - B) = \sin A \cos B - \cos A \sin B$$
$$\cos(A - B) = \cos A \cos B + \sin A \sin B$$

Geometrical Construction

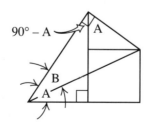

$90° - A$

continued

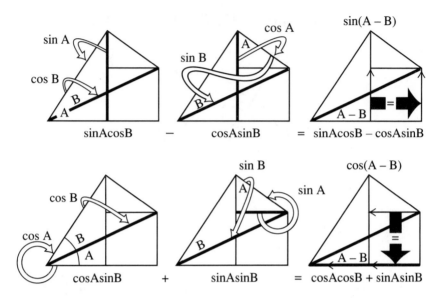

Sum and difference formulas

The second method of finding the formula for difference angles uses the sum formula already obtained, but makes *B* negative. From our investigation of the signs for various quadrants, negative angles from the 1st quadrant will be in the 4th quadrant. Making this substitution produces the same results that arrived geometrically in the previous section.

Finding the tangent formula follows the same method, either going through substitution into the sine and cosine formulas, or more directly, by making tan $(-B) = -\tan B$. Either way you get:

$$\tan(A - B) = [\tan A - \tan B]/[1 + \tan A \tan B]$$

$\boxed{\sin (A - B)}$	$= \sin (A + [-B])$	$= \sin A\cos [-B] + \cos A\sin [-B]$
		$= \sin A\cos B + \cos A (-\sin B)$
		$= \sin A\cos B - \cos A\sin B$

$\boxed{\cos (A - B)}$	$= \cos (A + [-B])$	$= \cos A\cos [-B] - \sin A\sin [-B]$
		$= \cos A\cos B - \sin A (-\sin B)$
		$= \cos A\cos B + \sin A\sin B$

$\boxed{\tan (A - B)}$	$= \tan (A + [-B])$	$= \dfrac{\tan A + \tan [-B]}{1 - \tan A\tan[-B]}$
		$= \dfrac{\tan A - \tan B}{1 - \tan A (-\tan B)}$
		$= \dfrac{\tan A - \tan B}{1 + \tan A\tan B}$

$$\boxed{\sin 15°} \quad = \sin (45° - 30°) \quad = \sin 45° \cos 30° - \cos 45° \sin 30°$$

$$= \frac{1}{\sqrt{2}} \times \frac{\sqrt{3}}{2} - \frac{1}{\sqrt{2}} \times \frac{1}{2} = 0.259$$

$$\boxed{\cos 15°} \quad = \cos (45° - 30°) \qquad \cos 45° \cos 30° + \sin 45° \sin 30°$$

$$= \frac{1}{\sqrt{2}} \times \frac{1}{2} + \frac{1}{\sqrt{2}} \times \frac{\sqrt{3}}{2} = 0.966$$

$$\boxed{\tan 15°} \quad = \tan (45° - 30°) \quad = \frac{\tan 45° - \tan 30°}{1 + \tan 45° \tan 30°} = \frac{1 - \dfrac{1}{\sqrt{3}}}{1 + \dfrac{1}{\sqrt{3}}}$$

$$= \frac{\sqrt{3} - 1}{\sqrt{3} + 1} = 2 - \sqrt{3} = \underline{\underline{0.268}}$$

Ratios through the four quadrants

You can deduce a few more ratios with the sum and difference formulas. You already did ratios for 75 degrees. Now, do those for 15 degrees. These formulas give ratios for angles at 15-degree intervals through the four quadrants. Plotting them out for the full 360 degrees, you can see how the three ratios change as the vector sweeps through the four quadrants.

Both the sine and cosine "wave" up and down between +1 and −1. Notice that the "waves" are displaced by 90 degrees, one from the other. This fact becomes important later.

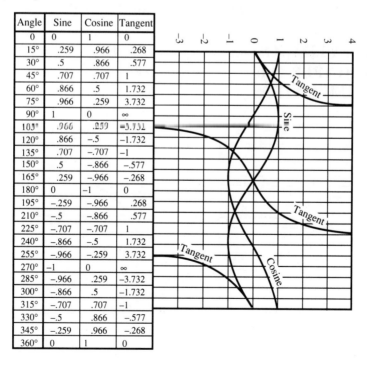

Angle	Sine	Cosine	Tangent
0	0	1	0
15°	.259	.966	.268
30°	.5	.866	.577
45°	.707	.707	1
60°	.866	.5	1.732
75°	.966	.259	3.732
90°	1	0	∞
105°	.966	.259	−3.732
120°	.866	−.5	−1.732
135°	.707	−.707	−1
150°	.5	−.866	−.577
165°	.259	−.966	−.268
180°	0	−1	0
195°	−.259	−.966	.268
210°	−.5	−.866	.577
225°	−.707	−.707	1
240°	−.866	−.5	1.732
255°	−.966	−.259	3.732
270°	−1	0	∞
285°	−.966	.259	−3.732
300°	−.866	.5	−1.732
315°	−.707	.707	−1
330°	−.5	.866	−.577
345°	−.259	.966	−.268
360°	0	1	0

The tangent starts out like the sine curve, but quickly it sweeps up to reach infinity at 90 degrees. Going "off scale" in the positive direction, it "comes on" from the negative direction on the other side of 90 degrees. Going through the 180-degree point, the tangent curve duplicates what it does going through 0 or 360 (whichever you view it as). At 270 degrees, it repeats what it did at 90 degrees.

Pythagoras in trigonometry

A formula can often be simplified, as was found by deriving the tangent formulas from the sine and cosine formulas, and changing it from terms using one ratio to terms using another ratio. In doing this, the Pythagorean theorem, expressed in trigonometry ratios, is very handy.

Assume that a right triangle has a hypotenuse of 1 unit long. Then one of the other sides will have a length of sin A and the other of cos A. From that, the Pythagorean theorem shows that: $\cos^2 A + \sin^2 A = 1$. This statement is always true, for any value of A.

A little thing here about the way it's written. $\cos^2 A$ means $(\cos A)^2$. If you wrote it cos A^2, the equation would mean something else. A is a number in some angular notation that represents an angle. A^2 would be the same number squared. Its value would depend on the angular notation used, so it's not a good term to use. What is meant is the angle's sine or cosine squared, not the angle itself.

The Pythagoras formula can be transposed. For instance, two other forms are: $\cos^2 A = 1 - \sin^2 A$, and $\sin^2 A = 1 - \cos^2 A$.

The PYTHAGORAS PRINCIPLE and ANGLE RATIOS

$$\cos^2 A + \sin^2 A = 1$$

$$\cos^2 A = 1 - \sin^2 A$$

$$\sin^2 A = 1 - \cos^2 A$$

$$\cos A = \sqrt{1 - \sin^2 A}$$

$$\sin A = \sqrt{1 - \cos^2 A}$$

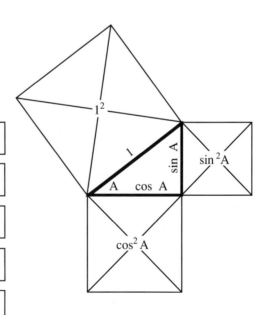

Multiple angles

The sum formulas, along with the Pythagorean theorem, are used for angles that are 2, 3, or a greater exact multiple of any original angle. Here, give formulas for $2A$ and $3A$. The same method is pursued further in Parts 3 and 4 of this book.

The sum formula works whether both angles are the same or different: $\sin(A + B)$ or $\sin(A + A)$. However, $\sin(A + A)$ is really $\sin 2A$. So, $\sin 2A$ is $\sin A \cos A + \cos A \sin A$. They are both the same product, in opposite order, so this statement can be simplified to $\sin 2A = 2 \sin A \cos A$.

Similarly, $\cos 2A = \cos A \cos A - \sin A \sin A$, which also can be written: $\cos 2A = \cos^2 A - \sin^2 A$. Using the Pythagorean theorem, change that to: $\cos 2A = 2 \cos^2 A - 1$. Finally, $\tan 2A = 2 \tan A/[1 - \tan^2 A]$.

Now, the triple angle ($3A$) is used just to show how further multiples are obtained. Basically, it's as simple as writing $3A = 2A + A$ and reapplying the sum formulas. But then, to get the resulting formula in workable form, you need to subtitute for the $2A$ part to get everything into terms of ratios for the simple angle A.

Work your way through the three derivations shown here. You can see that it will get more complicated for $4A$ and more (in Parts 3 and 4 of this book).

MULTIPLE ANGLES   Derived from Sum Formulas

$$\boxed{\sin 2A} = \sin (A + A) = \sin A \cos A + \cos A \sin A$$
$$= 2\sin A \cos A$$
$$= 2\sin A \sqrt{1 - \sin^2 A}$$

$$\boxed{\cos 2A} = \cos (A + A) = \cos A \cos A - \sin A \sin A$$
$$= \cos^2 A - \sin^2 A = \cos^2 A - (1 - \cos^2 A)$$
$$= 2\cos^2 A - 1$$

$$\boxed{\tan 2A} = \tan (A + A) = \frac{\tan A + \tan A}{1 - \tan A \tan A} = \frac{2\tan A}{1 - \tan^2 A}$$

or

$$\boxed{\tan 2A} = \frac{\sin 2A}{\cos 2A} = \frac{2 \sin A \cos A}{\cos^2 A - \sin^2 A} = \frac{2 \dfrac{\sin A \cos A}{\cos^2 A}}{\dfrac{\cos^2 A}{\cos^2 A} - \dfrac{\sin^2 A}{\cos^2 A}}$$

$$= \frac{2 \tan A}{1 - \tan^2 A}$$

MULTIPLE ANGLES   Ratios for 3A

$\boxed{\sin 3A}$ $= \sin(A + 2A)$ $= \sin A \cos 2A + \cos A \sin 2A$

$= \sin A (\cos^2 A - \sin^2 A) + 2\cos A \sin A \cos A$

$= \sin A (1 - 2\sin^2 A) + 2(1 - \sin^2 A) \sin A$

$= \sin A (3 - 4\sin^2 A)$ or

$= 3\sin A - 4\sin^3 A$

$\boxed{\cos 3A}$ $= \cos(A + 2A)$ $= \cos A \cos 2A - \sin A \sin 2A$

$= \cos A (2\cos^2 A - 1) - 2\sin^2 A \cos A$

$= \cos A (2\cos^2 A - 1) - 2(1 - \cos^2 A)\cos A$

$= \cos A (4\cos^2 A - 3)$ or

$= 4\cos A - 3\cos A$

$\boxed{\tan 3A}$ $= \tan(A + 2A)$ $= \dfrac{\tan A + \tan 2A}{1 - \tan A \tan 2A}$

$$= \frac{\tan A + \dfrac{2\tan A}{1 - \tan^2 A}}{1 - \dfrac{2\tan^2 A}{1 - \tan^2 A}}$$

$$= \frac{\tan A (1 - \tan^2 A) + 2\tan A}{1 - \tan^2 A - 2\tan^2 A}$$

$$= \frac{3\tan A - \tan^3 A}{1 - 3\tan^2 A}$$

Properties of the isosceles triangle

You have already seen that a right triangle is a useful building block for other shapes. An isosceles triangle has slightly different uses. But the fact on which these uses are based is that an isosceles triangle has two equal sides and two equal angles opposite those two sides. A perpendicular from the third angle (not one of the equal ones) to the third side (not one of the equal ones) bisects that third side. That is, it divides it into two equal parts, making the whole triangle into mirror-image right triangles.

With isosceles triangles, any triangle, except a right triangle, can be divided into three adjoining isosceles triangles, by dividing each side into two equal parts and erecting perpendiculars from the points of bisection. Where any two of these bisecting perpendiculars meet, if lines are drawn to the corners of the original triangle, the three lines must be equal, because two of them form the sides of an isosceles triangle. So, the perpendicular from the third side of the original triangle must also meet in the same point.

This statement is true, as we show here, whether the original triangle is acute or obtuse. The difference with an obtuse-angled triangle is that the meeting point is outside the original triangle, instead of inside.

What does a right triangle do? Perpendiculars from the mid-point of the hypotenuse to the other two sides will bisect those two sides—you get two out of three! The meeting point happens to sit on the hypotenuse.

$$\cos A = \sqrt{1 - \sin^2 A}$$

$$\text{and} \quad \sin A = \sqrt{1 - \cos^2 A}$$

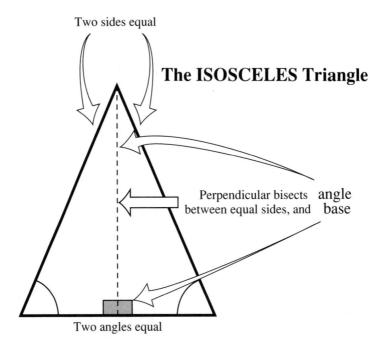

Two sides equal

The ISOSCELES Triangle

Perpendicular bisects angle
between equal sides, and base

Two angles equal

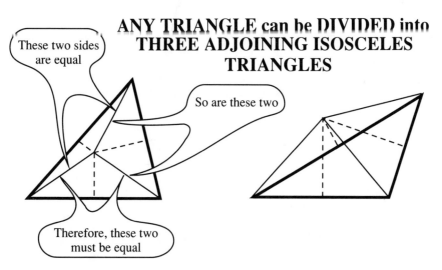

These two sides
are equal

ANY TRIANGLE can be DIVIDED into THREE ADJOINING ISOSCELES TRIANGLES

So are these two

Therefore, these two
must be equal

Angles in a circle

A basic property of a circle is that its center is at an equal distance from every point on its circumference. This equal distance is the radius of the circle.

If you draw any triangle inside a circle, the perpendiculars from the mid-points of its sides will meet at the circle's center and radii from the corners of the triangle will divide it into 3 isosceles triangles.

Now, if you name the equal pairs of angles in each isosceles triangle, A, A, B, B, C, C, you find that the original triangle has one angle $A + B$, one angle $B + C$, and one angle $A + C$. The three angles total $2A + 2B + 2C$. This, you know, adds up to 180 degrees.

In any isosceles triangle, the angle at the apex is 180 degrees minus twice each base angle. Because of the fact deduced in the previous paragraph, $180 - 2A$ must be the same as $2B + 2C$, for example.

Consider the angles that are opposite from the part of the circle, against which the top left side of the triangle sits. The angle at the center is $2B + 2C$, as just deduced. The angle at the circumference is $B + C$. You will find that, for any segment of a circle, the angle at the center is always twice the angle at the circumference.

The proof on the previous page leads to an interesting fact about angles in circles. Instead of identifying the angles with a side of a triangle, use an arc (portion of the circumference) of the circle. The important thing is the angle that corresponds to the arc at the center. A part of the circumference of a circle that is identified by the angle at the center is called the *chord* of the circle.

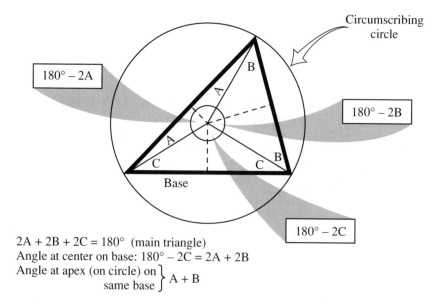

$2A + 2B + 2C = 180°$ (main triangle)
Angle at center on base: $180° - 2C = 2A + 2B$
Angle at apex (on circle) on $\left.\begin{array}{l} \\ \text{same base} \end{array}\right\}$ $A + B$

The angle at the center is twice the angle at the circumference

Any angle drawn touching the circumference, using this chord as termination for the lines bounding the angle, must be just half the angle at the center. Thus, all the angles in a circle, based on the same chord, must be equal. Suppose that the chord has an angle of 120 degrees. The angles at the circumference will all be exactly 60 degrees.

A special case is the semicircle (an exact half circle). The angle at the center is a straight line (180 degrees). Every angle at the circumference of a semicircle is exactly 90 degrees (a right angle). Any triangle in a semicircle is a right triangle.

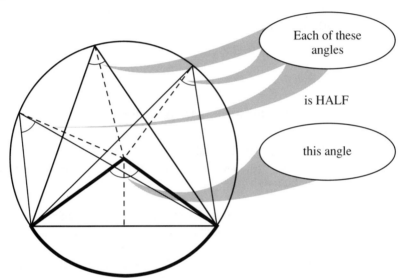

Each of these angles

is HALF

this angle

Every angle on the same *chord* is equal

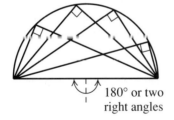

180° or two right angles

Every angle in a semicircle is a right angle

Definitions

The previous pages have often used angles that add up to either a right angle (90 degrees) or to two right angles (180 degrees). When two angles add up to 180 degrees (two right angles), they are called *supplementary*. When two angles add up to 90 degrees (one right angle), they are called *complementary*.

DEFINITIONS:

Supplementary angles	Complementary angles

Two angles that make up 180°,
or two right angles,
are **supplementary**

Two angles that make up 90°,
or one right angle,
are **complementary**

Questions and problems

1. The sine of angle A is 0.8 and the sine of angle B is 0.6. From the various relationships obtained so far, find the following: tan A, tan B, sin($A + B$), cos($A + B$), sin($A − B$), cos($A − B$), tan($A + B$), and tan($A − B$), without using tables or a calculator's trig buttons.

2. At the equator, Earth has a radius of 4000 miles. Angles around the equator are measured in meridians of longitude, with a north-to-south line through Greenwich, England as the zero reference. Two places are used to observe the moon: one is Mt. Kenya, on the equator at 37.5 east of Greenwich; the other is Sumatra, on the equator, at 100.5 east. How far apart are these two places, measured by an imaginary straight line through Earth's crust?

3. If sights were made horizontally from the observation points in question 2 (due east from the first, due west from the second), at what angle would the lines of sight cross?

4. At a certain time, exactly synchronized at both places, the moon is observed. In Kenya, the elevation of a line of sight, centered on the center of the moon's disc, is 58 degrees above horizontal, eastward. In Sumatra, the elevation is 58 degrees above horizontal, westward. How far away is the moon? Use the distance between the points calculated in question 2.

5. The cosine of a certain angle is exactly twice the sine of the same angle. What is the tangent of this angle? You don't need either tables or calculator for this question.

6. The sine of a certain angle is exactly 0.28. Find the cosine and tangent without tables or the trig functions on your calculator.

7. The sine of a certain angle is 0.6. Find the sine of twice this angle and three times this angle.

8. Find the sine and cosine of an angle exactly twice that of question 6.

9. Using 15 degrees as a unit angle, and the formulas for ratios of $2A$ and $3A$, find the values of sine, cosine, and tangent for 30 and 45 degrees. Do this calculation using exact values with surds and simplifying. Then try the problem with decimal values (with your calculator, if you wish).

10. Using 30 degrees as a unit angle, find the values for sine, cosine, and tangent for angles of 60 and 90 degrees to confirm those derived directly.

11. Using 45 degrees as a unit angle, find values for sine, cosine, and tangent for angles 90 and 135 degrees. Confirm your results from the quadrant information on page 237.

12. Using 60 degrees as a unit angle, find values for sine, cosine, and tangent for angles 120 and 180 degrees. Confirm your results from the quadrant information on page 237.

13. Using 90 degrees as a unit angle, find values for sine, cosine, and tangent for angles 180 and 270 degrees. Check your results from the quadrant boundary information on page 237.

14. Prove that the sine of an angle that is twice a certain original angle is equal to the cosine of an angle that consists of the difference between the original angle and its complement.

15. Using the tangent formulas for multiple angles and the tables, find the tangents for three times 29, 31, 59, and 61 degrees. Account for the changes in sign between three times 29 and 31 degrees and between 59 and 61 degrees.

16. The sine of an angle is 0.96. Find the sine and cosine for twice the angle.

17. A problem leads to an algebraic expression of the form $8\cos^2 A + \cos A = 3$. Solve for $\cos A$, and state in which quadrant the angle representing each solution will come. Give approximate values from tables or your calculator.

18. A problem leads to simultaneous trigonometry equations:

and
$$\cos^2 A + 0.1168 \cos A = 0.809472$$
$$\sin^2 A + 0.2576 \sin A = 0.150528$$

By solving each of these quadratics, find the possible solutions of $\cos A$ and $\sin A$, deduce which quadrants the possible answers are in, and find approximate values (in degrees).

Part 3

Developing algebra, geometry, trigonometry, and calculus

15

CHAPTER

Systems of counting

Degrees of accuracy

The system of counting that you use affects the accuracy of some calculations in what might seem to be an erratic way. One example can be seen in taking the square root of 3, with the decimal system. Follow the method from part 2 for taking the square root of 2.

APPROXIMATIONS

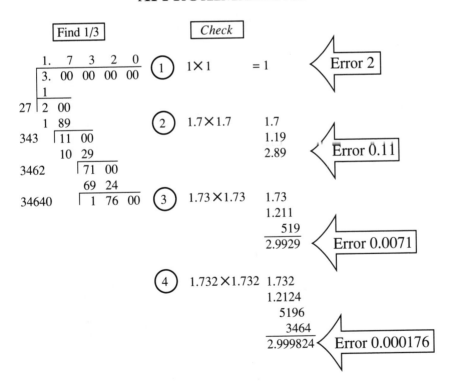

Find 1/3					Check			

| 1. | 7 | 3 | 2 | 0 | | | | |
| 3. | 00 | 00 | 00 | 00 | ① | 1×1 | = 1 | Error 2 |

```
        1
27 | 2  00
    1  89                    ② 1.7×1.7    1.7
343 | 11  00                              1.19
      10  29                              2.89     Error 0.11
3462  | 71  00
        69  24
34640 | 1  76  00           ③ 1.73×1.73   1.73
                                          1.211
                                           519
                                         2.9929    Error 0.0071

                           ④ 1.732×1.732  1.732
                                          1.2124
                                           5196
                                           3464
                                         2.999824  Error 0.000176
```

249

Notice how each "place" in the decimal system, yields a closer approximation to the square root of 3. To test it, see how close squaring the root brings you to the square that you started with: 3.

The first place is 1, which squared is only 1—an error of 2 from the true square of 3. Had 2 been used, the answer would have been closer: a square of 4—an error of 1. But our rule is to stay below the true value. Another method could use the closest value before going to the next step.

The second place comes closer quickly. 1.7 squared is 2.89, reducing the error to 0.11. The third place, 1.73 yields a square of 2.9929—an error of 0.0071. The fourth place, 1.732 comes a lot closer, making a square of 2.999824—an error of 0.000176.

Fractions in extended system counting

If you used a *septimal* (7s) *system*, a fraction of 1/7 would be 0.1—completely accurate with only one place beyond the point (not a decimal point, if this system is a septimal). In the decimal system, the fraction that results from dividing by 7 isn't so easy.

Follow the same error-noting procedure. Though noting the progressive reduction of error is similar, of more interest is the kind of decimal fractions that repeat.

FRACTIONS and DECIMALS

Decimal equivalent of: $\dfrac{1}{7}$

$$0.1 = \frac{1}{10} \qquad \frac{1}{7} \qquad = \frac{3}{10}$$

$$0.14 = \frac{14}{100} = \frac{7}{50} \qquad \frac{7}{49} \qquad \frac{1}{50}$$

$$0.142 = \frac{142}{1000} = \frac{71}{500} \qquad \frac{71}{497} \qquad \frac{3}{500}$$

$$0.1428 = \frac{1428}{10000} = \frac{357}{2500} \qquad \frac{357}{2499} \qquad \frac{1}{2500}$$

$$0.14285 = \frac{14285}{100000} = \frac{2857}{20000} \qquad \frac{2857}{19999} \qquad \frac{1}{20000}$$

$$0.142857 = \frac{142857}{1000000} \qquad \frac{142857}{999999} \qquad \frac{1}{1000000}$$

Error

These problems should make you ask how accurate or reliable the figures are. What does an error of 1 part in 1 million mean? Whether you happen (which is very unlikely) to be using a septimal system instead of a decimal one, just how precise is 1/7?

Orders of magnitude

The orders of magnitude begin another whole new concept in mathematics. To show another angle of this concept, suppose you are approaching an area that consists of a perfect square. To get the area you need more accurately, you add or subtract a little bit to or from both dimensions. Starting with a square of dimension L each way, you either add or subtract small piece S to or from each dimension. The change in area consists of two small, long slices (dimensions L by S) and one very much smaller piece that measures S both ways. The smaller S is, relative to L, the smaller S squared, relative to SL.

You could extend this concept to a similar adjustment on a cubic volume. Now, starting with a big cube, L each way, you add or subtract 3 slabs that are L square and S thick, three sticks that are L long by S square, and one very tiny cube that is S cubed. If S is 1/10 of L (and it might be much smaller), then S cubed is 1/1000 of L cubed.

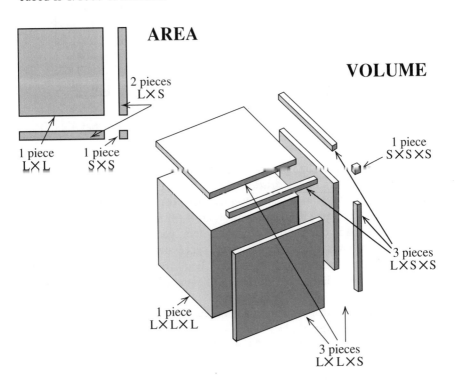

AREA

2 pieces
L×S

1 piece
L×L

1 piece
S×S

VOLUME

1 piece
S×S×S

3 pieces
L×S×S

1 piece
L×L×L

3 pieces
L×L×S

ORDERS of MAGNITUDE

$$1 + a$$
$$1 + a$$
$$1 + a$$
$$\underline{ a + a^2}$$
$$1 + 2a + a^2$$

$$(1 + a)^2 = 1 + 2a + a^2 \quad \Longleftarrow \quad$$

$$1 + a$$
$$\underline{1 + 2a + a^2}$$
$$ a + 2a^2 + a^3$$
$$1 + 3a + 3a^2 + a^3$$

$$(1 + a)^3 = 1 + 3a + 3a^2 + a^3 \quad \Longleftarrow \quad$$

$$1 + a$$
$$\underline{1 + 3a + 3a^2 + a^3}$$
$$ a + 3a^2 + 3a^3 + a^4$$
$$1 + 4a + 6a^2 + 4a^3 + a^4$$

$$(1 + a)^4 = 1 + 4a + 6a^2 + 4a^3 + a^4 \quad \Longleftarrow \quad$$

If a is	$(1 + a)^2$ is	$(1 + a)^3$ is	$(1 + a)^4$ is
0.1	1.21	1.331	1.4641
0.01	1.0201	1.030301	1.04060401
0.2	1.44	1.$\boxed{7}$28	$\boxed{2.073}$6
	$1 + 2a + a^2$	$1 + 3a + 3a^2 + a^3$	$1 + 4a + 6a^2 + 4a^3 + a^4$

You can show the same progression algebraically. To do this, if *a* is a small fraction, then powers of *a*, a^2, a^3, a^4, etc., consist of a descending series of orders of magnitude. Notice that successive powers of *a* have a series of coefficients which, if you take the fourth power, are 1, 4, 6, 4, and 1.

Still lingering in our familiar decimal system, you substitute different values for *a*, and show how changing it changes the successive powers of $(1 + a)$. If *a* is 0.1, successive powers begin to "spill over" into earlier "places". Up to the 4th power, the first two digits are 1.1, 1.2, 1.3, but at the 4th power, 1.5 would be nearer.

If *a* is 0.01, higher powers of *a* do not interfere with the first term, which is now in the second decimal place. Drop what follows the second place, the first two places are now 1.01, 1.02, 1.03, and 1.04. Further terms in that 4th power only make it 1.0406 in the 4th place.

However if *a* is 0.2, the later terms much sooner intrude into earlier ones. The blocked figures show this intrusion.

Systems of counting

Before electronic digital devices were invented, we used counters with little wheels that carried numbers. The numbers that showed through the front window were like those that electronic digital devices display. If you took the cover off the rest of the wheel, you could see how it worked, which helped you understand number systems.

The right-most wheel counted from 0 to 9 on a decimal system. When it came to 9, it would move from 9 to 0, and move the next wheel from 0 to 1. Every time the first wheel passed from 9 to 0, the next wheel would advance 1 more, until it returned to 9. Then, two wheels would read 99. As the first wheel moved from 9 to 0 this time, the next one would also move from 9 to 0, and the third wheel would move from 0 to 1, making it read 100.

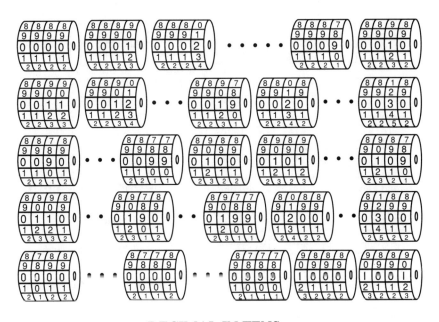

DECIMAL IN TENS

1

1 2 3 4 5 6 7 8 9 0

Duodecimal system

The decimal (base ten) system is not the only system that you could use. Years ago, some cultures used the *duodecimal system*—counting to twelve instead of ten. To use a wheel-counter system, you would need two more numbers on each wheel. In the wheels shown here, the extra symbols are *t* and *e* for ten and eleven. Modern digital systems more often use base 16, called the hexadecimal system.

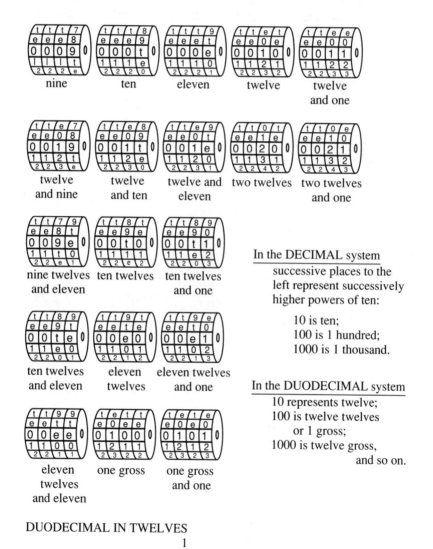

nine   ten   eleven   twelve   twelve
                                                    and one

twelve   twelve   twelve and   two twelves   two twelves
and nine   and ten   eleven   and one

nine twelves   ten twelves   ten twelves
and eleven   and one

ten twelves   eleven   eleven twelves
and eleven   twelves   and one

eleven   one gross   one gross
twelves   and one
and eleven

In the DECIMAL system

successive places to the left represent successively higher powers of ten:

10 is ten;
100 is 1 hundred;
1000 is 1 thousand.

In the DUODECIMAL system

10 represents twelve;
100 is twelve twelves
   or 1 gross;
1000 is twelve gross,
            and so on.

DUODECIMAL IN TWELVES
1
1 2 3 4 5 6 7 8 9 t e 0

The first 6 letters of the alphabet complete the single digit numbers up to what would be called 15.

Decimal
0 1 2 3 4 5 6 7 8 9 10 11 12 13 14 15

Hexadecimal
0 1 2 3 4 5 6 7 8 9 A B C D E F

In the decimal system, "10" (one zero) means ten. In the duodecimal system, "10" means twelve. In hexadecimal, "10" means sixteen. To get some exercise in different systems, use duodecimal for a bit. You will see why calculators or computers use hexadecimal on the inside—they usually read out in decimal.

Conversion from decimal to duodecimal

Why work in duodecimal when it's never used? Because something unfamiliar makes you think, it makes it easier to understand what is used. Hexadecimal is based on binary (base two), which is not as easy for systems that use a larger number base, because it's difficult to see something that is only two state (like yes or no) as counting. So, look at conversion from decimal to duodecimal.

To find how many times a number counts up to twelve, you divide the number by 12 in the familiar decimal system. The remainder at the bottom is the number of ones left over after a number of complete twelves in the quotient have been passed on to the twelves counter. Then, divide by 12 again. This time the remainder is eleven. In duodecimal, all numbers up to eleven must use a single digit, so *e* is used. You can follow through the rest of this conversion. The duodecimal equivalent of decimal 143131 is 6*t*9*e*7.

Convert decimal 143131 to duodecimal

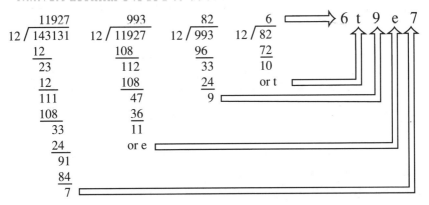

Conversion from duodecimal to decimal

How do you convert from duodecimal to decimal? Simply reverse the process. Using duodecimal, divide the duodecimal number by ten however many times is necessary. You need at least the tens' column of the duodecimal multiplication table. You were probably familiar with the twelve times column—enough to do it fairly easily. However, this way you need to use the ten times column in the twelve system. This system is unfamiliar and it makes you think.

Go down the ten times column. Ten times two are 18. That means 1 twelve and 8, which you would more normally call twenty. Twelve and eight make twenty, don't they? Next, ten times 3 are 26, meaning 2 twelves and 6. Two twelves are 24, and six make what is normally called 30. Finish to the end of the column.

DUODECIMAL MULTIPLICATION TABLE

x	2	3	4	5	6	7	8	9	t	e
2	4	6	8	t	10	12	14	16	18	1t
3	6	9	10	13	16	19	20	23	26	29
4	8	10	14	18	20	24	28	30	34	38
5	t	13	18	21	26	2e	34	39	42	47
6	10	16	20	26	30	36	40	46	50	56
7	12	19	24	2e	36	41	48	53	5t	65
8	14	20	28	34	40	48	54	60	68	74
9	16	23	30	39	46	53	60	69	76	83
t	18	26	34	42	50	5t	68	76	84	92
e	1t	29	38	47	56	65	74	83	92	t1

Convert duodecimal 6+9e7 to decimal

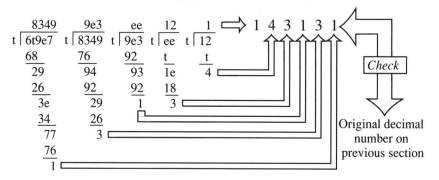

Binary counting

The difficulty about working in binary is that each place has only two "states," which are 0 and 1. You don't count "up to" something and then move to the next place. If you already have 1, the next 1 puts it back to 0 and passes a 1 to the next place. If you have a row of 1s, then adding another 1 shifts them all back to 0, and passes a 1 to the next place (from right to left).

In the window panel here, the decimal number equivalent replaces the binary numbers. In the binary system, every place would be either a 1 or a 0.

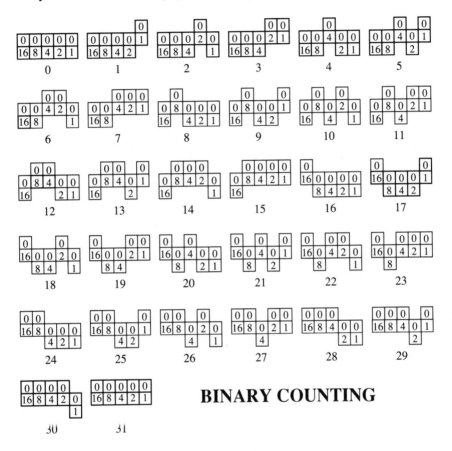

Converting decimal to binary

Here, at the top, values of places in binary that have a 1 instead of a 0, are listed as a decimal. Start with the number in decimal form, 1546. First, the 11th column of binary, is 1024. That puts a 1 in the 11th column of binary. Subtract 1024 from 1546, leaving 522. Next, the 10th column in binary is 512, so subtract 512 from 522, leaving 10, and put a 1 in the 10th column of binary. With 10 left, the next binary digit that you can use is the 4th column, which is 8. So we pass over the columns from 9th to 5th, put a 1 in the 4th column, and subtract 8 from 10 (leaving 2). The 2 puts a 1 in the 2nd column of binary, which finishes the conversion.

To complete what the previous section began, the table on the next page lists the binary equivalents for decimal numbers from 1 to 30.

11	10	9	8	7	6	5	4	3	2	1	Column
1024	512	256	128	64	32	16	8	4	2	1	Value

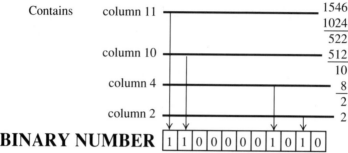

DECIMAL NUMBER

Contains   column 11 ——————————— 1546
1024
———
522
column 10 ——————— 512
———
10
column 4 —————————— 8
—
2
column 2 —————————————— 2

BINARY NUMBER | 1 | 1 | 0 | 0 | 0 | 0 | 0 | 1 | 0 | 1 | 0 |

Decimal Number	Binary Number	Decimal Number	Binary Number	Decimal Number	Binary Number
1	1	11	1011	21	10101
2	10	12	1100	22	10110
3	11	13	1101	23	10111
4	100	14	1110	24	11000
5	101	15	1111	25	11001
6	110	16	10000	26	11010
7	111	17	10001	27	11011
8	1000	18	10010	28	11100
9	1001	19	10011	29	11101
10	1010	20	10100	30	11110

Binary multiplication

Although you enter data into your calculator or computer in the familiar decimal notation, they all use binary to perform all of the mathematical functions that they perform. Try running a sample multiplication, basically as your calculator does it. Suppose you multiply 37 by 27. First, it must convert each number to binary, which it does as you enter the numbers. I'll simplify it a bit by converting it to true binary, instead of one of the biquinary conversions that make it easier for the calculator, but more difficult for you to understand. That comes later.

On the facing page are the conversions of 37 and 27 to pure binary.

Here is multiplication in binary, set out as you would set out ordinary long multiplication, but in a system where no numbers above 1 are "allowed." Every digit must be either 1 or 0. What it really amounts to is adding together the sequence of digits that represent 37 at every "place" where a 1-digit is in 27.

Four 1-digits are in 27, so the three 1-digits in 37 (with 0s interspersed) are entered 4 times in the proper places (to represent "27 times") and added. You can show them all added at once. However, the calculator does it. Every two 1s returns that place to 0 and passes a 1 to the next place to the left.

Working from the right, the first three places each have only one 1, which appears in the sum. The fourth place has two 1s, which make a 0 in that place and pass 1 to the fifth place, which already has a 1 of its own, so it becomes 0 and passes a 1 to the sixth place. This place already has two 1s, so that place goes to 1 again and passes a 1 to the seventh place, where again two 1s are. This place now has a 1 and it passes a 1 to the eighth place. The eigth place has no 1s, so the 1 passed is entered and that's the end of the "passing left." The remaining two places each have a single 1, which gets "brought down." The product, in binary, is 1111100111.

Convert the binary number back to decimal, by putting the decimal equivalent of each binary place where a 1 is. Adding the decimal equivalents comes to 999. To check, multiply 37 by 27, the old fashioned long way.

"Which is the long way?" you might ask. The binary way seems long to you. The only reason a calculator does it so quickly is that it performs millions of "operations" per second. It goes the long way around and calculates more quickly than you can via the short way that you are familiar with.

Multiply 37 × 27 by Binary

Convert to binary:

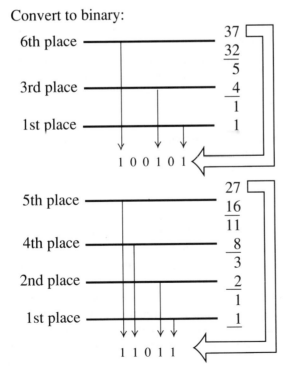

Place	Decimal Number
1	1
2	2
3	4
4	0
5	16
6	32
7	64
8	128
9	256
10	512

BINARY MULTIPLICATION

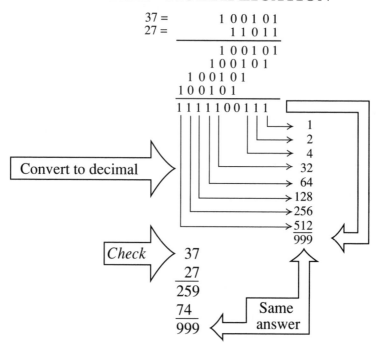

Alternative binary conversion

Here is another way to convert decimals to binary. It uses a table of binary equivalents for numbers from 1 to 9 in each decimal place. To illustrate its use, the two numbers for division on the next page are converted to binary below the table.

Notice that the binary equivalents for a particular digit bear no relationship to one another—from one column to the next. You cannot shift a decimal point or multiply by ten by making a similar shift in binary. I'll return to what calculators or computers do about this problem in a minute.

Binary division

Binary division shows what you learned in part 1 of this book in a rather dramatic way: division is really repetitive subtraction. Subtracting the binary for 37, which is 100101, in the top places of the dividend is exact with no remainder. What is left is the binary for 37 in the last place. So, the quotient, in binary, is 1000001.

To interpret the binary number back to decimal, use some more subtraction in binary and apply the table from the previous section. The first subtraction is the binary for 100, which leaves 11101. For the binary of 20, which leaves 1001, subtract the binary for 9. So working through binary, dividing 4773 by 37 leaves 129 as the quotient.

ALTERNATIVE BINARY CONVERSION

Conversion Table

1000	1111101000	100	1100100	10	1010	1	1
2000	11111010000	200	11001000	20	10100	2	10
3000	101110111000	300	100101100	30	11110	3	11
4000	111110100000	400	110010000	40	101000	4	100
5000	1001110001000	500	111110100	50	110010	5	101
6000	1011101110000	600	1001011000	60	111100	6	110
7000	1101101011000	700	1010111100	70	1000110	7	111
8000	1111101000000	800	1100100000	80	1010000	8	1000
9000	10001100101000	900	1110000100	90	1011010	9	1001

Divide 4773 by 37

4773:

4000	111110100000
700	1010111100
70	1000110
3	11
4773	1001010100101

37:

30	11110
7	111
37	100101

DIVIDE 4773 by 37
1001010100101 by 100101

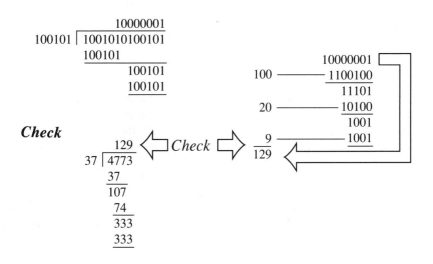

```
            10000001
100101 | 1001010100101
        100101
            100101
            100101
```

Check

```
        129
37 | 4773
     37
     107
      74
     333
     333
```

Check

```
              10000001
100 ───────── 1100100
              11101
 20 ───────── 10100
              1001
  9 ───────── 1001
129
```

Special calculator binary

You noticed that binary digits for various digits in decimal changing with each decimal place makes conversion complicated. When you enter a digit on a calculator, the first digit appears on the right. When you enter the next digit, the first digit moves left and the new one appears to its right. If the calculator had to convert the digit to the new binary sequence for the next place, the system would become very complicated.

So, the calculator allocates 4 binary places for each decimal place, which requires very little more "room" in the calculator's memory than pure binary would. In effect, the calculator now "works" in decimal, but uses 4 "bits" of binary to convey each place of decimal.

You've heard of "bytes" in computer technology. A byte is a sequence of bits that goes together in a "bunch," like the 4 shown here. However a byte contains more than 4 bits. For a long while, a byte consisted of 8 bits, which is why computer technology began using the octal (base 8) system of numbers. Later, a byte was doubled in size to 16 bits, which is why more recent computer technology uses the hexadecimal (base 16) system of numbers.

Decimal Numeral	True Binary	Place Biquinary
1	0001	0001
2	0010	0010
3	0011	0011
4	0100	0100
5	0101	1000
6	0110	1001
7	0111	1010
8	1000	1011
9	1001	1100

Indices

In any system of numbers, binary, octal, decimal, or hexadecimal (or even some others that are not in common use), the place of a number indicates a power of the number on which the system is based. In the binary system, according to where the 1 appears, it represents some power of 2. In the 4th place, it is the 3rd power of 2, which is 8. Here is a comparison between powers of 2 and powers of 10.

From this example, you can see some rules for using indices that help us take further short cuts in multiplication and division. First, remember that multiplication and division are short-cut methods for performing repeated addition and subtraction. Now, *indices* are short-cut methods for repeated multiplication and division.

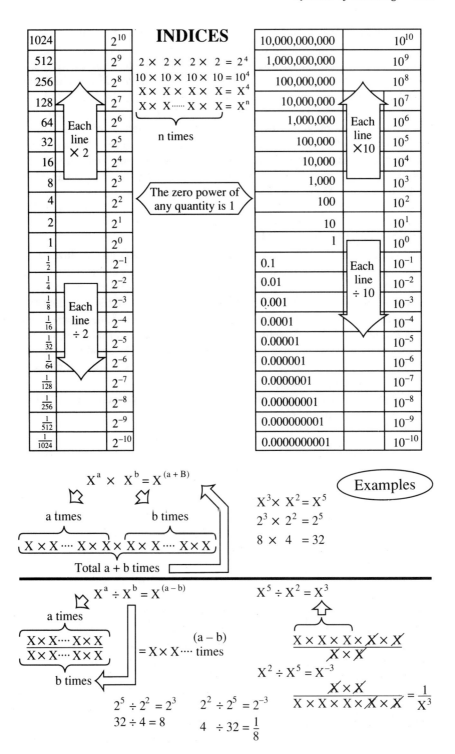

Suppose you have to multiply x^a by x^b. The product is $x^{(a+b)}$. You can easily see it if you write x multiplied by itself a times, then multiply the product by x multiplied by itself b times. The total number of times that you multiply x by itself is $a + b$ times. To illustrate, assume that a is 3 and b is 2; x^3 multiplied by x^2 makes x^5. Numerically, 2^3 is 8, 2^2 is 4, and 2^5 is 32. $8 \times 4 = 32$. It checks.

Now try division. Dividing x^a by x^b, the quotient is $x^{(a-b)}$. You can check this answer by multiplying x by itself a times the numerator of a fraction and using x times itself b times for the denominator. You can cancel b times the number of x's in the numerator and leave a residue of x's in the numerator that is $(a - b)$ times. To illustrate, make $a = 5$ and $b = 2$. x^5 divided by x^2 equals x^3. If you used 2 for x, x^5 is 32, x^2 is 4, and x^3 is 8. 32 divided by 4 is 8.

Roots: inverse of powers

Here, you must distinguish between the inverse of a number and the inverse of a power. A *minus index* is the inverse or reciprocal of the number raised to the power, indicated by the index. *Roots* are the opposite of powers. For example, because 2^2 is 4, $4^{1/2}$ is 2; 2^3 is 8, so $8^{1/3}$ is 2; 2^4 is 16, so $16^{1/4}$ is 2, and so on.

Fractional indices represent roots. The 3/2 power of 4 is 8, the square root of $4^{1/2}$ is 2, and 2^3 is 8. Reversing this process, $8^{2/3}$ is 4. You can find other numbers in roots by the process of square root. For example, $2^{1/2}$ (the square root of 2) is 1.414 etc.; $8^{1/2}$ is twice this. Why? Since $4^{1/2}$ is 2 and $2^{1/2}$ is 1.414, (2 times 4) $^{1/2}$ is $8^{1/2}$ (twice 1.414), which is 2.828.

You are not restricted to square roots, or indeed to any specific roots. Now, a whole new field of numbers is opened.

$$2^2 = 4$$
$$2^3 = 8$$
$$4^2 = 16$$
$$4^{\frac{3}{2}} = 8$$
$$2^{\frac{1}{2}} = 1.414....$$
$$8^{\frac{1}{2}} = 2.828....$$
$$2^{\frac{3}{2}} = 2.828....$$
$$2^5 = 32$$

$$4^{\frac{1}{2}} = 2$$
$$8^{\frac{1}{3}} = 2$$
$$16^{\frac{1}{2}} = 4$$
$$8^{\frac{2}{3}} = 4$$

$$32^{\frac{1}{5}} = 2$$
$$32^{\frac{2}{5}} = 4 \qquad 32^{\frac{1}{2}} = 5.656....$$
$$32^{\frac{3}{5}} = 8$$
$$32^{\frac{4}{5}} = 16$$

What about other fractions?

Surds and indices

Introducing surds is actually going back to a virtually obsolete way of writing roots. Before the fractional index notation introduced in the previous section came into vogue, it was customary to use a surd in front of a number to indicate its square root. Thus, a surd in front of x represented the square root of x, the same as $x^{1/2}$. Putting a 3 in front of the surd indicated the cube root of x, instead of the square root. Putting a small n or any other letter or number in front of the surd likewise represented a specific root. If the number under the surd has a power b and an a in front of the surd, the expression can be written as: $x^{b/a}$. A surd followed by a *vinculum* over (line over the top of) $a^2 + b^2$ is the root of the whole expression. This expression can be written: $(a^2 + b^2)^{1/2}$.

$$\sqrt{x} = x^{\frac{1}{2}}$$

$$\sqrt[3]{x} = x^{\frac{1}{3}}$$

$$\sqrt[n]{x} = x^{\frac{1}{n}}$$

$$\sqrt[a]{x^b} = x^{\frac{b}{a}}$$

$$\sqrt{a^2 + b^2} = (a^2 + b^2)^{\frac{1}{2}}$$

Questions and problems

NOTE: The questions and problems on these pages are not in graded order; they assume a knowledge of the earlier parts of this book. If you have difficulty with a question, try others first and return to the one that is difficult. These questions are designed to exercise initiative in applying the mathematical principles that have been introduced so far.

1. Find the decimal equivalent of the fraction 1/37 and determine the fractional error in only the first three decimal places.

2. Arrange a table of the recurring groups of figures in the decimals for the 36 fractions from 1/37 to 36/37. Then, rearrange the table with the decimals that use the same group of digits in the same column. How many columns do you need?

3. The following three calculations are correct, each in a different, unfamiliar notation, or system of numbers (base). Find, in each case, what number base is used, in place of the usual 10:

$$
\begin{array}{r}
8137 \\
420 \\
332 \\
\hline
10000
\end{array}
\qquad
\begin{array}{r}
6543 \\
15 \\
\hline
46011 \\
6543 \\
\hline
144441
\end{array}
\qquad
\begin{array}{r}
4567 \\
31)\overline{166237} \\
144 \\
\hline
222 \\
175 \\
\hline
253 \\
226 \\
\hline
257 \\
257 \\
\hline
\end{array}
$$

4. Check the working of the second two examples in the previous question (a) by the opposite calculation (division and multiplication respectively) and (b) by converting the numbers to the decimal system.

5. In the duodecimal system, divide $1t294$ by 1708 and check the result (a) by multiplying the result by 1708 and (b) by converting all the numbers to the decimal system.

6. Using the binary system, multiply 15 by 63 and convert back to the decimal system. Check your result by simple multiplication.

7. Using the binary system, divide 1922 by 31 and convert back to the decimal system. Check your result by simple division.

8. Find the values of the following:

$16^{3/4}$, $243^{0.8}$, $25^{1.5}$, $64^{2/3}$, $343^{4/3}$
$98^{1/2}$, $243^{0.4}$, $243^{0.5}$,   and $243^{0.6}$.

9. Convert the following numbers from decimal to binary. As a check, convert them back:

62,   81,   111,   49,   98,   222,   2000,   2345,   670

10. Convert the following numbers from binary to decimal then, as a check, convert them back:

1011011101,   110111,   10000011,   10100011,   1010011000,   1111111111

11. Why is the binary system used in calculators instead of the decimal system? To illustrate, multiply 129 by 31, both by the binary and the decimal systems. Now, assume an error is made in the tens' figure so that 129 is multiplied by 41 instead of 31. A similar error in binary would cause the second place in the

binary equivalent of 31 to be advanced by 1. Compare the error made in the decimal system with that in the binary system.

12. Evaluate $(a^2 + b^2)^{1/2}$, when a and b have the following pairs of values: 4 and 3, 12 and 5, 24 and 7, 40 and 9, 60 and 11, 84 and 13, and 112 and 15. What does each pair have in common? Can you explain it?

13. Evaluate $(a^2 + b^2)^{1/2}$, when a and b have the following pairs of values: 3 and 4, 8 and 6, 15 and 8, 24 and 10, 35 and 12, 48 and 14, and 63 and 16. What does each pair have in common? Expand the reason for the relationship in question 12 to explain those in this one.

14. Find the first 6 pairs of numbers that satisfy the expression: $a^2 + b^3 = (a + b)^2$ by any method you choose. Having found them, show a reason from the algebra for the pattern formed by successive values of a and b.

15. In a certain number system, multiplying 84 by 2 produces 148; by 3 produces 210; by 4 produces 294; by 5 produces 358; by 6 produces 420. What will 7 times 84 be in this number system?

16. Make a conversion table from duodecimal to binary, such as the one for the decimal system from this chapter. Use it to convert the numbers in question 15 to the binary system. Convert the same numbers to the decimal system.

17. Write as simple decimal numbers, without fractions, the following quantities: 100^2, $100^{1/2}$, 100^{-2}, and $100^{-1/2}$. From these four values, verify values of $100^{3/2}$, $100^{5/2}$, $100^{-3/2}$, and $100^{-5/2}$ by the method of subtracting and adding indices.

18. Evaluate, to three places of decimals, $100^{1/4}$.

19. Find values, correct to 3 places of decimals (where necessary) for: $32^{0.1}$, $32^{0.2}$, $32^{0.3}$, $32^{0.4}$, $32^{0.5}$, $32^{0.6}$, $32^{0.7}$, $32^{0.8}$, $32^{0.9}$. What is the ratio between each quantity and the next?

20. Evaluate the following expressions:

$$(10^2 - 2^6)^{1/2} \quad (36^2 - 8^3)^{1/2} \quad (28^2 - 21^2)^{1/3}$$
$$(5^2 - 3^2)^{1/4} \quad (17^2 - 15^2)^{1/6} \quad (9^2 - 7^2)^{-0.2}$$
$$(2^2 + 3^3 + 4^3 + 5)^{1/2} \quad 998001^{1/2}.$$

16
CHAPTER

Progressions

Arithmetic progression

Whatever system you use in arithmetic or in algebraic numbers, a series of numbers can show some kind of regular pattern, which will help you to develop and check that series in various ways. Such a pattern is unlike the ones in multiplication tables. They depended on the system or notation that was used: decimal, octal, hexadecimal, binary, or whatever. These patterns exist independently of the base in which you write them.

Here are some arithmetic progressions. Four are numerical. First is the counting number system itself. Second are the even numbers. Odd numbers would be similar, but starting with 1 and adding 2 for each successive number. Third are the numbers that are divisible by 3. Fourth are the numbers that also add 3 to each term, but begins with 1 instead of 3.

$$1 \ 2 \ 3 \ 4 \quad 5 \quad 6 \quad 7 \ .... \quad \text{ARITHMETIC PROGRESSION}$$
$$2 \ 4 \ 6 \ 8 \quad 10 \ 12 \ 14 \ ....$$
$$3 \ 6 \ 9 \ 12 \ 15 \ 18 \ 21 \ ....$$
$$1 \ 4 \ 7 \ 10 \ 13 \ 16 \ 19 \ ....$$

$$a, \quad a + d, \quad a + 2d, \quad a + 3d, \quad a + 4d \ ....$$

(a) is 1st term

(d) is difference between one term and the next

| 1st term | 2nd term | 3rd term | 4th term | 5th term | 6th term | 7th term |

The algebraic series gives the general form of an arithmetic progression, or series of numbers, where *a* is the first term and each term is *d* more than the previous one. If the numbers diminish, instead of increase, *d* would have a minus sign.

The geometric construction shows the relationship of a series of terms that are represented by a series of equally spaced vertical lines. Each line is longer than its neighbor to the left by the same amount.

Geometric progression

On the previous page, each term differs from the previous term by the same amount, added or subtracted. In geometric progression, each term is multiplied by the same amount to get the next term. Just as the difference in arithmetic progression can either add or subtract, so the ratio of one term to the next in geometric progression can either expand or contract successive terms.

Here are some samples of geometric progression. In the first one, each term is twice the previous one. In the second, it is 3 times the previous one. In the third, each term is 1.5 times the previous one. In the fourth, the process is reversed: each term is half of the previous one. In the fifth, each term is 2/3 of the previous one.

$1, \ 2, \ 4, \ 8, \ 16 \dots$

$1, \ 3, \ 9, \ 27, \ 81 \dots$ $a, \ ar, \ ar^2, \ ar^3, \ ar^4$

$16, 24, 36, 54, 81 \dots$ (a) is 1st term

$1, \ \dfrac{1}{2}, \ \dfrac{1}{4}, \ \dfrac{1}{8}, \ \dfrac{1}{16} \dots$ (r) is ratio between one term and the next

$9, \ 6, \ 4, 2\dfrac{2}{3}, 1\dfrac{7}{9} \dots$

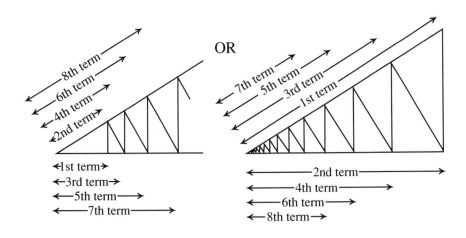

The algebra gives the general form for all geometric progressions. *a* is the first term and *r* is the ratio by which each term is multiplied to find the next term.

The geometric construction shows the two types. At the left is an expanding series. At the right is a contracting, or as mathematicians call it, a converging series. Notice that the expanding series quickly "runs away" (goes off scale), while the converging series gets smaller and smaller, indefinitely. The construction in each case uses similar right triangles to represent the changing ratio.

Harmonic progression

A third kind of progression has been taught in schools. It is not used as often as arithmetic and geometric progressions. Other progressions are none of these three. You can learn something about progressions in general by studying these three types.

The easy way to understand harmonic progression is as a reciprocal of arithmetic progression. Instead of each term increasing by *d* (a constant amount), it is diminished by being divided by an arithmetic series. If the first term is *a*, the second term is *a* divided by 1 + *d*, the third divided by 1 + 2*d*, and so on.

In the first numerical example, *d* is 1. In the second, *d* is 1/2. In the third, *d* is 1 again, but the first term is 1, not 60 (as in the first series).

These patterns in numbers serve a variety of purposes that will be developed as you study them. They form basis for series that calculate trig ratios for any angle and many other things.

$$60, \ 30, \ 20, \ 15, \ 12, \ 10, \ \ldots.$$

$$60, \ 40, \ 30, \ 24, \ 20, \ 17\tfrac{1}{7}, \ \ldots.$$

$$1, \ \frac{1}{2}, \ \frac{1}{3}, \ \frac{1}{4}, \ \frac{1}{5}, \ \frac{1}{6}, \ \ldots.$$

$$a, \ \frac{a}{1+d}, \ \frac{a}{1+2d}, \ \frac{a}{1+3d}, \ \frac{a}{1+4d}, \ \frac{a}{1+5d}, \ \ldots.$$

Patterns in Numbers will help

such as ⟶

ARITHMETIC
GEOMETRIC
HARMONIC

PROGRESSION

① Check working
② Find more direct methods
③ Find answers that eluded other methods

Sum of an arithmetic series

Often, to find the answer, you must sum a series by adding all its terms. The long way would be to write down all the terms and add them up. Where many terms are, you have two ways of getting the sum.

If you know the first term, a, the last term, 1, and the number of terms, n, then the total is n times the average term. Because the terms increase (or decrease) uniformly, the average term is midway between first and last. The midpoint can be found by adding together the first and last and dividing by 2. So the sum is $2(a + 1)/2$.

If, on the other hand, you don't know the last term, but know the first term, a, the difference, d, and the number of terms, the last term is $a + (n - 1)d$. So, sum of the first and last is: $a + a + (n - 1)d$ or $2a + (n - 1)d$. Now, use the first method and simplify: $n[2a + (n - 1)d]/2$. This equation can be simplified to: $na + n(n - 1)d/2$. The symbol universally used for the sum of a series is a Greek capital sigma, Σ.

First term: a
Second " a + d
Third " a + 2d
Fourth " a + 3d
Nth " a + (n − 1)d

$$a, a + d, a + 2d, a + 3d, ..., a + (n - 1)d$$

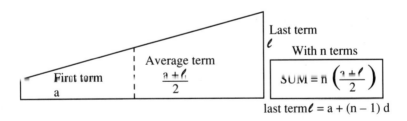

Last term
ℓ
With n terms

$$\text{SUM} = n \left(\frac{a + \ell}{2}\right)$$

last term $\ell = a + (n - 1)\,d$

First term
a

Average term
$\frac{a + \ell}{2}$

$\displaystyle\sum_{1}^{n}$ $\left(\begin{array}{c}\text{Greek capital}\\ \text{sigma}\end{array}\right)$

means

Sum of terms 1 through n

$$\boxed{\text{SUM}} = n \left(\frac{2a + (n - 1)\,d}{2}\right)$$

$$\sum_{1}^{n} = na + \frac{n\,(n - 1)\,d}{2}$$

continued

1 Add together 3, 5, 7, 9, 11, 13, 15

Check

First term: $a = 3$
Last term: $\ell = 15$
Number of terms: $n = 7$

$$\sum_1^n = n\left(\frac{a+\ell}{2}\right)$$

$$\sum_1^7 = 7\left(\frac{3+15}{2}\right) = 7\left(\frac{18}{2}\right) = 7 \times 9$$

$$= \underline{63}$$

3
5
7
9
11
13
15
———
63

2 Find the sum of −7, −4, −1, +2,

+5 .... to 12 terms

Check

+ 2	− 7
5	− 4
8	− 1
11	−12
14	
17	
20	
23	
26	

$126 - 12 = 114$

First term: $a = -7$
Difference: $d = 3$
Number of terms: $n = 12$

$$\sum_1^n = na + \frac{n(n-1)d}{2}$$

$$\sum_1^{12} = -12 \times 7 + \frac{12 \times 11 \times 3}{2}$$

$$= -84 + \frac{396}{2} = -84 + 198$$

$$= \underline{114}$$

Sum of a geometric series

This problem is not so simple, but the trick that you learn here is helpful with series that you will study later.

Because a is a factor common to all the terms, it can be put outside parentheses with a list of all the multipliers inside. Now, if you multiply every term by ratio r, they all move one to the right. In one line, you have r times the sum. However, it begins with r times a and ends with r^n times a. The original sum has, inside the parentheses, terms that begin with 1 and finish with r^{n-1}.

Assuming that r is greater than 1, subtract the original sum from r times the sum. All the middle terms disappear, because they are all the same. Inside the parentheses, -1 and r^n is left. The sum on the other side of the equation is multiplied by $r - 1$. So, the sum can now be found by dividing both sides by $r - 1$.

GEOMETRIC PROGRESSION $\quad a, ar, ar^2, ar^3, ar^4, \ldots.$

First term: a

Ratio: r

The nth term: ar^{n-1}

Sum of terms 1 to n:

$$\sum_1^n = a + ar + ar^2 + ar^3 + \ldots + ar^{n-2} + ar^{n-1}$$
$$= a\,[1 + r + r^2 + r^3 + \ldots + r^{n-2} + r^{n-1}\,]$$

How to simplify this ⬆ ?

Multiply both sides by r:

$$r \times \sum_1^n = ar\,[1 + r + r^2 + r^3 + \ldots + r^{n-2} + r^{n-1}\,]$$
$$= a\,[\quad r + r^2 + r^3 + r^4 \ldots + r^{n-1} + r^n\,]$$

Subtract original

$$\sum_1^n = a\,[1 + r + r^2 + r^3 + r^4 \ldots + r^{n-1} \qquad]$$

$$(r-1)\sum_1^n = a\,[-1 \qquad\qquad + r^n\,]$$

$$= a\,[r^n - 1]$$

All the middle terms disappear

Divide both sides by (r – 1)

$$\sum_1^n = \frac{a\,(r^n - 1)}{r - 1}$$

---

① **Find the sum of 10 terms $\quad$ 1, 2, 4, 8, 16 ….**

a = 1
r = 2
n = 10

$$\sum_1^n = \frac{a\,(r^n - 1)}{r - 1}$$

$$\sum_1^{10} = \frac{2^{10} - 1}{2 - 1} = \frac{1024 - 1}{1}$$

$$= 1023$$

Check

1
2
4
8
16
32
64
128
256
512
1023

continued

---

② **Find the sum of 7 terms**   **4, 12, 36, 108 ....**

$a = 4$
$r = 4$
$n = 7$

$$\sum_1^n = \frac{a\,(r^n - 1)}{r - 1}$$

$$\sum_1^7 = \frac{4 \times (3^7 - 1)}{3 - 1}$$

$3^3 = 27$
$3^4 = 81$
      27
     216
$3^7 = 2187$

Check
4
12
36
108
324
972
2916
4372

$$= \frac{4 \times (2187 - 1)}{3 - 1} = \frac{4 \times 2186}{2}$$

$$= 4372$$

Here, you sum a couple of geometric series, one to 10 terms, the other to 7 terms. In each case, verify the result by doing it "the long way." Actually, you probably won't have to use this formula very often, but it's worth hanging on to for those few occasions when it is useful.

Look on the checking as a sort of reversible process. If you haven't used the formula for some time, you might not feel sure about it. Do it "the long way," then by formula, which will prove two things: that you didn't make a mistake in doing it the long way and that the formula "works!"

Converging series

The series you have considered so far have been expanding or diverging: each term is larger than the one before it. This occurs because r is greater than 1. In a converging series, each term is smaller than the one before it.

A story exists about an Eastern potentate who offered a philosopher a reward for some work. He offered him a chess board with grains of wheat on each of its 64 squares. He would put 1 grain on the first, 2 grains on the second, 4 on the third, 8 on the fourth, and so on, until he got to the 64th square. It didn't sound like much, until he figured it. The grand total is $2^{64} - 1$, which is 18,446,744,073,709,551,615 grains! That was more wheat than he expected!

When r is less than 1, r^n is less than 1, and r is less than 1. If n is large—especially if you sum the series to infinity, r^n is zero. So, the sum of the series reduces to a very simple expression.

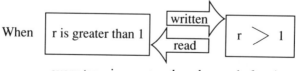

When   r is greater than 1   written / read   r > 1

every term is *greater* than the one before it.

For example: r = 2
    1, 2, 4, 8, 16, 32, 64 .... This is a DIVERGING SERIES

When   r is less than 1   written / read $r < 1$

              every term is *smaller* than the one before it.

For example: r = $\dfrac{1}{2}$

8, 4, 2, 1, $\dfrac{1}{2}$, $\dfrac{1}{4}$, $\dfrac{1}{8}$ .... This is a CONVERGING SERIES

Sum of a converging series

Because the order of both the numerator and the denominator are reversed by making the series converge, the result is still positive.

An interesting thing happens if you sum to infinity. You continue until the series disappears into nothingness. When *r* is less than 1, r^n becomes zero by making *n* infinity. So, the formula becomes very simple: $a/(1 - r)$.

$$\sum_{1}^{n} = a\left[1 + r + r^2 + r^3 + \ldots + r^{n-1}\right]$$

Multiply by r

$$r\sum_{1}^{n} = a\left[\ \ r + r^2 + r^3 + \ldots + r^{n-1} + r^n\right]$$

Lower line is smaller, so subtract

$$(1-r)\sum_{1}^{n} = a\left[1 \qquad\qquad - r^n\right]$$

Divide by $(1-r)$

$$\sum_{1}^{n} = \frac{a(1 - r^n)}{1 - r}$$

In an endless series

TO INFINITY $n = \infty$ ⟹ n is infinity

$r < 1$ ⟹ So increasing powers of r become smaller, and $r^\infty = 0$

$$\sum_{1}^{\infty} = \frac{a(1 - r^\infty)}{1 - r} = \frac{a(1 - 0)}{1 - r}$$

$$= \frac{a}{1 - r}$$

Rate of convergence

You can see how the use of different series for the same calculation can often make the work easier. The sum of three different series produces a value of 4 at infinity.

The first one is 1, 3/4, 9/16, 27/64, etc. Notice that the sum to 6 terms is 3 295/1024: still quite far from 4, although it seems logical that it will eventually get there.

The second one is 2, 1, 1/2, 1/4, etc. Notice that the sum to 6 terms now is 3 15/16. It's already only 1/16 short of its ultimate value of 4.

The third one is 3, 3/4, 3/16, 3/64, etc. Notice that the sum to 6 terms is now 3 1023/1024—only 1/1024 short of its ultimate value.

Each of these series has a successively greater rate of convergence than the one before it.

(1) Sum of series $1, \dfrac{3}{4}, \dfrac{9}{16}, \dfrac{27}{64} \dots$ to infinity

$a = 1$

$r = \dfrac{3}{4}$

$n = \infty$

$$\sum_{1}^{\infty} = \frac{a}{1-r} = \frac{1}{1-\dfrac{3}{4}} = \frac{1}{\dfrac{1}{4}} = 4$$

PROGRESSIVE TOTALS ⇨

$$1 + \frac{3}{4} + \frac{9}{16} + \frac{27}{64} + \frac{81}{256} + \frac{243}{1024} +$$

$$1\frac{3}{4} \quad 2\frac{5}{16} \quad 2\frac{47}{64} \quad 3\frac{13}{256} \quad 3\frac{295}{1024}$$

Eventually ⇨ reaches 4

(2) Sum of series $2, 1, \dfrac{1}{2}, \dfrac{1}{4} \dots$ to infinity

$a = 2$

$r = \dfrac{1}{2}$

$n = \infty$

$$\sum_{1}^{\infty} = \frac{a}{1-r} = \frac{2}{1-\dfrac{1}{2}} = \frac{2}{\dfrac{1}{2}} = 4$$

PROGRESSIVE TOTALS ⇨

$$2 + 1 + \frac{1}{2} + \frac{1}{4} + \frac{1}{8} + \frac{1}{16} +$$

$$3 \quad 3\frac{1}{2} \quad 3\frac{3}{4} \quad 3\frac{7}{8} \quad 3\frac{15}{16}$$

Eventually ⇨ reaches 4

(3) Sum of series $3, \dfrac{3}{4}, \dfrac{3}{16}, \dfrac{3}{64} \dots$ to infinity

$a = 3$

$r = \dfrac{1}{4}$

$n = \infty$

$$\sum_{1}^{\infty} = \frac{a}{1-r} = \frac{3}{1-\dfrac{1}{4}} = \frac{3}{\dfrac{3}{4}} = 4$$

PROGRESSIVE TOTALS ⇨

$$3 + \frac{3}{4} + \frac{3}{16} + \frac{3}{64} + \frac{3}{256} + \frac{3}{1024} +$$

$$3\frac{3}{4} \quad 3\frac{15}{16} \quad 3\frac{63}{64} \quad 3\frac{255}{256} \quad 3\frac{1023}{1024}$$

Eventually ⇨ reaches 4

FOR EACH SERIES, SUM TO INFINITY IS 4 Notice difference in rate of convergence

Permutations

Permutations have many practical applications. If gamblers used it, they'd realize how great the real odds are and probably quit gambling. With 10 horses in a race, basically each one has a 1 in 10 chance of winning. If one or two are better runners than the others, the others have a less than 1 in 10 chance of winning. But why waste time? Gamblers will keep losing more money than they win!

In this example, 7 horses can pass the winning post. Any one of the 7 can come in first. For each winner, 6 are left that can come in second. So, the possible first and second place are 7 times 6 (42). For each first two, 5 choices are left for third place. This makes 210 possibilities for the first three places.

To find the order of the rest, the possibilities are: $7 \times 6 \times 5 \times 4 \times 3 \times 2 \times 1$. Maybe that 1 is redundant: he's the only one left! However, it means that 5040 orders are possible.

In how many orders can
7 articles be placed?

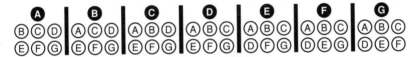

There are 7 choices for 1st place. Then ....

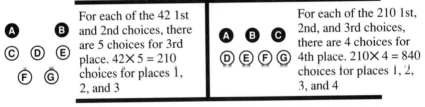

For each 1st, there are 6 choices for 2nd place
This makes $7 \times 6 = 42$ choices for 1st and 2nd place

Ⓐ Ⓑ ⒸⒹⒺ Ⓕ Ⓖ — For each of the 42 1st and 2nd choices, there are 5 choices for 3rd place. $42 \times 5 = 210$ choices for places 1, 2, and 3	Ⓐ Ⓑ Ⓒ ⒹⒺⒻⒼ — For each of the 210 1st, 2nd, and 3rd choices, there are 4 choices for 4th place. $210 \times 4 = 840$ choices for places 1, 2, 3, and 4

Ⓐ Ⓑ Ⓒ Ⓓ ⒺⒻⒼ — For each of the 840 1st, 2nd, 3rd, and 4th choices, there are 3 choices for 5th place.
$840 \times 3 = 2520$ choices for places 1, 2, 3, 4, and 5

Ⓐ Ⓑ Ⓒ Ⓓ ⒺⒻⒼ — For each of the 2520 1st, 2nd, 3rd, 4th, and 5th choices, there are 2 choices for 6th place.
$2520 \times 2 = 5040$ choices for places 1, 2, 3, 4, 5, and 6

When 6 places are filled, only one is left for the remaining place

7 articles can be arranged in 5040 different orders

Factorial notation!

How many ways can so many be taken from so many? The usual symbols for expressing this are k articles chosen from an available n. Examining the figures used so far, you could write that $n(n - 1)(n - 2) \ldots (n - k + 1)$. Mathematicians use a shortened way of writing these equations and calculators use the same method to make the calculation, factorial notation. The accepted symbol is an exclamation point after the number. *Factorial notation* is defined as the product of every integral number from 1 up to the number so marked.

The permutation of k articles taken from an available n is the product of all the numbers from n, down to the number: $n - k + 1$. That seems to be the simplest form. But calculators use another way to arrive at the answer. If you divide

HOW MANY PERMUTATIONS OF k ARTICLES TAKEN FROM AN AVAILABLE n?

There are n choices of 1st place
For each of these, there are n – 1 choices of 2nd place
For each of these, there are n – 2 choices of 3rd place

TOTAL: n (n – 1) (n – 2) choices for
first three places

[n – (k – 1)] or [n – k + 1] choices of k th position,
or a total of n (n – 1) x .... x (n – k + 1) choices for k places

FACTORIAL NUMBERS $1 \times 2 \times 3 \times 4 \ldots \times n$ is called
FACTORIAL n
written $\lfloor n$ or n!

$$n(n - 1) \times \ldots \times (n - k + 1) \text{ is } \frac{n!}{(n - k)!}$$

$$\frac{n(n-1) \times \ldots \times (n-k+1)(n-k)(n-k-1) \times \ldots \times 3 \times 2 \times 1}{(n-k)(n-k-1) \times \ldots \times 3 \times 2 \times 1}$$

How many permutations of 3 articles taken from an available 10?

$$10 \times 9 \times 8 = 720 \qquad \frac{10 \times 9 \times 8 \times 7 \times 6 \times 5 \times 4 \times 3 \times 2 \times 1}{7 \times 6 \times 5 \times 4 \times 3 \times 2 \times 1}$$

factorial *n* (written *n*!) by factorial (*n* − *k*) (written (*n* − *k*)!), by canceling the shorter string of numbers from the longer, you are left with a product of all the numbers from (*n* − *k* + 1) to *n*.

Pocket calculators provide a *x*! button that will factorially read out whatever number was entered when that button was pressed. It won't work if you enter a fractional number or if you enter a number show factorial expansion is too big for the calculator to handle.

Combinations

Calculating permutations considers the order in which the articles were chosen. In combinations, order is not important. If you must pick 3 letters from the first 10 letters of the alphabet, A through J, you have 10 choices for your first pick, 9 choices for second, and 8 for third. That multiplies out to 720 possible choices. Suppose your choice picks the first 3 letters, A, B, and C. These could have been picked in any order. 6 different orders exist for which you could pick just the first 3 letters.

So, permutations take order into account. Combinations only deal with which ones are picked, in any order. Pursuing the 10-letter choice from which you picked 3: 720 possible choices exist (taking sequence into account). However, if you only want to know which 3, any order, divide 720 by 6, which is 120.

PERMUTATIONS of 3 taken from 10 includes

$$
\left.\begin{array}{l}
\text{ABC} \\
\text{ACB} \\
\text{DAC} \\
\text{BCA} \\
\text{CAB} \\
\text{CBA}
\end{array}\right\} \text{6 different orders of each combination}
$$

How many COMBINATIONS of 3 taken from 10?

$$\frac{10 \times 9 \times 8}{3 \times 2 \times 1} = \frac{720}{6} = 120$$

How many COMBINATIONS of 7 taken from 10?

Each 3 taken leaves 7, so it should be the same answer

$$\frac{10 \times 9 \times 8 \times \cancel{7} \times \cancel{6} \times \cancel{5} \times \cancel{4}}{\cancel{7} \times \cancel{6} \times \cancel{5} \times \cancel{4} \times 3 \times 2 \times 1} = \frac{10 \times 9 \times 8}{3 \times 2 \times 1} = 120$$

COMBINATIONS OF k TAKEN FROM n

PERMUTATIONS of k taken from n are $\dfrac{n!}{(n-k)!}$

This is written $_nP_k = \dfrac{n!}{(n-k)!}$

COMBINATIONS of k taken from n are $\dfrac{_nP_k}{k!} = \dfrac{n!}{(n-k)!\,k!}$

This is written $_nC_k = \dfrac{n!}{(n-k)!\,k!}$

BECAUSE $_nC_k = \dfrac{n!}{(n-k)!\,k!}$ and $_nC_{(n-k)} = \dfrac{n!}{k!\,(n-k)!}$

Therefore $_nC_k = {}_nC_{(n-k)}$

The symbol for permutations is P, with two subscripts, one in front of the P and one after it. The one before it says how many the choice is from. The one after it says how many are chosen. The formula for permutations of k taken from n is: $n!/(n - k)!$

The symbol for combinations is C, written in the same style. The formula for combinations of k taken from n is: $n!/(n - k)!k!$ Notice that the same formula also gives the number of combinations of $(k - n)$ taken from n. What this says is that the number of ways that a certain number, k, can be taken from a bigger number, n, is equal to the combinations of those left in the same case. This fact might seem obvious if you think about it. Still, it is a useful fact when calculating.

Powers of a binomial

A binomial is any expression that consists of two terms, with either a plus or a minus sign between them. For a general form of a binomial, write: $(a + b)$. When such an expression is raised to successive powers (squared, cubed, fourth, etc.), it generates a successively more complicated series of terms, each of which consists of a power of a, of b, or a product of both powers.

You can multiply successive expressions of the general binomial $(a + b)$ to get the pattern of terms that form. In any of the powers of the binomial, start with a raised to that power, followed by terms that consist of successively lower powers a, multiplied by successively higher powers of b, until you get to the last term, which is b raised to the same power. Each of the product terms, with powers of both a and b, has a numerical coefficient.

First, investigate the pattern of these coefficients by forming a pyramid, as is shown here.

$$a + b \longleftarrow \textbf{is a binomial}$$

multiply by .... $a + b$

$$a^2 + 2ab + b^2 \qquad \text{is } (a + b)^2$$

multiply by .... $a + b$ again

$$a^3 + 3a^2b + 3ab^2 + b^3 \qquad \text{is } (a + b)^3$$

$$a + b$$

$$a^4 + 4a^3b + 6a^2b^2 + 4ab^3 + b^4$$

Power Index	Coefficients
1	1 1
2	1 2 1
3	1 3 3 1
4	1 4 6 4 1
5	1 5 10 10 5 1
6	1 6 15 20 15 6 1
7	1 7 21 35 35 21 7 1
8	1 8 28 56 70 56 28 8 1
9	1 9 36 84 126 126 84 36 9 1
10	1 10 45 120 210 252 210 120 45 10 1

Binomial expansion

Write out the expansion: $(a + b)^n$. It is: $(a + b) (a + b) (a + b) (a + b) \ldots n$ times. Now, multiply all those terms together. First, is simple: the first term of each is a^n. Next, is almost as easy: take $n - 1$ of a's and 1 b. You have n of those. Now, it begins to get involved: the combinations in which you take $n - 2$ a's and 2 b's.

After going through this step, write the general term in the whole expansion. It has 3 parts: the coefficient, which is the combination of k [you are at the $(k + 1)$th term] a's taken with $n - k$ b's; then, a^{n-k}; and finally b^k; all multiplied together.

$(a + b)^n$ is

$(a + b)(a + b) (a + b) (a + b) (a + b)$ .... n times

Coefficeint of a^n is 1. All first terms multiplied together.

Terms in $a^{n-1}b$ are each $(n - 1)$ terms a, multiplied by one term b.

There are n combinations like this.

Terms in $a^{n-2}b^2$ are each $(n - 2)$ terms a, multiplied by two terms b.

Combinations are $_nC_2$

Each term in the EXPANSION has three parts or factors

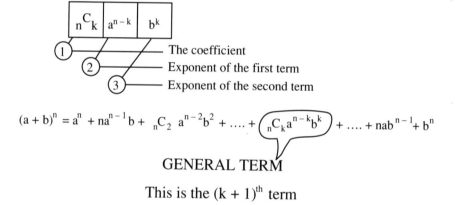

$(a + b)^n = a^n + na^{n-1}b + {}_nC_2 a^{n-2}b^2 + ... + \left({}_nC_k a^{n-k}b^k \right) + ... + nab^{n-1} + b^n$

GENERAL TERM

This is the $(k + 1)^{th}$ term

Binomial series

Before going into uses of the series that do new things, check it by using it for something you can check with a familiar method. First, show the general forms of the series with *C*, then interpret it with extended notation (partly factorial). Then, substitute an expansion for 13^5, in the form $(10 + 3)^5$. Each term is evaluated and added, then multiply 13 by itself directly 5 times to show the same answer.

Notice something here about the writing. Up till now, to indicate multiplication, the "times" sign (an $\times$ has been used). Now, use a dot, like a period. To avoid making it look like a decimal point, keep the dot above the line. This measure saves a lot of space when you have a lot to write.

BINOMIAL SERIES

Expand $(a + b)^7$

$$(a + b)^n = a^n + na^{n-1}b + {}_nC_2 a^{n-2}b^2 + \cdots + {}_nC_k a^{n-k}b^k + \cdots$$

$$= a^n + na^{n-1}b + \frac{n(n-1)}{2}a^{n-2}b^2 + \cdots \frac{n(n-1)\cdots(n-k+1)}{k!}a^{n-k}b^k + \cdots$$

$$(a + b)^7 = a^7 + 7a^6b + \frac{7\cdot6}{2}a^5b^2 + \frac{7\cdot6\cdot5}{3\cdot2}a^4b^3 + \frac{7\cdot6\cdot5\cdot4}{4\cdot3\cdot2}a^3b^4 + \cdots$$

These two the same; from here on repeat ⟶

$$= a^7 + 7a^6b + 21a^5b^2 + 35a^4b^3 + 35a^3b^4 + 21a^2b^5 + 7ab^6 + b^7$$

Find 13^5 by expanding $(10 + 3)^5$

$$
\begin{array}{lll}
10^5 & & \to 100,000 \\
+5\cdot10^4\cdot3 & & \to 150,000 \\
+\frac{5\cdot4}{2}\cdot10^3\cdot3^2 & \to 10\cdot10^3\cdot9 & \to 90,000 \\
+\frac{5\cdot4}{2}\cdot10^2\cdot3^3 & \to 10\cdot10^2\cdot27 & \to 27,000 \\
+5\cdot10\cdot3^4 & \to 50\cdot81 & \to 4,050 \\
+3^5 & & \to 243
\end{array}
$$

$$= \boxed{371,293}$$

$$
\begin{array}{l}
13 \\
\times 13 \\
\hline
13 \\
39 \\
\hline
13^2 = 169 \\
\times 13 \\
\hline
169 \\
507 \\
\hline
13^3 = 2197
\end{array}
\qquad
\begin{array}{l}
2197 \\
\times\ 13 \\
\hline
2197 \\
6591 \\
\hline
28561 = 13^4 \\
\times\ 13 \\
\hline
28561 \\
85683 \\
\hline
371293 = 13^5
\end{array}
$$

CHECK

Completing some patterns

For some pages now, strange "ground" has been covered. When you do that, it is always good to check your bearings to see where you're going. First, find out if the binomial series will work when the value of n is such that the series continues forever—a nonterminating series.

Try expanding $(4 + 1)^{-1}$. Obviously, $4 + 1 = 5$ and the minus -1 exponent takes the reciprocal of 5. The result should be 1/5. Following the rules, the expansion reduces to $1/4 - 1/16 + 1/64 - 1/1024\ldots$ In this geometrical series, r is $-1/4$. The sum to infinity is $a/(1 - r)$. So, 1/4 is divided by 5/4, which reduces to 1/5. As a geometrical series, it "works." As far as you've taken it (only 4 terms) the value is 205/1024. 205/1025 would be exactly 1/5. Pretty close!

Another example is $(4 - 1)^{-1}$, which should be 1/3. In 4 terms, it is 341/1024; 341/1023 is exactly 1/3.

BINOMIAL EQUIVALENT of INFINITE GEOMETRIC

$\boxed{\text{Expand } (4 + 1)^{-1}}$ $a = 4, b = 1, n = -1$ $\boxed{4 + 1 = 5, 5^{-1} = \frac{1}{5}}$

$(4 + 1)^{-1} = 4^{-1} + (-1)\, 4^{-2} \cdot 1 + \left(\frac{-1 \cdot -2}{2}\right) 4^{-3} \cdot 1^2 + \left(\frac{-1 \cdot -2 \cdot -3}{3 \cdot 2}\right) 4^{-4} \cdot 1^3 + \left(\frac{-1 \cdot -2 \cdot -3 \cdot -4}{4 \cdot 3 \cdot 2 \cdot 1}\right) 4^{-5} \cdot 1^4$

$= \frac{1}{4} - \frac{1}{16} + \frac{1}{64} - \frac{1}{256} + \frac{1}{1024}$ .... (to infinity)

$\longrightarrow \frac{1}{5}$

$\frac{3}{16} \quad \frac{13}{64} \quad \frac{51}{256} \quad \frac{205}{1024}$

$\frac{3}{15} \quad \frac{13}{65} \quad \frac{51}{255} \quad \frac{205}{1025}$

$\boxed{\text{Expand } (4 - 1)^{-1}}$ $a = 4, b = -1, n = -1$ $\boxed{4 - 1 = 3, 3^{-1} = \frac{1}{3}}$

$(4 - 1)^{-1} = 4^{-1} + (-1)\, 4^{-2}(-1) + \left(\frac{-1 \cdot -2}{2}\right) 4^{-3}(-1)^2 + \left(\frac{-1 \cdot -2 \cdot -3}{3 \cdot 2}\right) 4^{-4}(-1)^3 + ...$

$= \frac{1}{4} + \frac{1}{16} + \frac{1}{64} + \frac{1}{256} + \frac{1}{1024}$ .... (to infinity)

$\longrightarrow \frac{1}{3}$

$\frac{5}{16} \quad \frac{21}{64} \quad \frac{85}{256} \quad \frac{341}{1024}$

$\frac{5}{15} \quad \frac{21}{63} \quad \frac{85}{255} \quad \frac{341}{1023}$

$\boxed{\text{Expand } \left(\frac{1}{a} - \frac{r}{a}\right)^{-1}}$ $\boxed{\frac{1}{a} - \frac{r}{a} = \frac{1-r}{a}, \left(\frac{1-r}{a}\right)^{-1} = \frac{a}{1-r}}$

$\left(\frac{1}{a} - \frac{r}{a}\right)^{-1} = \left(\frac{1}{a}\right)^{-1} + (-1)\left(\frac{1}{a}\right)^{-2}\left(-\frac{r}{a}\right) + \left(\frac{-1 \cdot -2}{2}\right)\left(\frac{1}{a}\right)^{-3}\left(-\frac{r}{a}\right)^2 + \left(\frac{-1 \cdot -2 \cdot -3}{3 \cdot 2}\right)\left(\frac{1}{a}\right)^{-4}\left(-\frac{r}{a}\right)^3 + ...$

$= a + \frac{a^2 r}{a} + \frac{a^3 r^2}{a^2} + \frac{a^4 r^3}{a^3} + ...$

$= a + ar + ar^2 + ar^3 + ...$ Sum to infinity $= \frac{a}{1-r}$

Using binomial to find roots

The previous pages might seem like mathematical exercises. Actually, you were becoming familiar with a new technique—the use of series for making calculations. Now, you will see how it can be used to shorten the method for finding square roots.

Here, use three different expansions to find the square root of 2. The first one is $(1 + 1)^{1/2}$. The second is $(9/4 - 1/4)^{1/2}$. The third is $(49/25 + 1/25)^{1/2}$. Notice that each is a method of writing root 2.

Look at the results. The first looks as if it might get there eventually, but it hasn't got very close after 5 terms. The second has 4 terms to get it correct to 4 places of decimals; certainly an improvement. The third has 4 terms to get it correct to 6 places of decimals.

FIND $\sqrt{2}$ BY BINOMIAL SERIES

$$\text{(1)}\quad (1+1)^{1/2} = 1^{1/2} + \left(\frac{1}{2}\right)1^{-1/2}(1) + \left(\frac{\frac{1}{2}\cdot-\frac{1}{2}}{2}\right)1^{-3/2}(1)^2 + \left(\frac{\frac{1}{2}\cdot-\frac{1}{2}\cdot-\frac{3}{2}}{3\cdot2}\right)1^{-5/2}(1)^3 + \left(\frac{\frac{1}{2}\cdot-\frac{1}{2}\cdot-\frac{3}{2}\cdot-\frac{5}{2}}{4\cdot3\cdot2}\right)1^{-7/2}(1)^4 + \dots$$

$$= 1 \quad + \frac{1}{2} \quad - \frac{1}{8} \quad + \frac{1}{16} \quad - \frac{5}{128} \quad +\dots$$

$$1.5 \qquad \rightarrow 1.375 \qquad \rightarrow 1.4375 \qquad \rightarrow 1.3984375$$
$$-0.125 \qquad +0.0625 \qquad -0.0390625$$

$$\text{(2)}\quad \left(\frac{9}{4}-\frac{1}{4}\right)^{\frac{1}{2}} = \left(\frac{9}{4}\right)^{\frac{1}{2}} + \left(\frac{1}{2}\right)\left(\frac{9}{4}\right)^{-\frac{1}{2}}\left(-\frac{1}{4}\right) + \left(\frac{\frac{1}{2}\cdot-\frac{1}{2}}{2}\right)\left(\frac{9}{4}\right)^{-\frac{3}{2}}\left(-\frac{1}{4}\right)^2 + \left(\frac{\frac{1}{2}\cdot-\frac{1}{2}\cdot-\frac{3}{2}}{3\cdot2}\right)\left(\frac{9}{4}\right)^{-\frac{5}{2}}\left(-\frac{1}{4}\right)^3 + \dots$$

$$= \frac{3}{2} \quad -\frac{1}{2}\cdot\frac{2}{3}\cdot\frac{1}{4} \quad -\frac{1}{8}\cdot\frac{8}{27}\cdot\frac{1}{16} \quad -\frac{1}{16}\cdot\frac{32}{243}\cdot\frac{1}{64} \quad -\dots$$

$$\rightarrow 1.5$$
$$= \frac{3}{2} \quad -\frac{1}{12} \quad -\frac{1}{432} \quad -\frac{1}{7776} \qquad \begin{matrix}0.083333\\1.416667\end{matrix}$$
$$\rightarrow 0.002315$$
$$1.414352$$
$$\rightarrow 0.000129$$
$$1.414223$$

$$\text{(3)}\quad \left(\frac{49}{25}+\frac{1}{25}\right)^{\frac{1}{2}} = \left(\frac{49}{25}\right)^{\frac{1}{2}} + \left(\frac{1}{2}\right)\left(\frac{49}{25}\right)^{-\frac{1}{2}}\left(\frac{1}{25}\right) + \left(\frac{\frac{1}{2}\cdot-\frac{1}{2}}{2}\right)\left(\frac{49}{25}\right)^{-\frac{3}{2}}\left(\frac{1}{25}\right)^2 + \left(\frac{\frac{1}{2}\cdot-\frac{1}{2}\cdot-\frac{3}{2}}{3\cdot2}\right)\left(\frac{49}{25}\right)^{-\frac{5}{2}}\left(\frac{1}{25}\right)^3 + \dots$$

$$= \frac{7}{5} \quad +\frac{1}{2}\cdot\frac{5}{7}\cdot\frac{1}{25} \quad -\frac{1}{8}\cdot\frac{125}{343}\cdot\frac{1}{625} \quad +\frac{1}{16}\cdot\frac{3125}{16807}\cdot\frac{1}{15625} \quad -\dots$$

$$\rightarrow 1.4$$
$$= \frac{7}{5} \quad +\frac{1}{70} \quad -\frac{1}{13720} \quad +\frac{1}{1344560} \quad -\dots \qquad \begin{matrix}0.0142857\\1.4142857\end{matrix}$$
$$\rightarrow 0.0000729$$
$$1.4142128$$
$$\rightarrow 0.0000007$$
$$1.4142135$$

Making a series converge

Notice how you went about making the series in the previous section. You wanted a square root, so you made the first term of the binomial a perfect square. That way the first term of the series would be a simple root. The second term was also a square. Computing the successive terms was relatively simple. The better convergence was obtained by making b much smaller than a.

You have proved that the binomial "works" for finding square root with an infinite series, and you can find a result to a given accuracy more quickly. Next, you can use it for other things, as you did with the square root.

Making a Series Converge

$(1+1)^{\frac{1}{2}}$	1	$\frac{1}{2}$	$\frac{1}{8}$	$\frac{1}{16}$	$a=b$	$a^{\frac{1}{2}}=1$
$\left(\frac{9}{4}-\frac{1}{4}\right)^{\frac{1}{2}}$	$\frac{3}{2}$	$\frac{1}{12}$	$\frac{1}{432}$	$\frac{1}{7776}$	$a=9b$	$a^{\frac{1}{2}}=\frac{3}{2}$
$\left(\frac{49}{25}+\frac{1}{25}\right)^{\frac{1}{2}}$	$\frac{7}{5}$	$\frac{1}{70}$	$\frac{1}{13720}$	$\frac{1}{1344560}$	$a=49b$	$a^{\frac{1}{2}}=\frac{7}{5}$

$\boxed{\text{Find } 10^{\frac{1}{3}}}$

$8\frac{1}{3}=2$

$a=8$

$b=2$

$n=\frac{1}{3}$

$$(8+2)^{\frac{1}{3}} = 8^{\frac{1}{3}} + \frac{1}{3}\cdot 8^{-\frac{2}{3}}\cdot 2 + \left[\frac{\frac{1}{3}\cdot-\frac{2}{3}}{2}\right]8^{-\frac{5}{3}}\cdot 2^2 + \left[\frac{\frac{1}{3}\cdot-\frac{2}{3}\cdot-\frac{5}{3}}{3\cdot 2}\right]8^{-\frac{8}{3}}\cdot 2^3 + \left[\frac{\frac{1}{3}\cdot-\frac{2}{3}\cdot-\frac{5}{3}\cdot-\frac{8}{3}}{4\cdot 3\cdot 2}\right]8^{-\frac{11}{3}}\cdot 2^4 + ...$$

$$= 2 + \frac{1}{3}\cdot\frac{1}{4}\cdot 2 \quad -\frac{1}{9}\cdot\frac{1}{32}\cdot 4 + \frac{5}{81}\cdot\frac{1}{256}\cdot 8 - \frac{10}{243}\cdot\frac{1}{2048}\cdot 16 + ...$$

$$= 2 \quad +\frac{1}{6} \quad -\frac{1}{72} \quad +\frac{5}{2592} \quad -\frac{5}{15552} \quad +....$$

```
2.00000
+ 0.16667
2.16667
- 0.01389
2.15278
+ 0.00193
2.15471
- 0.00032
2.15439
+ 0.00006
2.15445
- 0.00001
2.15444
```

$\boxed{2.1544}$

$\times\frac{1}{3}\cdot\frac{1}{4} \qquad \times\frac{1}{3}\cdot\frac{1}{4} \qquad \times\frac{5}{9}\cdot\frac{1}{4} \qquad \times\frac{2}{3}\cdot\frac{1}{4} \qquad \times\frac{11}{15}\cdot\frac{1}{4} \qquad \times\frac{7}{9}\cdot\frac{1}{4}$

Ratio between successive terms

Next coefficients

$$\frac{\frac{1}{3}\cdot-\frac{2}{3}\cdot-\frac{5}{3}\cdot-\frac{8}{3}\cdot-\frac{11}{3}}{5\cdot 4\cdot 3\cdot 2}$$

$$\cdot-\frac{14}{3}$$

$$6\cdot$$

Questions and problems

1. In each of the following series, identify its kind: arithmetic, geometric, or harmonic. State the first term and the value of *d* or *r*. Finally, check your conclusion by reconstructing the series from the general form for that series.

1, 5, 9, 13, 17	16, 12, 9, 6 3/4, 5 1/16
−9, −3, +3, +9, +15	−81, +54, −36, +24, −16
10, 11, 12.1, 13.31, 14.461	9, 7, 5, 3, 1
1512, 1260, 1080, 945, 840, 756	105, 120, 140, 168, 210
16, 25, 34, 43, 52, 61	12, 23, 34, 45, 56, 67

2. Find the sum of the following series:

1, 3, 5, 9 . . . to 42 terms

23 . . ., 73 (last term in arithmetic progression of 26 terms)

5, 10, 20, 40 . . . to 11 terms

  5, −10, 20, −40 . . . to 10 terms
  5, −19, 20, −40 . . . to 11 terms
  100, −50, 25, −12.5 . . . to 6 terms
  100, −50, 25, −12.5 . . . to 7 terms
  6, 17, 28, 39, 50, 61 . . . to 19 terms
  14.5, 23.5, 32.5, 41.5 . . . 50.5, to 20 terms
  15625, 3125, 625, 125, 25 . . . to 10 terms

3. Sum the following series to infinity:

  10, 6, 3.6, 2.16, 1.296   10, −6, 3.6, −2.16, 1.296
  20, 16, 12.8, 10.24, 8.192   180, −144, 115.2, −92.16
  190, −171, 153.9, −138.51   10, 9, 8.1, 7.29, 6.561
  900, 90, 9, 0.9, 0.09   1100, −110, 11, −1.1, 0.11
  1300, −390, 117, −35.1, 10.53   700, 210, 63, 18.9, 5.67

4. Evaluate the following permutations: $_{50}P_3$, $_{10}P_5$, $_{12}P_6$, $_{10}P_4$, $_7P_6$.

5. Evaluate the following combinations: $_{50}C_3$, $_{12}C_4$, $_{12}C_8$, $_{10}C_4$, $_7C_6$.

6. For a new system of automatic dialing, a man wishes to reserve telephone numbers using seven digits, none of which uses the same figure twice. For example, 035−2468 is ok, but 355−9876 is not because it uses 5 twice. How many such numbers is it theoretically possible to find?

7. The existing system reserves certain 3-digit prefixes for area codes so that they can't be used for station prefixes. The ones used for area codes have a second digit of either 0 or 1; the first and third digits are not either 0 or 1. How many numbers that fit the specification of question 6 does this leave?

8. Twelve horses run in a race. Without any knowledge of their previous form, what are the odds against (a) naming the correct winner: (b) naming the first 2 places in order; (c) naming the first 3 places in order; (d) naming the first 3 places, but not in any particular order?

9. In the same race, what are the odds against a particular horse coming in among the first 3?

10. Use binomial expansion to find series for the following:

  $(0.1 - 0.06)^{-1}$ $(0.1 + 0.06)^{-1}$ $(0.05 - 0.04)^{-1}$
  $(1/180 + 1/225)^{-1}$ $(1/190 + 9/1000)^{-1}$ $(0.1 - 0.09)^{-1}$
  $(1/900 - 1/9000)^{-1}$ $(1/1100 + 1/11,000)^{-1}$
  $(1/1300 + 3/13,000)^{-1}$ $(1/700 - 3/7000)^{-1}$

To check your answers, verify that the series are the same as those in question 3.

11. Use binomial expansion to find the square root of 3: first, with the expansion of $(2.25 + 0.75)^{1/2}$; then, with the expansion of $(2.89 + 0.11)^{1.5}$. Use 5 terms of the first series and approximate evaluation to find the root to as many places as the accuracy justifies. Use 4 terms of the second series with approximate evaluation to 6 places of decimals.

12. Use the binomial expansion of $(27 + 3)^{1/3}$ to find the cube root of 30, using 5 terms of the series to find the approximate evaluation, correct to 4 places of decimals.

13. Use the first 4 terms in the expansion of $(32 - 2)^{1/5}$ to find the 5th root of 30, correct to 5 places of decimals.

14. A tossed coin is equally likely to land heads or tails. Assume the first throw is heads. In any two throws, the possibilities are: two throws alike or two throws opposite. Based on this statement, the second throw is also equally likely to be heads or tails. Any 3 throws has two chances of 3 throws alike, with 6 other possibilities for different combinations. Based on this fact, if the first 2 throws are heads, the probability is 3 to 1 that the third throw will be tails. Extend this reasoning to find the probability of a 4th throw being tails, if 3 throws have been heads.

15. A simple gambling game consists of taking a stack of randomly mixed coins (for example, pennies) and offering odds for guessing the order of the first so many coins that are uncovered in succession. For example, the first 3 coins might be predicted as: heads, heads, tails. If the odds offered are 6 to 1, is the guesser likely to win or lose over a large number of guesses?

16. An alternative offer is 10 to 1 for guessing the order of the first 4 coins. Is this a better or worse offer than 6 to 1 for three guesses?

17

CHAPTER

Putting progressions to work

Rates of change

The relationship between a quantity and its derivative (rate of change) can provide short cuts in calculating. In fact, as well as being relatively simple, these rates are sometimes the only way to find an accurate result. To understand the principle involved, consider a car travelling along a highway.

The rate at which it moves along the highway is speed (velocity). A stationary car doesn't move. Change from being at rest to moving, or from moving to moving more rapidly, is acceleration. By means of acceleration, velocity is increased. So the rate at which velocity is increased is acceleration, as was described in chapter 12.

Before radar speed guns were invented, police timed the movement of a car between two points. If the time was less than what travelling at the legal speed would require, the driver got a ticket. A smart driver might see the first cop and slow down before he got to the second. The time check would show that he wasn't speeding, even though he was when he saw the first cop. Radar speed guns stopped this practice by reading speed at an instant, instead of averaging it over a distance.

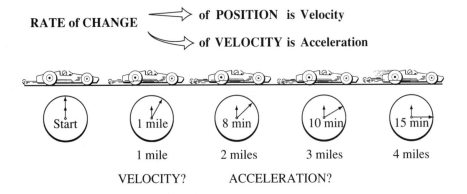

RATE of CHANGE ⟹ of **POSITION** is Velocity

⟹ of **VELOCITY** is Acceleration

Start	1 mile	8 min	10 min	15 min
	1 mile	2 miles	3 miles	4 miles

VELOCITY? ACCELERATION?

Weight and Spring Resonance

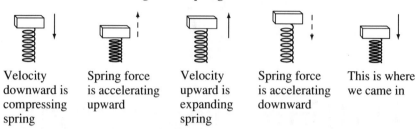

| Velocity downward is compressing spring | Spring force is accelerating upward | Velocity upward is expanding spring | Spring force is accelerating downward | This is where we came in |

But how do you find   POSITION
VELOCIITY
ACCELERATION
at any particular instant?

The weight and spring resonance system, also described in chapter 12, shows how velocity and acceleration change during its movement. How do you check those facts? *Infinitesimal calculus*, which is really quite easy, despite its imposing name, helps study all these problems.

Infinitesimal Changes

Relationship between quantities x and y of form $\boxed{y = x^n}$

Infinitesimal increase in x: $x + dx$

Corresponding increase in y: $y + dy$

$$y + dy = (x + dx)^n$$

Binomial

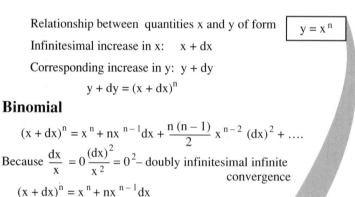

$$(x + dx)^n = x^n + nx^{n-1}dx + \frac{n(n-1)}{2} x^{n-2} (dx)^2 + \ldots.$$

Because $\dfrac{dx}{x} = 0$ $\dfrac{(dx)^2}{x^2} = 0^2$ – doubly infinitesimal infinite
convergence

$$(x + dx)^n = x^n + nx^{n-1}dx$$
$$y + dy = x^n + nx^{n-1}dx$$
$$\underline{y \qquad\quad = x^n \qquad\qquad\qquad}$$
$$dy = nx^{n-1}dx \quad \text{or} \quad \boxed{\frac{dy}{dx} = nx^{n-1}}$$

The rate of change of y with respect change in x at instant $y = x^n$ is:
nx^{n-1}

Infinitesimal changes

Any relationship can be plotted as a graph. Any graph can have an algebraic equation to express the relationship that was plotted. Some equations are simple, some are complex. Those that relate to the real world have two types of variables: independent and dependent. In an equation, such as $y = x^n$, x is the independent variable, on which the value of y depends, so the dependent is also variable.

Increasing x causes y to increase. Looking at one point on a graph doesn't tell you the rate of y's increase. A stretch of the graph, like timing the car over a distance, can tell you the average rate of y's increase.

Infinitesimal calculus lets you calculate speed at a point, like the radar gun measures it. The idea is simple. To measure slope at a point, measure change over an infinitesimally small "piece" of the graph.

The very small change in y, divided by the very small change in x that caused it, gives the slope at that point. Use the symbols dx and dy to represent infinitesimally small changes. Compared to x and y, they are too small to be measurable.

DERIVATIVE of a LINEAR EQUATION

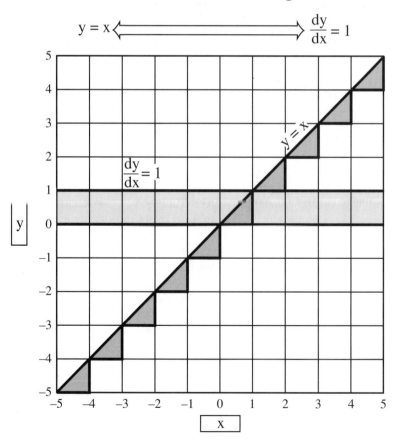

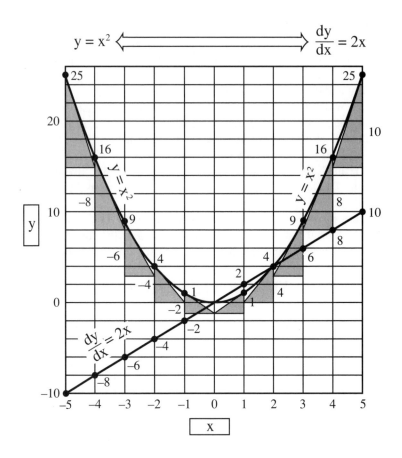

Here is the general solution for the slope of a graph for $y = x^n$. The binomial expansion is used to expand $(x + dx)^n$. Then, $y = x^n$ is subtracted from the expression for $y + dy = (x + dx)^n$, to leave an equation for dy in terms of dx.

The smaller you make a change, the smaller the higher powers become. If $dx/x = 0$ (but is comparable with dy), $(dx)^2/x^2$ completely vanishes.

Now apply these principles to a few cases and see how it "works." Apply it to the simplest possible equation, $y = x$. Dividing dy by dx, after subtracting $y = x$, we have $dy/dx = 1$. This statement is easily seen to be correct, because for every change of 1 in x, y also changes by 1.

Take the next power of x, $y = x^2$. From the formula in this section, $dy/dx = 2x$. Plotting the curve for $y = x^2$, for values of x from -5 to $+5$, draw a smooth curve through the 11 points.

If your curve represents the true curve for $y = x^2$, you can draw triangles under the curve with a base length of 1, and a slanting side that just touches the curve at the same slope as the curve. In each case, the vertical height of the triangle is $2x$ for the particular point.

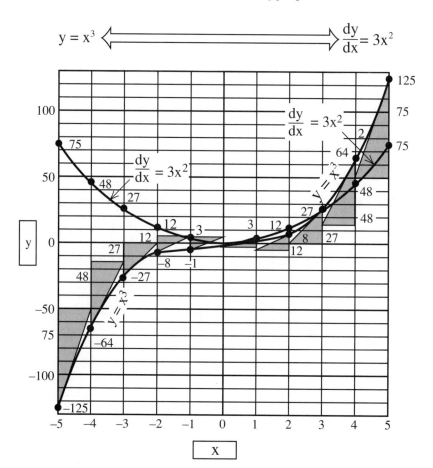

Values of *dy/dx* can be plotted as another graph (this time a straight line), which shows the slope of the first curve at every point.

This time, try the graph of y   x³. From the formula, dy/dx — 3x². For higher powers of x, the scale for y has to be changed to get the graph on the page. Draw tangent triangles for the unit horizontal base and measure the height to verify the formula. Then, plot a graph for *dy/dx*.

Notice that the cubic curve, unlike the square curve or parabola, curves upward for positive values of x, and downward for negative values of x. Its slope is always upwards, except when x = 0, where it is momentarily horizontal, the zero slope. The slope is equally positive for the same values of x, positive and negative.

Take one more power of x for the time being: y = x⁴. Changing scales again, plot from x = −5 to x = +5. This curve is similar to that for y = x², in that values of y are positive for both positive and negative values of x. Of course, the curvature is quite different. Correspondingly, the curve for *dy/dx* is again negative when x is negative, because the y = x⁴ slope is downward (from left to right). Once again, the triangle constructions verify the formula result.

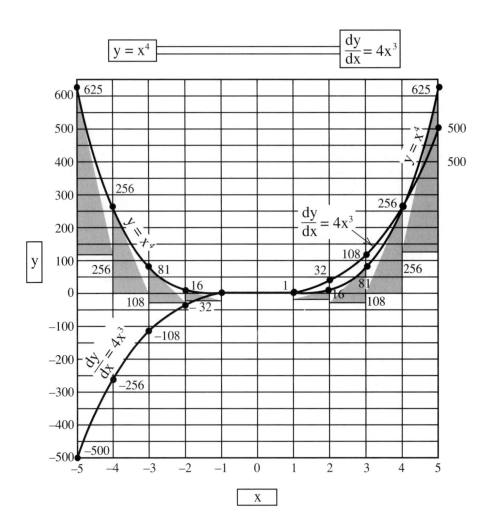

Successive differentiation

In the preceding pages, the slope of the curve was found by a single differentiation, *dy/dx*. Some problems require *successive differentiation*. For example, velocity is a rate of change of position. Acceleration is a rate of change of velocity. So, acceleration is a second differentiation of position; the first gives velocity, the second gives acceleration (or deceleration), and the third gives the rate of change of acceleration. Because acceleration changes, it must have a rate of change.

Starting with $y = x^4$, successively differentiate to find these *successive derivatives* (as mathematicians call them). Previously the simplest equation, $y = x^n$ was used. It seems obvious that a constant will transfer to the derivative. The work here confirms it:

$$\text{If } y = ax^n, \text{ then } dy/dx = anx^{n-1}$$

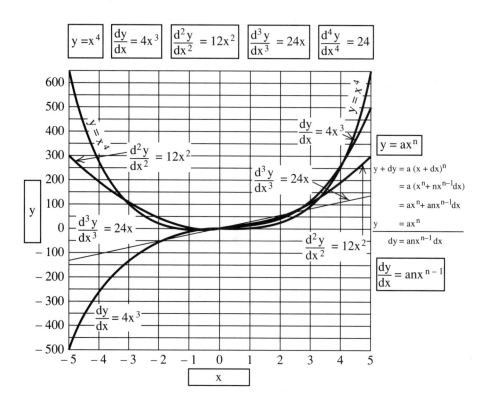

Differentiating a complete expression

Suppose an equation for y in terms of x contains a whole expression, with several terms that involve different powers of x. In general form, this might be $y = ax^m + bx^n + \ldots$ It could continue to any number of terms, so only two are taken here.

Increasing x by dx, substitute $x + dx$ for x, each time it occurs in the original equation and increase y by dy (to make $y + dy$). Using the same method as on previous pages, you have:

$$y + dy = ax^m + amx^{m-1}dx + bx^n + dnx^{n-1}dx$$

Subtracting the parts that correspond to y, the parts that correspond with dy are left:

$$dy = amx^{m-1}dx + bnx^{n-1}dx$$

Dividing through by dx gives an expression for dy/dx that shows that dy/dx is equal to the sum of differentiations of individual terms in the original expression.

Put in sample numbers: $y = x^5 - 50x^3 + 520x$. The derivative is: $dy/dx = 5x^4 - 150x^2 + 520$. Notice that where the original term has a minus sign (as does $50x^3$) the term in the derivative also has a minus sign.

Work through the example in the previous section with tabulated values of y and dy/dx, term by term, from $x = -6$ to $x = +6$. Study the curves. Where the curve for y reaches a maximum or minimum, the curve for dy/dx passes through zero value. Momentarily, y is neither increasing nor decreasing.

Where the curve for y crosses the zero line, except at the ends, where both curves go "off scale," the dy/dx curve is at a maximum positive or negative. At those points, the slope of y reaches a maximum, either up or down.

$$y = ax^m + bx^n +$$

$$
\begin{aligned}
y + dy &= a\,(x + dx)^m + b\,(x + dx)^n \\
&= a\,[x^m + mx^{m-1}\,dx] + b\,[x^n + nx^{n-1}\,dx] \\
&= ax^m + amx^{m-1}dx + bx^n + bnx^{n-1}dx \\
y &= ax^m \qquad\qquad\quad + bx^n \\
\hline
dy &= \quad amx^{m-1}\,dx \qquad + bnx^{n-1}dx
\end{aligned}
$$

$$\frac{dy}{dx} = amx^{m-1} + bnx^{n-1}$$

Example

$$y = x^5 - 50x^3 + 520x$$

$a = 1$	$a = 50$	$a = 520$
$n = 5$	$n = 3$	$n = 1$

$$\frac{dy}{dx} = 5x^4 - 150x^2 + 520\,(x^0)$$

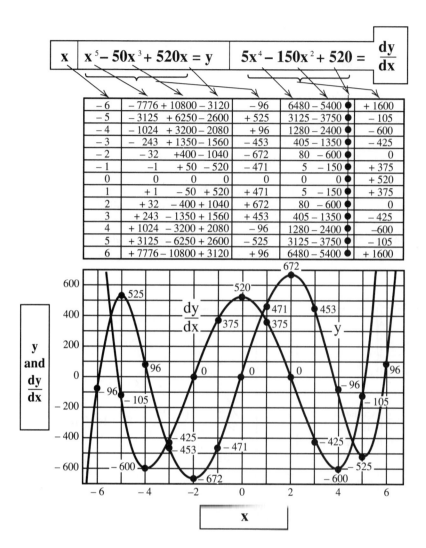

Successive differentiation of movement

The successive differentiation of movement can be applied to movement (for example, a travelling car). Time is the *independent variable* because whatever happens, time continues. Distance is the *dependent variable* because it depends on time, speed, and other things. Time never depends on anything, so it is always the independent variable when considered in relation to other variables.

From the variables of time and distance, which usually have either the symbols t and d or x and y, the successive differentiations are called *derivatives*. The first derivative, distance measured against time, is *velocity*. The second derivative is *acceleration*, the rate of change of velocity. Rate of change of acceleration has no name. Its units would be feet per second cubed! These relationships can be demonstrated during 5 successive time intervals.

At first, everything is stationary. Distance (from any other point) is fixed. Velocity and acceleration are both zero.

For the second interval acceleration increases, assume a steady rate, shown by the straight slanting line. Velocity will increase with a quadratic curve. Distance will begin increasing with a cubic curve.

For the third interval, acceleration holds constant, so velocity increases steadily in a straight line segment. That part of the distance curve will be quadratic.

For the fourth interval, acceleration decreases back to zero. Velocity follows an inverted quadratic curve and distance approaches a straight line.

For the fifth interval, acceleration is again zero. Velocity is constant (a horizontal straight line) and distance is a steeply sloping straight line.

You can learn quite a bit about derivatives by studying these relationships.

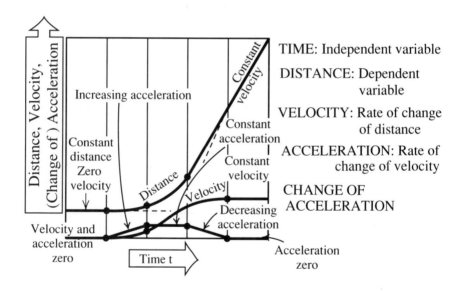

Circular measure of angles

Angles can be measured in a variety of ways, but all are virtually ratios. Degree measure divides a complete circle, (rotation) into 360 degrees. The number of degrees shows how much of a complete rotation the angle is. Advocates of metric wanted to divide the circle into 100 parts, which they call *grades*. This measure is sometimes used, but the principle is the same as the others.

Dividing a circle into quadrants is quite basic. All methods of measuring angles are essentially a way to take the circumference, divided by the radius, as a ratio. By this definition of an angle, none of the well-known measures result in convenient numbers. Many parts of mathematics define an angle as the ratio of the arc length around the circumference, divided by the radius.

The circular measure of angles, as this is called, fits with definitions of trigonometric ratios. Sine, cosine, and tangent are all ratios that identify an angle. None of them is conveniently proportional to all angles. Sine and tangent begin at small angles, in direct proportion to the angle, but this proportionality breaks down long before the first right angle. The cosine begins at 1 and decreases, slowly at first, reaching zero at the first right angle.

Arc length, measured along the circumference and divided by radius, is always proportional to the angle that the ratio identifies. The circular measure of an angle is stated in *radians*.

Measuring distance around the circumference, the first semicircle (180 degrees) accommodates a little over 3 radii. A whole circle accommodates twice as many, a little over 6 radii. The ratio of the length of a semicircular arc to its radius has the universal symbol π (the Greek letter pi). For the present, use its approximate value of 3.14. On this basis, a right angle is half of π radians. Using this equivalent, assign radian values to angles to correspond to degree measure for 60, 45, 30, and 15 degrees.

MEASURES of ANGLES

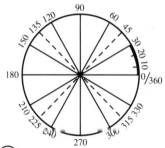

(1) By degrees: 360 to a circle or complete revolution

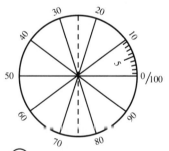

(2) Dividing a revolution or circle into 100 parts

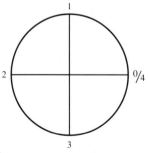

(3) By quadrants: 4 to a circle

(4) By RADIANS

$$\text{Angle in radians} = \frac{\text{length of arc}}{\text{radius}}$$

Tabulating the radian equivalents for these angles, as well as the values for sine and tangent that you calculated in Part 2, notice how close the three figures are at 15 degrees. For very small angles, the figures are even closer. As the angle increases, the divergence becomes greater.

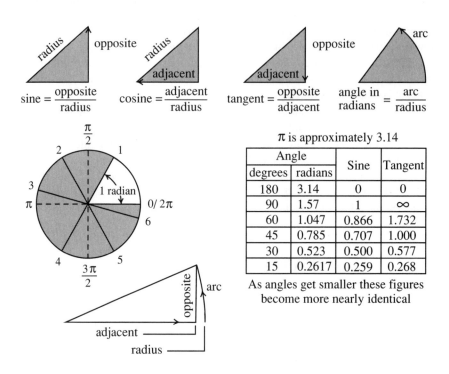

$$\text{sine} = \frac{\text{opposite}}{\text{radius}} \qquad \text{cosine} = \frac{\text{adjacent}}{\text{radius}} \qquad \text{tangent} = \frac{\text{opposite}}{\text{adjacent}} \qquad \frac{\text{angle in}}{\text{radians}} = \frac{\text{arc}}{\text{radius}}$$

π is approximately 3.14

Angle		Sine	Tangent
degrees	radians		
180	3.14	0	0
90	1.57	1	∞
60	1.047	0.866	1.732
45	0.785	0.707	1.000
30	0.523	0.500	0.577
15	0.2617	0.259	0.268

As angles get smaller these figures become more nearly identical

Differential of angles

Circular measure allows you to apply the principles of differential calculus, concerning infinitesimal changes. As already pointed out, for very small angles, the ratio for the sine is almost the same as its circular measure (in radians). So, for nearly zero angle, sin dx = dx. Because the adjacent equals the hypotenuse at zero angle and $1/1$ = 1, cos dx = 1. These values are true, regardless of the value of x, because they concern only the infinitesimally small angle, dx.

Suppose y = sin x. Apply the sum formula to the right-hand part of $y + dy$ = sin ($x + dx$). Substituting sin dx = dx and cos dx = 1 gives: $y + dy$ = sin x + cos $x \cdot dx$

Take away the original part:	y = sin x
This leaves:	dy = cos $x \cdot dx$
Dividing both sides by dx:	dy/dx = cos x

Similarly, if y = cos x, again use the sum formula and substitute for cos dx and sin dx: dy/dx = $-$ sin x.

INFINITESIMAL CHANGES of ANGLE

For infinitely small angles, using radian measure

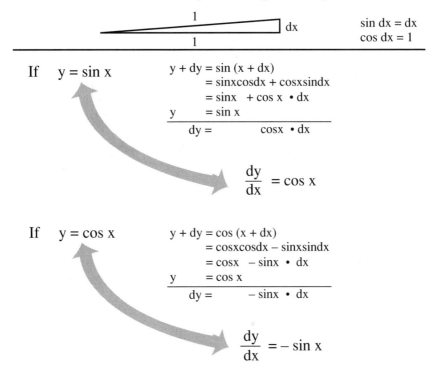

sin dx = dx
cos dx = 1

If   y = sin x

y + dy = sin (x + dx)
= sinxcosdx + cosxsindx
= sinx   + cos x • dx
y   = sin x
_____
dy =   cosx • dx

$$\frac{dy}{dx} = \cos x$$

If   y = cos x

y + dy = cos (x + dx)
= cosxcosdx – sinxsindx
= cosx   – sinx • dx
y   = cos x
_____
dy =   – sinx • dx

$$\frac{dy}{dx} = -\sin x$$

Successive differentiation of sine wave

These facts about trigonometrical ratios help draw the sine-wave curve (for both sine and cosine); drawing the curves also provides a better picture of this relationship. Here, values between 0 and 2π radians (0 and 360 degrees) for sin, *dy/dx* of sin *x*, cos *x*, and so on, are tabulated and the curves are drawn. Notice that differentiating the sine yields the cosine, differentiating the cosine yields minus the sine, differentiating minus the sine yields minus the cosine, and differentiating minus the cosine yields plus the sine.

The angle base shows both degree measure and radian measure. A radian is slightly less than 60 degrees. If you marked off 1 radian, a straight line from zero would reach a value of 1 at 1 radian. Also, at the 60-degree point, the slope is precisely half that starting at zero.

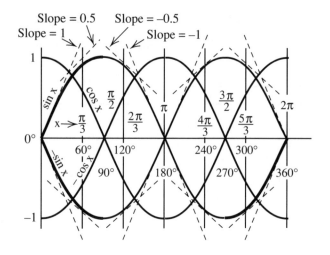

When y is		Values of x								
		0	$\frac{\pi}{3}$	$\frac{\pi}{2}$	$\frac{2\pi}{3}$	π	$\frac{4\pi}{3}$	$\frac{3\pi}{2}$	$\frac{5\pi}{3}$	2π
sin x	y =	0	0.866	1	0.866	0	–0.866	– 1	–0.866	0
	$\frac{dy}{dx}=$	1	0.5	0	–0.5	– 1	–0.5	0	0.5	1
cos x	y =	1	0.5	0	–0.5	– 1	–0.5	0	0.5	1
	$\frac{dy}{dx}=$	0		–1		0		+ 1		0
– sin x	y =	0	–0.866	–1	–0.866	0	0.866	1	0.866	0
	$\frac{dy}{dx}=$	– 1	–0.5	0	0.5	1	0.5	0	–0.5	– 1
– cos x	y =	– 1	–0.5	0	0.5	1	0.5	0	–0.5	– 1
	$\frac{dy}{dx}=$	0		1		0		– 1		0
sin x	y =	0	0.866	1	0.866	0	–0.866	– 1	–0.866	0
	$\frac{dy}{dx}=$	1	0.5	0	–0.5	– 1	0.5	0	0.5	1

Finding series for sine

This fact about sine and cosine curves helps relate the quantities that are represented by these ratios to actual angles (measured in radians). Although better ways show up later, this basis will calculate ratios for all angles, not just the simple angles in Part 2. Notice the style of writing that is used to save repeating the dependent variable.

Instead of writing, "If $y = \sin x$, the $dy/dx =$" or whatever the variables are, substitute what it is that you make y equal to. So, write "$d/dx \sin x =$," which is read as "*d* by *dx* sine *x* equals."

$$y = \sin x, \frac{dy}{dx} = \quad\Longrightarrow\quad = \frac{d}{dx}\sin x \quad \textbf{Shortened method of writing}$$

Suppose $\quad \sin x = ax + bx^2 + cx^3 + dx^4 + ex^5 + fx^6 + gx^7 + \ldots.$

then $\dfrac{d}{dx}\sin x = a + 2bx + 3cx^2 + 4dx^3 + 5ex^4 + 6fx^5 + 7gx^6 + \ldots.$

or $\cos x =$

then $\dfrac{d}{dx}\cos x = 2b + 3{\cdot}2cx + 4{\cdot}3dx^2 + 5{\cdot}4ex^3 + 6{\cdot}5fx^4 + 7{\cdot}6gx^5 + \ldots.$

or $-\sin x =$

so $\sin x =$ $\quad -2b \quad -3\cdot 2cx \quad -4\cdot 3dx^2 \quad -5\cdot 4ex^3 \quad -6\cdot 5fx^4 \quad -7\cdot 6gx^5 \quad -\ldots.$

also $= \qquad\qquad\qquad ax \;+\; bx^2 \;+\; cx^3 \;+\; dx^4 \;+\; ex^5 \;+\; \ldots.$

When x = 0

$\dfrac{d}{dx}\sin x$ or $\cos x = 1$; so $a = 1$ $\quad\Longrightarrow\quad$ $\dfrac{d}{dx}\cos x$ or $-\sin x = 0$; so$(2)b = 0$

$-3{\cdot}2c = a$	so $c = -\dfrac{1}{3{\cdot}2}$	$-4{\cdot}3\,d = b = 0$	$d = 0$
$-5{\cdot}4e = c$	so $e = \dfrac{1}{5{\cdot}4{\cdot}3{\cdot}2}$	$-6{\cdot}5\,f = d = 0$	$f = 0$
$-7{\cdot}6g = e$	so $g = -\dfrac{1}{7{\cdot}6{\cdot}5{\cdot}4{\cdot}3{\cdot}2}$		

$$a = 1,\; b = 0,\; c = -\frac{1}{3!},\; d = 0,\; e = \frac{1}{5!},\; f = 0,\; g = -\frac{1}{7!}$$

$$\sin x = x - \frac{x^3}{3!} + \frac{x^5}{5!} - \frac{x^7}{7!} + \ldots$$

What helps derive this series is that the 4th derivative is the same quantity repeated, or the second derivative is minus the same quantity. So, first write sin x as a series of powers of *x* with different coefficients, which you need to find values for.

Differentiate the series for sin *x*, term by term, to yield *d/dx* sin *x*, which must also be cos *x*, as already shown. Pursuing this, differentiate it term by term to yield *d/dx* cos *x*, which must also be −sin *x*. Reversing signs, you have two expressions (series) for sin *x*. Both expressions consist of a series of powers of *x*. So, the respective coefficients must be the same, which lets you calculate them.

The last form begins with −2*b*, which has no counterpart in the first one. When *x* is 0, sin *x* is 0, so −2*b* must be 0, too. The term −3 · 2*c* is equal to *a*. You know that for small values of *x*, sin *x* = *x*, so *a* must be 1 and *c* must be −1/3! As *b* was zero, *d* must also be zero. Next, −5 · 4*e* = *c*, which you already know is equal to −3 · 2*a*, so *e* must be +1/5! (minus times minus makes a plus). You have developed a series for sin *x*, in terms of powers of *x*, when *x* is the angle in radians.

Finding series for cosine

To find a similar series for cosine, the procedure is the same. When $x = 0$, $\cos x$ = 1. The first term is 1 without a power of x (x^0). Following through, you get another series, which uses the terms that the sine series left out.

To try out these series, in the sine series, put $x = 0.5$. That number is slightly less than 30 degrees. This measure converges rapidly, allowing $\sin 0.5$ to be evaluated to 6 places of decimals after only 4 terms.

Suppose

$$\cos x = a + bx + cx^2 + dx^3 + ex^4 + fx^5 + gx^6 + \ldots$$

then
$$\left.\begin{array}{l} \dfrac{d}{dx}\cos x \\ \text{or} - \sin x \end{array}\right\} = \quad b + 2cx + 3dx^2 + 4ex^3 + 5fx^4 + 6gx^5 + \ldots$$

$$\left.\begin{array}{l} \dfrac{d}{dx} - \sin x \\ \text{or} - \cos x \end{array}\right\} = \quad 2c + 3{\cdot}2d\,x + 4{\cdot}3e\,x^2 + 5{\cdot}4f\,x^3 + 6{\cdot}5g\,x^4 + \ldots$$

so $\cos x$ = $-2c\ -3{\cdot}2d\,x\ -4{\cdot}3e\,x^2\ -5{\cdot}4f\,x^3\ -6{\cdot}5g\,x^4\ -\ldots$
also = $a\ +\ b\,x\ +\ c\,x^2\ +\ d\,x^3\ +\ e\,x^4\ +\ldots$

When x = 0

$$\cos x = 1 \qquad\qquad \text{so } a = 1$$
$$\dfrac{d}{dx}\cos x \text{ or} - \sin x = 0 \qquad \text{so } b = 0$$
$$-3{\cdot}2d = b = 0, \quad d = 0$$

$$-2c = a, \text{ so } c = -\frac{1}{2}$$

$$-4{\cdot}3e = c, \text{ so } e = \frac{1}{4{\cdot}3{\cdot}2} \text{ or } \frac{1}{4!}$$

$$-6{\cdot}5g = e, \text{ so } g = -\frac{1}{6{\cdot}5{\cdot}4{\cdot}3{\cdot}2} \text{ or } -\frac{1}{6!}$$

$$\boxed{\cos x = 1 - \frac{x^2}{2!} + \frac{x^4}{4!} - \frac{x^6}{6!} + \ldots}$$

$\boxed{\text{4th term is less than } \dfrac{1}{500,000} \text{ or } .000002}$

① **Suppose** $x = 1.5$ (slightly less than 30°)

$$\sin x = \frac{1}{2} - \frac{1}{6}{\cdot}\frac{1}{8} + \frac{1}{120}{\cdot}\frac{1}{32} - \frac{1}{5040}{\cdot}\frac{1}{128} +$$

② **Suppose** $x = 1.5$ (slightly less than 90°)

$$\cos x = 1 - \frac{1}{2}{\cdot}\frac{9}{4} + \frac{1}{24}{\cdot}\frac{81}{16} - \frac{1}{720}{\cdot}\frac{729}{64}$$

$$= 1 - \frac{9}{8} + \frac{27}{128} - \frac{81}{5120} + \ldots$$

1st	0.500000
2nd	− 0.020833
	0.479167
3rd	+ 0.000260
	0.479427
4th	− 0.000002
	0.479425

$\sin 0.5 = 0.479425$ correct to six places

$\boxed{\text{4th term is } 0.01582}$ So many more terms are needed for corresponding accuracy

For the cosine series, try an angle of $x = 1.5$, which is slightly less than 90 degrees, so its cosine is less than 0.1. The 4th term evaluates to 0.01582. . . (in decimals). Many more terms are necessary to find an answer that is correct to 6 places.

Questions and problems

1. A curve takes the form: $y = a(x^3 + bx^2 + cx) + d$, where a, b, c, and d are constants, which might have positive or negative signs. The first fact known about the curve is that level points occur for values of $x = -3$ and $+2$. From this fact, evaluate constants b and c. Values of y that correspond to $x = -3$ and $+2$, respectively, are $+65$ and -60. Evaluate constants a and d.

2. For 6 seconds from standstill a car increases its rate of acceleration uniformly (units: ft/sec³), covering a distance of 36 feet. What is the rate of acceleration for these 6 seconds and what are the acceleration and velocity at the end of the 6 seconds?

3. After the first 6 seconds, the car maintains constant acceleration for 8 more seconds. What will be the speed at the end of the 14 seconds (in ft/sec or miles/hr) and how far will it have travelled?

4. For 6 more seconds, the acceleration is uniformly decreased to zero. What is the final speed and distance from the start at the end of 20 seconds?

5. Calculate, correct to 7 places of decimals, the sine and cosine ratios for each of the following angles, given in radians: 0.1, 0.2, 0.3, 0.4, and 0.5.

6. If $y = \tan x$, find dy/dx. Use the same method as for sine and cosine, and substitute $\tan dx = dx$. The right side of the equation will be a fraction with terms in dx in both the numerator and the denominator. Simplify by multiplying both sides by the denominator and collecting terms both with and without dy or dx. A term with the product $dydx$ can be ignored.

7. Factor the expressions: $\sin A + \sin B$; $\sin A + \sin B + \sin(A + B)$. Prove that:
$$[\sin A + \sin B + \sin(A + B)][\sin A - \sin B - \sin(A - B)]$$
$$= \sin 2A \sin B - \sin 2B \sin A$$

8. From the sum and difference formulas:
$$\sin(A + B) = \sin A \cos B + \cos A \sin B$$
$$\sin(A - B) = \sin A \cos B - \cos A \sin B$$
$$\cos(A + B) = \cos A \cos B - \sin A \sin B$$
$$\cos(A - B) = \cos A \cos B + \sin A \sin B$$

Find formulas using ratios of the angles $(A + B)$ and $(A - B)$ to express the following ratio products as the sum or difference of two ratios:

(a) $\sin A \sin B$ (b) $\sin A \cos B$ (c) $\cos A \sin B$ (d) $\cos A \cos B$

9. By reversing the process from the formulas already obtained for ratios of multiple angles, find expressions in terms of ratios of the angle A for $\sin 1/2\ A$, $\cos 1/2\ A$, and $\tan 1/2\ A$.

10. By successive differentiation of the expression:

$$y = \sin x + 2 \cos x$$

find a series for y, in the form:

$$y = a + bx + cx^2 + dx^4$$

11. A man standing due east of a monument on a hill notices that its elevation, measured carefully with a sextant, is 5 degrees 42 1/2 minutes. After walking due north for 1 mile, he takes a bearing on the monument with a compass and finds that it is 10 degrees 26 minutes south of west from his new position. How high is the monument from his first viewpoint point as a reference level? Take the tangent of 5 degrees 42 1/2 minutes as 0.1000 and 18 degrees 26 minutes as 0.3333.

12. Show that $\sqrt{2} - \sin\left(A + \dfrac{\pi}{4}\right) = 2 - \sin + \cos A$.

13. If $\tan B = 0.5$, show that $\sqrt{5} - \sin(A + B) = 2 - \sin A + \cos A$.

14. If $\tan B = 0.75$, show that $5 - \sin(A + B) = 4 - \sin A + 3 - \cos A$

15. Mass (m), velocity (v), power (p), and time (t), are related during acceleration from standstill at constant power, by $\dfrac{1}{2}mv^2 = pt$. Because acceleration is given by: $a = \dfrac{dv}{dt}$, find a formula for acceleration with time under this condition.

16. Find the derivatives of:

$y = 45x^2$ $s = 4t^4 - 2.5t^3 + 1.6t^2 + 8t$

$v = 17x^{1.5}$ $y = 3.5x^{2.5} + 5x^{1.5}$

$v = 6t^{1/2}$ $f = \dfrac{5.3}{t^{1/2}}$

$y = 5x^4 + 4x^3 + 3x^2 + 2x$

$z = 29.3y^5 - 14.5y^2 + 27$ $h = f^2 - f^{-1}$

$g = 2.1f - 17.5f^3$

17. Find a series in the form: $a + bx + cx^2 + dx^3 + 3x^4 + fx^5 + \ldots$ so that:

$$\frac{d^2y}{dx^2} = y.$$

18. Find a series of the same form so that:

$$\frac{dy}{dx} = y.$$

19. Find a series of the same form so that:

$$\frac{dy}{dx} = 2y.$$

20. Plot curves for $y = \frac{1}{x}$ for both positive and negative values of x, with values of x and y from -5 to $+5$. Differentiate $y = \frac{1}{x}$, to show that both parts of the curve always have a negative slope.

18
CHAPTER

Putting differentiation to work

Differential of sine waves

The previous chapter showed that successive differentiation of sine waves results in the same equation with its starting point shifted. Starting at sin x, successive derivatives are cos x, $-\sin x$, $-\cos x$ and back to sin x. It was particularly easy, taking $y = \sin x$, with x in radians. The angle might be proportional to x in radians, but often it is not directly x radians. If the angle is in degrees, it has to be converted into radians for this method to work. Sometimes the quantity is not really an angle at all, but is something that trigonometry can represent as an angle.

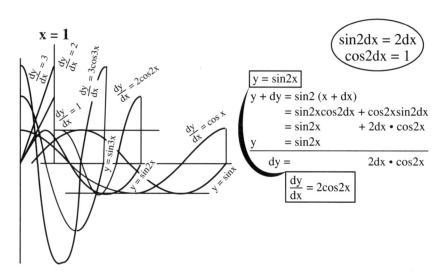

$$\sin 2dx = 2dx$$
$$\cos 2dx = 1$$

$$y = \sin 2x$$
$$y + dy = \sin 2\,(x + dx)$$
$$= \sin 2x \cos 2dx + \cos 2x \sin 2dx$$
$$= \sin 2x + 2dx \cdot \cos 2x$$
$$\underline{\quad y \qquad\qquad = \sin 2x \qquad\qquad\qquad\qquad\qquad}$$
$$dy = \qquad\qquad\qquad 2dx \cdot \cos 2x$$

$$\frac{dy}{dx} = 2\cos 2x$$

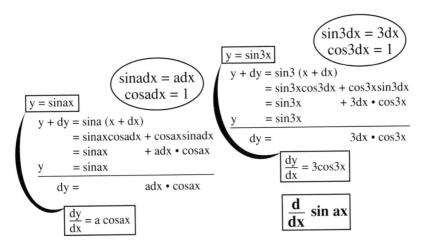

Assume the equation is $y = \sin 2x$, instead of $y = \sin x$. Following the same method, $dy/dx = 2 \cos 2x$. Next, take $y = \sin 3x$. The derivative is $dy/dx = 3 \cos 3x$. Finally, using a general multiplier, a, if $y = \sin ax$, $dy/dx = a \cos ax$. Using the shortened form of writing: $d/dx \sin ax = a \cos ax$.

Sinusoidal motion

Using the simple equation: $y = \sin x$, the slope of the curve at the zero (starting) point is equal to its magnitude at maximum, which is why the cosine has the same amplitude as the sine wave. When the multiple constant a was introduced, a times as many waves were in the basic period, so the slope of the original wave is a times as steep. The amplitude of the derivative is multiplied by a.

Now assume the equation $y = A \sin bt$ represents some motion with passage of time, t. A is the maximum movement from its average position and y is the distance from this reference position at time t. b is a constant rate that shows how fast the thing moves every time it passes through the zero (reference) position and thus, how many times it will make its complete excursion back and forth in a given time.

Velocity is the first derivative, given by dy/dt. It figures to $dy/dt = Ab \cos bt$. Acceleration is the next derivative, given by the equation:

$$d^2y/dt^2 = -Ab^2 \sin bt$$

Notice that maximum velocity occurs every time that the object passes through zero (reference) position and zero velocity occurs at each extreme. Zero acceleration occurs at the zero position, when velocity is a maximum, and is a maximum at each extreme.

Notice that the zero position and the zero acceleration coincide. Maximum excursion and maximum acceleration also coincide. Acceleration is b^2 times position (in whatever units are used) and it is of the opposite sign. When position is maximum upwards, acceleration is maximum downwards, and vice versa.

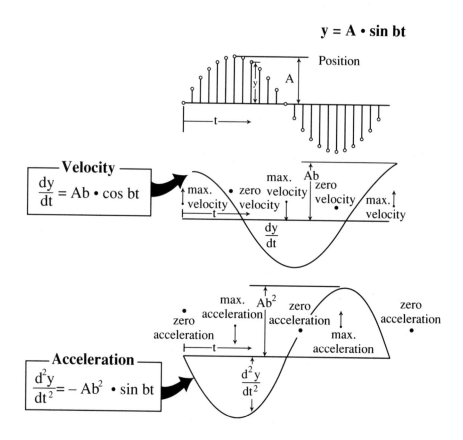

$$y = A \cdot \sin bt$$

Velocity

$$\frac{dy}{dt} = Ab \cdot \cos bt$$

Acceleration

$$\frac{d^2y}{dt^2} = -Ab^2 \cdot \sin bt$$

Harmonic motion

This fact about systems in mechanics, electric circuits, acoustics—in fact every branch of science—explains the cyclic interchange of energy, called *harmonic motion*. Part 2 showed that such a system has a characteristic period (oscillation time) regardless of the amplitude of movement. There the movement was called resonance.

Harmonic motion is the name given to the movement it makes during the cyclic period, which is sinusoidal. In this particular example, the pressure and movement are sinusoidal. In an electrical system, the voltage and current would be sinusoidal. In an acoustic system, the air flow and pressure variation would be sinusoidal, and so on.

The natural relationship is fixed by quantity b. Only at one frequency, which makes $bt = 2\pi$, will this natural relationship hold, where energy interchanges with no external force applied. The value of b^2 is fixed by the stiffness of the spring and by the mass of the moving weight. Change either one and b^2 changes, which results in a different natural resonance frequency.

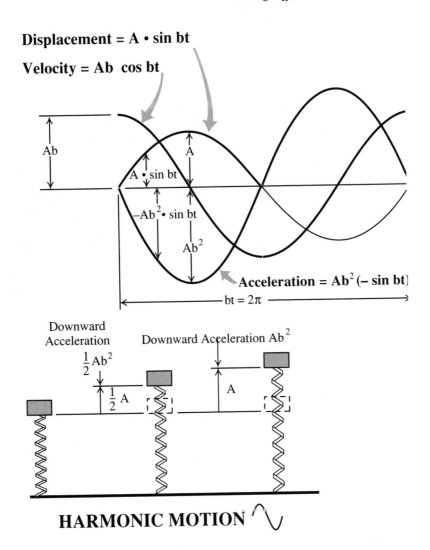

Displacement = A • sin bt

Velocity = Ab cos bt

Ab

A • sin bt

A

$-Ab^2$ • sin bt

Ab^2

Acceleration = Ab^2 (– sin bt)

bt = 2π

Downward Acceleration

$\frac{1}{2}Ab^2$

$\frac{1}{2}A$

Downward Acceleration Ab^2

A

HARMONIC MOTION

Linear or nonlinear relationship

Pure harmonic motion at a specific frequency occurs only if the spring is *linear*—
the relationship between the amount of compression or tension and the force produced is a straight line graph. In a linear spring, if every inch results in a force of
15 poundals, the force for successive inches will increase in uniform steps, 15,
30, 45, 60, 75, etc., both ways.

However, a spring might not be linear. The extra force might increase as the spring becomes more fully compressed. Instead of increasing 15, 30, 45, 60, and 75 for successive inches, the figures might run 16, 34, 54, 76, and 100. In tension, the effect might be reversed, so successive forces of tension, at inch intervals, read 14, 26, 36, 44, and 50 poundals. Such a spring is *nonlinear*, because the force is not proportional to displacement.

LINEAR SPRING

15 poundals/ in

Neutral – No Force	Compression					Tension				
	15 poundals	30 poundals	45 poundals	60 poundals	75 poundals	15 poundals	30 poundals	45 poundals	60 poundals	75 poundals

Neutral – No Force	16 poundals	34 poundals	54 poundals	76 poundals	100 poundals	14 poundals	26 poundals	36 poundals	44 poundals	50 poundals
	Compression					Tension				

NONLINEAR SPRING

Nonlinear relationships

Plotting the force/displacement relationship tabulated in the previous section as a graph, you find that it can be resolved in two components. The linear part is the same as the linear spring—15 poundals for every inch. Then, the square-law part is proportional to the square of displacement. Because displacement is in opposite directions, one is considered positive, the other negative. So, one way the square-law component will add to the linear force. The other way, it will partially cancel the linear force.

Now, assume the movement is somehow made sinusoidal so that the force produced is determined by the sinusoidal variation in position. You can show this position by plotting motion and force separately, each against time. By plotting each, you are using the curve at top left as a "transfer characteristic." For each point in time on the motion sinusoid, you project the corresponding point on the transfer curve horizontally onto the corresponding time point on the force curve.

On that last graph, the dashed line is a true sinusoid, and the solid line curve is the force produced by this nonlinear spring. The top part is more pointed and the bottom part is flattened.

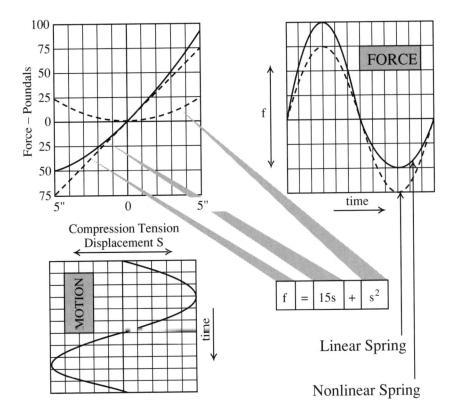

Analysis of nonlinear relationships

By replotting the curve on the right of the previous section, you analyze it. The equation for the transfer characteristic can be written: $f = 15s + s^2$, where s is in inches and f is in poundals. On the transfer curve, s is the independent variable time.

Substituting the movement equation into the transfer characteristic, you have a term in $\sin^2 at$. Transposition in the bottom left panel converts the $\sin^2 at$ term to a form that contains $\cos 2at$, a double-frequency sinusoid. This component is usually called a *second harmonic*, because its frequency is twice that of the basic (fundamental) movement frequency.

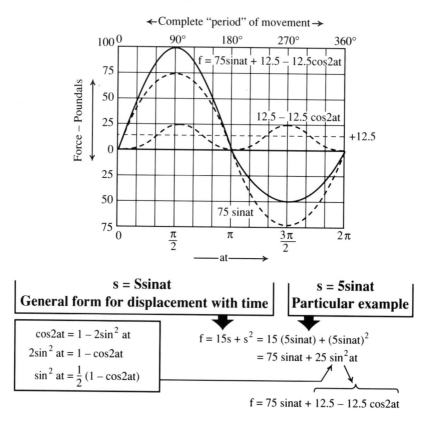

s = Ssinat	s = 5sinat
General form for displacement with time	**Particular example**

$$\cos 2at = 1 - 2\sin^2 at$$
$$2\sin^2 at = 1 - \cos 2at$$
$$\sin^2 at = \frac{1}{2}(1 - \cos 2at)$$

$$f = 15s + s^2 = 15(5\sin at) + (5\sin at)^2$$
$$= 75\sin at + 25\sin^2 at$$

$$f = 75\sin at + 12.5 - 12.5\cos 2at$$

Symmetrical nonlinearity

The nonlinearity considered in the last two sections wasn't symmetrical. The top of the wave was stretched and the bottom was compressed. This relationship is asymmetrical. Consider a symmetrical nonlinearity, in which both top and bottom are compressed. You might think of it as a spring, but it can apply to many things.

Both ways, for greater displacement, force ceases to be proportional. You find that this nonlinearity is equivalent to adding a cubic term to the transfer characteristic. The equation takes the form: $f = as + bs^3$. In the example shown, constants a and b are 125 and -1, respectively.

Making a similar transposition, the $\sin^3 at$ term gives a combination of a sin at term and a sin $3at$ term. When substituted into the main equation, the sin at

term reduces the amplitude of the fundamental, but the sin 3*at* term helps it follow the original amplitude as it leaves the zero line, and flattens it as it approaches maximum amplitude. Study this concept carefully.

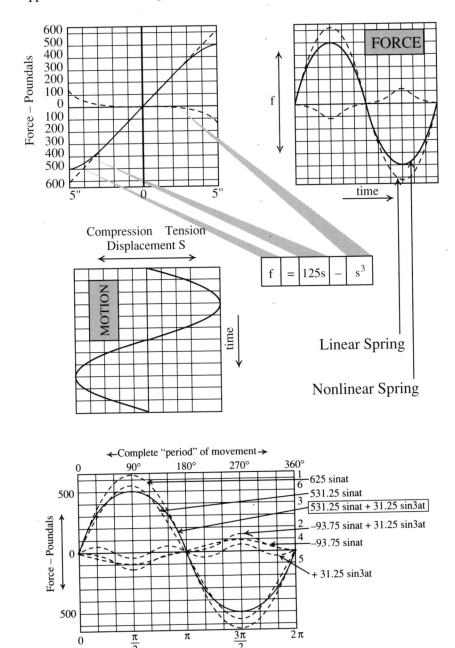

Compression   Tension
Displacement S

MOTION

$$f = 125s - s^3$$

Linear Spring

Nonlinear Spring

FORCE

time

←Complete "period" of movement→

$\frac{1}{6}$ 625 sinat

531.25 sinat

3 531.25 sinat + 31.25 sin3at

2 −93.75 sinat + 31.25 sin3at

4 −93.75 sinat

5 + 31.25 sin3at

Force – Poundals

—at—→

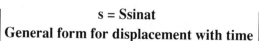

s = Ssinat	s = 5sinat
General form for displacement with time	**Particular example**

$$f = 125s - s^3 = 125\,(5\text{sinat}) - (5\text{sinat})^3$$

$$= 625\text{sinat} - 125\,\sin^3 at$$

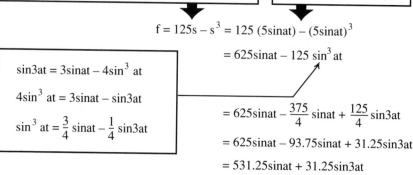

$$\sin3at = 3\text{sinat} - 4\sin^3 at$$

$$4\sin^3 at = 3\text{sinat} - \sin3at$$

$$\sin^3 at = \frac{3}{4}\text{sinat} - \frac{1}{4}\sin3at$$

$$= 625\text{sinat} - \frac{375}{4}\text{sinat} + \frac{125}{4}\sin3at$$

$$= 625\text{sinat} - 93.75\text{sinat} + 31.25\sin3at$$

$$= 531.25\text{sinat} + 31.25\sin3at$$

Multiple components of power sinusoids

The previous pages have discussed the simplest form that asymmetrical and symmetrical nonlinearity can take. However, the departure from the linear might not be exactly a square or cube term—or even a combination of both. It might have higher power terms. Any curve can be resolved into a power series that involves successively higher powers of *at* (the independent variable).

sin^{4}at

$$\cos4at = \cos^2 2at - \sin^2 2at$$

$$= 2\cos^2 2at - 1$$

$$\cos2at = 1 - 2\sin^2 at$$

$$\cos^2 2at = (1 - 2\sin^2 at)^2$$

$$= 1 - 4\sin^2 at + 4\sin^4 at$$

$$\cos4at = 2\cos^2 2at - 1$$

$$= 1 - 8\sin^2 at + 8\sin^4 at$$

$$8\sin^2 at = 4 - 4\cos2at$$

$$\cos4at = -3 + 4\cos2at + 8\sin^4 at$$

$$8\sin^4 at = 3 - 4\cos2at + \cos4at$$

$$\boxed{\sin^4 at = \frac{1}{8}\left\{3 - 4\cos2at + \cos4at\right\}}$$

sin^{5}at

$$\sin4at = 2\sin2at\cos2at$$

$$= 4\,\text{sinatcosat}\,(\cos^2 at - \sin^2 at)$$

$$\cos4at = 1 - 8\sin^2 at + 8\sin^4 at$$

$$\sin5at = \text{sinatcos4at} + \text{cosatsin4at}$$

$$= \text{sinat}\,(1 - 8\sin^2 at + 8\sin^4 at)$$

$$+ 4\,\text{sinatcos}^2 at\,(\cos^2 at - \sin^2 at)$$

$$= \text{sinat}\,\{1 - 8\sin^2 at + 8\sin^4 at$$

$$+ 4\,(1 - \sin^2 at)\,(1 - 2\sin^2 at)\}$$

$$= \text{sinat}\,\{5 - 20\sin^2 at + 16\sin^4 at\}$$

$$= 5\,\text{sinat} - 20\sin^3 at + 16\sin^5 at$$

$$20\sin^3 at = 15\text{sinat} - 5\sin3at$$

$$\sin5at = -10\text{sinat} + 5\sin3at + 16\sin^5 at$$

$$16\sin^5 at = 10\,\text{sinat} - 5\sin3at + \sin5at$$

$$\boxed{\sin^5 at = \frac{1}{16}\left\{10\,\text{sinat} - 5\sin3at + \sin5at\right\}}$$

Look at 4th and 5th power terms. A pure 4th power term results in $\sin^4 at$. Follow down the substitutions that lead to $\sin^4 at = 1/8\{3 - 4 \cos 2at + \cos 4at\}$. The 4th power adds both second and fourth harmonics, as well as a zero line offset. To get a pure 4th (if for some reason you'd want it) added to the fundamental, you must add a 2nd power term as well.

The method is similar with 5th power. Follow down the substitutions that lead to:

$$\sin^5 at = 1/16\{10 \sin at - 5 \sin 3at + \sin 5at\}.$$

So, this modifies the fundamental, as well as adding (or subtracting from) the third harmonic and providing some of the fifth.

Fourth power term in transfer characteristic

To make the treatment more general, use x and y for the independent and dependent variables of the transfer curve. These variables could apply to any of the many things where harmonic motion can occur. Assume that the equation is: $y = 32x + 8x^4$, and take variations of x between $+1$ and -1.

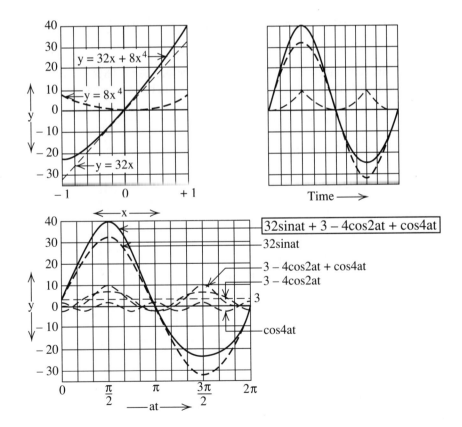

The sharpening of the upward peak and blunting of the downward one is even more abrupt than when nonlinearity was caused by just the square term. The resulting transfer curve "stays with" the straight part (fundamental) more closely for some distance, then leaves it more rapidly.

In place of the input sine wave, the resultant wave is analyzed in terms of harmonics, as obtained by the algebra. You should begin to see by now that algebra, geometry, trigonometry, and calculus are not separate subjects, as once taught, but are different "tools" in mathematics.

Combination of power terms

The last few pages show how to investigate successively higher power terms. By following the expression in "Multiple components of power sinusoids," the fifth power makes the combination of fundamental, third, and fifth harmonic almost cancel a region near the "zero line," and depart suddenly at the ends. Third and fifth harmonics are additive in their effects near the peaks—either accentuating or flattening them, according to the sign of the fifth power term.

Power terms do not usually come alone—especially the higher order ones. The fourth will usually have some second with it. The fifth will usually have some third with it. Sometimes odds and evens will combine.

To illustrate, assume a transfer curve that represents the equation:

$$y = 100x - 4x^2 + x^4.$$

Nearer to the zero line, the x^2 term causes a downward bend. Then, as the curve extends outwards, the x^4 term overtakes it, causing an upward bend at the ends.

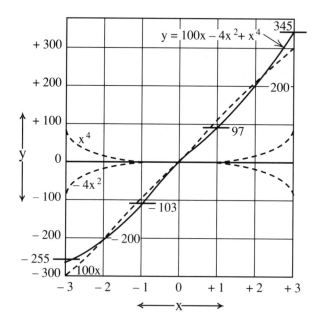

Now, substitute this transfer curve into a sinusoid of the form $x = A \sin at$. This equation gives a general expression for the resultant wave of the form shown in the box. Substitute different values for A into the coefficients and tabulate the results to show amplitudes of fundamental, offset, second, and fourth harmonics. Notice that the signs of the offset and the second harmonic change as the amplitude increases. Only the fourth harmonic stays in the same direction, however, it's very small at smaller amplitudes, so it has a negligible effect. At an amplitude of $A = 2$, the second disappears.

The points on the output, for values of A at 1, 2, and 3, are marked on the curve shown in the previous section. Later, this book defines maximum and minimum points more specifically.

$$y = 100x - 4x^2 + x^4 \Longleftarrow \qquad x = A\sin at$$

$$y = 100A\sin at - 4A^2 \sin^2 at + A^4 \sin^4 at$$

$$-4A^2 \sin^2 at = -2A^2 + 2A^2 \cos 2at$$

$$A^4 \sin^4 at = \frac{3A^4}{8} - \frac{A^4}{2}\cos 2at + \frac{A^4}{8}\cos 4at$$

$$y = 100A\sin at + \frac{3A^4}{8} - 2A^2 + \left(2A^2 - \frac{A^4}{2}\right)\cos 2at + \frac{A^4}{8}\cos 4at$$

A				
0.5	50	$-\frac{61}{128}$	$+\frac{15}{32}$	$\frac{1}{128}$
1.0	100	$-1\frac{5}{8}$	$+1\frac{1}{2}$	$\frac{1}{8}$
1.5	150	$-2\frac{75}{128}$	$+1\frac{31}{32}$	$\frac{81}{128}$
2.0	200	-2	0	2
2.5	250	$12\frac{19}{128}$	$7\frac{1}{32}$	$4\frac{113}{128}$
3.0	300	$+11\frac{5}{8}$	$-22\frac{1}{2}$	$10\frac{1}{8}$

Multiples and powers

Finding expressions for multiple angle functions in terms of powers of unit angle functions can be pursued systematically, as shown here. The first step is to extend the multiple angle formulas as far as you need—a step at a time. Using sum formulas, first find expressions for sin $2A$ and cos $2A$; regard $2A$ as $(A + A)$. Next, regarding $3A$ as $(2A + A)$, you find expressions for sin $3A$ and cos $3A$. Regarding $(n + 1)A$ as $(A + nA)$, substitute already found expressions for sin nA and cos nA.

The tabulation takes this as far as $6A$. The working is not shown. You might wish to work each step through to see how it "works."

MULTIPLES and POWERS

$\sin 2A = 2\sin A\cos A$

$\cos 2A = \cos^2 A - \sin^2 A$
$= 2\cos^2 A - 1$
$= 1 - 2\sin^2 A$

$\sin 3A = 3\sin A - 4\sin^3 A$

$\cos 3A = 4\cos^3 A - 3\cos A$

$\sin 4A = 4\sin A\,(2\cos^3 A - \cos A)$

$\cos 4A = 1 - 8\cos^2 A + 8\cos^4 A$
$= 1 - 8\sin^2 A + 8\sin^4 A$

$\sin 5A = 5\sin A - 20\sin^3 A + 16\sin^5 A$

$\cos 5A = 5\cos A - 20\cos^3 A + 16\cos^5 A$

$\sin 6A = \cos A\,(6\sin A - 32\sin^3 A + 32\sin^5 A)$

$\cos 6A = 32\cos^6 A - 48\cos^4 A + 18\cos^2 A - 1$
$= 1 - 18\sin^2 A + 48\sin^4 A - 32\sin^6 A$

$\sin^2 A = \dfrac{1}{2}\ (1 - \cos 2A)$

$\cos^2 A = \dfrac{1}{2}\ (1 + \cos 2A)$

$\sin^3 A = \dfrac{1}{4}\ (3\sin A - \sin 3A)$

$\cos^3 A = \dfrac{1}{4}\ (3\cos A + \cos 3A)$

$\sin^4 A = \dfrac{1}{8}\ (3 - 4\cos 2A + \cos 4A)$

$\cos^4 A = \dfrac{1}{8}\ (3 + 4\cos 2A + \cos 4A)$

$\sin^5 A = \dfrac{1}{16}\ (10\sin A - 5\sin 3A + \sin 5A)$

$\cos^5 A = \dfrac{1}{16}\ (10\cos A + 5\cos 3A + \cos 5A)$

$\sin^6 A = \dfrac{1}{32}\ (10 - 15\cos 2A + 6\cos 4A - \cos 6A)$

$\cos^6 A = \dfrac{1}{32}\ (10 + 15\cos 2A + 6\cos 4A + \cos 6A)$

For the even powers, both $\sin^n A$ and $\cos^n A$ use a form of the expression for $\cos nA$, and substitute in for the lower powers. For the odd powers, take the expression for $\sin nA$ to get $\sin^n A$ and the expression for $\cos nA$ to get $\cos^n A$, with similar substitutions. Here again, only the results are tabulated. Try a few to see how to do it.

Notice that the expressions for even powers all have a constant term, but the odd powers do not. This pattern is because the even powers cause an asymmetrical effect.

The substitutions used to derive the expressions in the previous section become involved in detail working. Nothing was difficult and the routine became familiar with practice, but the very number of substitutions made means that a mistake could creep in at any point. We need a simple means to check the results. Although the checks shown here are no absolute guarantee that an expression is correct, if they follow the pattern, and the numerical coefficients check out correctly, the answer is much more likely to be right.

Use two angles for each check: $A = 0$ and $A = \pi/2$, which is 90 degrees. Whatever the multiple, nA is always 0 when A is 0, so $\sin nA$ should always be 0 and $\cos nA$ always 1 in this column. For the 90-degree column, nA should always be n right angles. So, for $\sin nA$, the sequence will be $+1$, 0, -1, 0, and $+1$ . . . and $\cos nA$ will have the same sequence, beginning at 0 instead of 1.

In the powers' table $(0)^n$ is always 0 and $(1)^n$ is always 1. So, $\sin^n A$ will always be 0 for $A = 0$ and 1 for A as a right angle. Similarly, $\cos^n A$ will always be 1 for $A = 0$ and 0 for A as a right angle.

◀ CHECKS: A = 0 and A = $\frac{\pi}{2}$ (90°) ▶

Quantity	A = 0	A = $\frac{\pi}{2}$
$\sin 2A = 2\sin A\cos A$	$2{\cdot}0{\cdot}1 = 0$	$2{\cdot}1{\cdot}0 = 0$
$\cos 2A = \cos^2 A - \sin^2 A$	$1 - 0 = 1$	$0 - 1 = -1$
$\sin 3A = 3\sin A - 4\sin^3 A$	$3{\cdot}0 - 4{\cdot}0 = 0$	$3{\cdot}1 - 4{\cdot}1 = -1$
$\cos 3A = 4\cos^3 A - 3\cos A$	$4{\cdot}1 - 3{\cdot}1 = 1$	$4{\cdot}0 - 3{\cdot}0 = 0$
$\sin 4A = 4\sin A\,(2\cos^3 A - \cos A)$	$4{\cdot}0\,(2{-}1) = 0$	$4{\cdot}1\,(2{\cdot}0 - 1{\cdot}0) = 0$
$\cos 4A = 1 - 8\cos^2 A + 8\cos^4 A$	$1 - 8 + 8 = 1$	$1 - 8{\cdot}0 + 8{\cdot}0 = 1$
$\sin 5A = 5\sin A - 20\sin^3 A + 16\sin^5 A$	$5{\cdot}0 - 20{\cdot}0 + 16{\cdot}0 = 0$	$5{\cdot}1 - 20{\cdot}1 + 16{\cdot}1 = 1$
$\cos 5A = 5\cos A - 20\cos^3 A + 16\cos^5 A$	$5{\cdot}1 - 20{\cdot}1 + 16{\cdot}1 = 1$	$5{\cdot}0 - 20{\cdot}0 + 16{\cdot}0 = 0$
$\sin 6A = \cos A\,(6\sin A - 32\sin^3 A + 32\sin^5 A)$	$1(6{\cdot}0 - 32{\cdot}0 + 32{\cdot}0) = 0$	$0(6{\cdot}1 - 32{\cdot}1 + 32{\cdot}1) = 0$
$\cos 6A = 32\cos^6 A - 48\cos^4 A + 18\cos^2 A - 1$	$32 - 48 + 18 - 1 = 1$	$0 - 0 + 0 - 1 = -1$

$\sin^2 A = \frac{1}{2}(1 - \cos 2A)$	$\frac{1}{2}(1 - 1) = 0$	$\frac{1}{2}(1 + 1) = 1$
$\cos^2 A = \frac{1}{2}(1 + \cos 2A)$	$\frac{1}{2}(1 + 1) = 1$	$\frac{1}{2}(1 - 1) = 0$
$\sin^3 A = \frac{1}{4}(3\sin A - \sin 3A)$	$\frac{1}{4}(0 - 0) = 0$	$\frac{1}{4}(3 + 1) = 1$
$\cos^3 A = \frac{1}{4}(3\cos A + \cos 3A)$	$\frac{1}{4}(3 + 1) = 1$	$\frac{1}{4}(0 + 0) = 0$
$\sin^4 A = \frac{1}{8}(3 - 4\cos 2A + \cos 4A)$	$\frac{1}{8}(3 - 4 + 1) = 0$	$\frac{1}{8}(3 + 4 + 1) = 1$
$\cos^4 A = \frac{1}{8}(3 + 4\cos 2A + \cos 4A)$	$\frac{1}{8}(3 + 4 + 1) = 1$	$\frac{1}{8}(3 - 4 + 1) = 0$
$\sin^5 A = \frac{1}{16}(10\sin A - 5\sin 3A + \sin 5A)$	$\frac{1}{16}(0 - 0 + 0) = 0$	$\frac{1}{16}(10 + 5 + 1) = 1$
$\cos^5 A = \frac{1}{16}(10\cos A + 5\cos 3A + \cos 5A)$	$\frac{1}{16}(10 + 5 + 1) = 1$	$\frac{1}{16}(0 + 0 + 0) = 0$
$\sin^6 A = \frac{1}{32}(10 - 15\cos 2A + 6\cos 4A - \cos 6A)$	$\frac{1}{32}(10 - 15 + 6 - 1) = 0$	$\frac{1}{32}(10 + 15 + 6 + 1) = 1$
$\cos^6 A = \frac{1}{32}(10 + 15\cos 2A + 6\cos 4A + \cos 6A)$	$\frac{1}{32}(10 + 15 + 6 + 1) = 1$	$\frac{1}{32}(10 - 15 + 6 - 1) = 0$

All these check columns should have either 0 or 1 in the appropriate pattern. If one of the coefficients in the detail working has gone wrong, it will almost inevitably cause a different result in one or both of these checks.

Formulating expressions to specific requirements

These expressions for sin nA and cos nA and for powers of sin A and cos A, using algebra on the trig functions, can derive an expression that meets any requirement that you choose. Previously, the expression resulted in a zero coefficient of cos $2at$ when $A = 2$. Suppose you wanted an expression so that this term disappeared when $A = 5$?

You want an expression in the form: $y = x + ax^2 - bx^4$ so that by substituting the time variable, $x = A \sin ct$, the coefficient of cos $2ct$ disappears when $A = 5$. This equation leads to a ratio between the coefficients a and b. So long as the coefficients are in this ratio, the coefficient of cos $2ct$ will disappear when A

= 5. To tie the coefficients down to definite values, rather than just a ratio, when $A = 5$, the coefficient of cos $4ct$ must be 1.

For the coefficient of cos $2ct$ to disappear, $bA^{4/2}$ must equal $aA^{2/2}$ or $A^2 = a/b$. For this to occur when $A = 5$, a/b must be 25. For the coefficient of cos $4ct$ to be 1, $bA^{4/8}$ must be 1. That makes $b = 8/625$, in decimal form 0.0128. Because a/b is 25, 25 times 0.0128 is 0.32. The expression that you want is:

$$y = x + 0.32x^2 - 0.0128x^4$$

To check, substitute $x = 5 \sin ct$ into that equation and satisfy yourself that the cos $4ct$ coefficient is 1.

---

1 $y = x + ax^2 - bx^4$ **2** $x = A\sin ct$

3 Coefficient of cos2ct to be 0 when $A = 5$

4 Coefficient of cos4ct to be 1   Find a and b

$$y = A\sin ct + a\,(A\sin ct)^2 - b\,(A\sin ct)^4$$

$$= A\sin ct + \frac{aA^2}{2} - \frac{aA^2}{2}\cos 2ct - \frac{3bA^4}{8} + \frac{bA^4}{2}\cos 2ct - \frac{bA^4}{8}\cos 4\,ct$$

Required: $\dfrac{bA^4}{2} = \dfrac{aA^2}{2}$ $A^2 = \dfrac{a}{b} = 25$ **1**

$\dfrac{bA^4}{8} = 1$ $b = \dfrac{8}{A^4} = \dfrac{8}{625}$ **2**

$\underline{b}$ is $\dfrac{8}{625}$ or $\underline{0.0128}$ $\underline{a}$ is $\dfrac{8}{25}$ or $\underline{0.32}$

$$y = x + 0.32x^2 - 0.0128x^4$$

Check

$y = 5\sin ct + 4 - 4\cos 2\,ct - 3 + 4\cos 2\,ct - \cos 4\,ct$

$= 5\sin ct + 1 - \cos 4\,ct$

Combining algebra and trigonometry

Algebra, such as that in the previous section, often helps solve trig problems. Suppose a problem reduces to the trigonometric equation: $3\cos 2A + 8\sin A = 5$. You could solve this equation by hunting for a value of A that satisfies that equation. However, algebra gives us a more direct way.

First use the substitution $\cos 2A = 1 - 2\sin^2 A$ to bring all the functions of A into the form of $\sin A$ and its powers. Now, the equation is $3 + 8$:

$$\sin A - 6\sin^2 A = 5$$

Rearranging this as a quadratic, it is:

$$6\sin^2 A - 8\sin A + 2 = 0$$

Dividing through by 2 makes it simpler. Now, solve the equation as a quadratic. It doesn't matter that the variable is sin A instead of x or some more familiar variable. The formula method gives values of sin A as 1 or $1/3$. If the problem required it, you can now give angle A appropriately.

Check the answer(s), either in the original problem (which wasn't given here) or in the form. In this case, both answers check.

Solve 3cos2A + 8sinA = 5

$$\cos 2A = 1 - 2\sin^2 A$$

$$3\cos 2A = 3 - 6\sin^2 A$$

$$3\cos 2A + 8\sin A = 3 + 8\sin A - 6\sin^2 A = 5$$

$$6\sin^2 A - 8\sin A + 2 = 0 \qquad 3\sin^2 A - 4\sin A + 1 = 0$$

$$\sin A = \frac{2}{3} + \sqrt{\frac{4}{9} - \frac{1}{3}} = \frac{2}{3} + \frac{1}{3}$$

$$= 1 \text{ or } \frac{1}{3}$$

Check

If $\sin A = 1$, $\cos 2A = -1$

$$3\cos 2A + 8\sin A = -3 + 8 = 5$$

If $\sin A = \frac{1}{3}$, $\cos 2A = \frac{7}{9}$

$$3\cos 2A + 8\sin A = \frac{7}{3} + \frac{8}{3} = \frac{15}{3} = 5$$

Questions and problems

1. The compression of a spring by 1 foot produces a steadily increased force of 39.5 poundals. Tension produces an equal force in the opposite direction. If a 1-pound weight is on the end of the spring, find the period that is required for the spring to make a complete oscillation, neglecting any weight of the spring itself. Take $4\pi^2$ as 39.5.

2. What will be the period of oscillation if the weight is (a) increased to 4 pounds; (b) decreased to 1/4 pound?

3. A transfer characteristic is given by: $y = 100x + x^2$. If the sinusoid $x = A \sin bt$ is substituted for x, what is the coefficient of cos $2bt$ in the expression for y?

4. In the last question, express the coefficient of cos $2bt$ as a fraction of the coefficient of sin bt and deduce how the proportion of double frequency harmonic varies with A.

5. A transfer characteristic is given by $y = 100x + x^3$. Using the substitution $x = A \sin bt$, find the coefficients of $\sin bt$ and $\sin 3bt$ in the expression for y. Express the third harmonic as a fraction of the fundamental. Deduce how the proportion of third harmonic varies with A.

6. A transfer curve is given by: $y = 97x + x^2 + x^3$. Using the substitution $x = A \sin bt$, find an expression for y in terms $\sin bt$, a constant, $\cos 2bt$, and $\sin 3bt$. Find the value of A that makes the coefficients of $\cos 2bt$ and $\sin 3bt$ numerically equal and express this as a fraction of the coefficient of $\sin bt$ for the same value of A.

7. A transfer curve is given by: $y = 100x + 9x^2 - 4x^4$. Find expressions for the relative magnitude of fundamental, constant, second and fourth harmonics, in terms of A. Also, find the value of A for which the second harmonic is zero.

8. A transfer curve is given by: $y = 55x - 5x^3 + x^5$. Find expressions for relative magnitude of fundamental, third and fifth harmonics, in terms of A. For what value of A does third harmonic vanish and what percentage of the fundamental (total) is the fifth harmonic at this value of A?

9. $y = 30x + 9x^2 - 5x^3 - 4x^4 + x^5$ is the equation for a transfer curve. Find relative magnitudes of fundamental, constant, second, third, fourth, and fifth harmonics, in terms of A. Find what values of A make (a) the second and (b) the third harmonic disappear.

10. A transfer curve is given by: $y = 100x + bx^2 - cx^4$. The second harmonic disappears when $A = 5$ and the fourth harmonic is 6.25% of fundamental. Find coefficients b and c.

11. A transfer curve: $y = ax + bx^2 + cx^3 - dx^4 - ex^5$, has the second harmonic disappear when $A = 2$ and the fourth harmonic is 1.5% of fundamental, and the third harmonic when $A = 3$ and the fifth harmonic is 1% of fundamental. When $A = 3$, the fundamental coefficient is 300. Find values for a, b, c, d, and e.

12. By using successive substitutions into the $\sin (A + B)$ and $\cos (A + B)$ formulas, check the expressions in "Multiples and powers" for $\sin nA$ and $\cos nA$ to $n = 6$ and continue the series of expressions to $n = 8$. Check the coefficients by using substitutions at $A = 0$ and 90 degrees.

13. Using these expressions, check the expressions in "Multiples and powers" for $\sin^n A$ and $\cos^n A$ to $n = 6$ and continue the series to $n = 8$. Check these coefficients by making the same substitutions.

14. A problem yields equations, 4 sin² A − cos 4A = 3 and 4 cos² A − cos 4A = 3. Find four values for the angle A that satisfy these equations.

15. Find possible values of A that will satisfy equation:

$$\cos 6A - 6\cos 4A + 15\cos 2A = 6$$

16. What angles of A satisfy the equation:

$$\sin 5A - 5\sin 3A + 10\sin A = 1/2 \text{ ?}$$

17. A problem yields the equation:

$$\sin 7A - 7\sin 4A + 21\sin 3A - 34\sin A = 0$$

Find possible values of A to satisfy this equation. From 0 to 360 degrees, it should have 7 angles.

18. The equations: sin($A + B$) = 0.8 and sin($A - B$) = 0.352 are given. Find the values of sin A, sin B, cos A, cos B, cos ($A + B$), and cos ($A - B$).

19. A weight of 1 pound has its movement controlled by two separate compression springs. The spring it compresses in one direction produces a force of 5 poundals for every inch that is compressed. The spring it compresses in the other direction produces a force of 20 poundals for every inch that is compressed. It oscillates so as to compress these springs alternately. Find the relative time made by the weight compressing and decompressing each spring, and find the relative distances that the two springs will be compressed.

20. If the weight passes from one spring to the other with no free movement, will the period of oscillation vary as amplitude of movement varies? If a distance of several feet is between the springs over which the weight moves freely, will the time period change according to how much the springs compress each time? Explain why the greater compression results in a shorter time period.

19
CHAPTER

Developing
calculus theory

The concept of functions

An expression that contains a variable is a function of that variable. You have already considered several functions without calling them that. Powers of x, trig functions of x, and many more still to come are all functions of x.

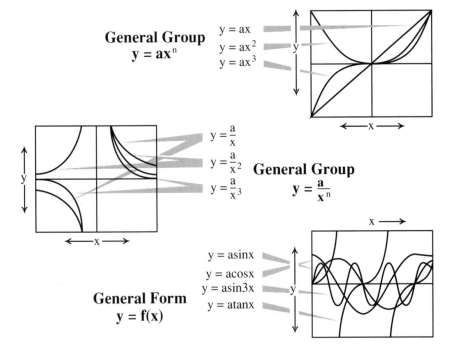

General Group
$$y = ax^n$$

$y = ax$
$y = ax^2$
$y = ax^3$

$y = \dfrac{a}{x}$
$y = \dfrac{a}{x^2}$
$y = \dfrac{a}{x^3}$

General Group
$$y = \dfrac{a}{x^n}$$

$y = a\sin x$
$y = a\cos x$
$y = a\sin 3x$
$y = a\tan x$

General Form
$$y = f(x)$$

You worked through some examples where functions could be differentiated, term by term. Here are examples of various functions in general groups. The family $y = ax^n$, for which the derivative is: $dy/dx = anx^{n-1}$. Trig functions can be assembled as a series of such terms. A related group has the general form: $y = a/x^n$. Another way of writing that, as in the section on indices, would be: $y = ax^{-n}$. Then, the general form gives the derivative as: $dy/dx = -an \cdot x^{-(n+1)}$, which can be switched back to its original form: $dy/dx = -an/x^{n+1}$.

To work with functions in general, since they can take various forms, a general form is: $y = f(x)$, which is read "y is a function of x" or "y equals f of x." This expression can mean any function of x, with the implication that it is a function for which you already can derive the derivative.

Two functions multiplied together

Assume that $y = uv$, where u and v are each (different) functions of x (they don't have to be different, but this procedure would be pointless if they were the same). You want to know dy/dx.

This general form assumes the independent variable x. If you increase x by dx to $x + dx$, then from derivatives that you already know, u will increase to $u + du$ and v will increase to $v + dv$. Saying increase could mean decrease if the sign happens to be negative. As $y = uv$, then: $y + dy = (u + du)(v + dv)$. Multiplying out, the right side has four terms. The first is uv, which corresponds to y. The next two are udv and vdu, and the fourth term is the product of two infinitely small changes, which makes it negligibly small—even in infinitely small terms. So, the fourth term is meaningless; throw it away. After taking away the finite part, $y = uv$, you are left with: $dy = udv + vdu$. By dividing through by dx, the equation is in its more complete form.

Derivative of Two Functions Multiplied Together

> If $u - f(x)$ and $v - f(x)$
>
> and $y = uv$
>
> Find $\dfrac{dy}{dx}$
>
> ───────────────────────────
>
> $y = uv$
>
> $y + dy = (u + du)(v + dv)$
>
> $\qquad = uv + udv + vdu + dudv$
>
> $y \qquad = uv$
>
> ───────────────────────────
>
> $dy = \qquad udv + vdu$

$$\frac{d}{dx}uv = u\frac{dv}{dx} + v\frac{du}{dx}$$

Checking the formula

A general proof like that on the previous page is rather vague and difficult to visualize. Does it really work? To begin with, try it on something you can check.

x^7 is equal to x^4 times x^3. You already know that d/dy y is $7x^6$. Try it as a product, x^4 times x^3. Make $u = x^4$ and $v = x^3$. $du/dx = 4x^3$ and $dv/dx = 3x^2$. The next step: $udv/dx = x^4$ times $3x^2$, which multiplies out to $3x^6$; $vdu/dx = x^3$ times $4x^3$, which multiplies to $4x^6$. Adding the two together is $7x^6$. The same answer.

Here's another one: $y = \sin 2x$. You know the direct way that $dy/dx = 2 \cos 2x$. Also $\sin 2x = 2 \sin x \cos x$. That's a product. Make $u = \sin x$ and $v = \cos x$. Leave 2 outside the parentheses. $du/dx = \cos x$ and $dv/dx = -\sin x$. So, $udv/dx = -\sin^2 x$ and $vdu/dx = \cos^2 x$, which, put together, is $\cos 2x = 2[\cos^2 x - \sin^2 x]$. You already established that $\cos 2x = \cos^2 x - \sin^2 x$. So, $dy/dx = 2 \cos 2x$, which again confirms the method.

PRODUCT FUNCTIONS
Checking ones we know

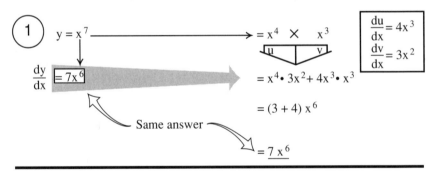

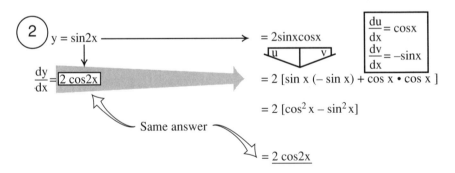

Using the product formula

In each of the previous problems, you left the answer in the simplest form that was derived from this method. Different forms could be used. The best form to use would probably depend on the rest of the problem.

The second example uses a reciprocal function of x with a trig function. Next, you derive a quotient formula. This one could be done either way. In the section "Checking quotient functions," it will be treated as a quotient to get the same answer.

USING FORMULA

① $y = x^3 \sin 2x$

$\boxed{u}\ \boxed{\ v\ }$

$\dfrac{dy}{dx} = x^3 \cdot 2\cos 2x + 3x^3 \cdot \sin 2x$

$\quad = 2x^3 \cos 2x + 3x^2 \sin 2x$

$\dfrac{du}{dx} = 3x^2$

$\dfrac{dv}{dx} = 2\cos 2x$

---

② $y = \dfrac{\sin 2x}{x^3}$

$\dfrac{dy}{dx} = -\dfrac{3\sin 2x}{x^4} + \dfrac{2\cos 2x}{x^3}$

or $\dfrac{2\cos 2x}{x^3} - \dfrac{3\sin 2x}{x^4}$

$u = \sin 2x \qquad \dfrac{du}{dx} = 2\cos 2x$

$v = x^{-3} \qquad \dfrac{dv}{dx} = -3x^{-4}$

One function divided by another

Turn to one function divided by another. $y = u/v$, where u and v are both functions of x. As usual, add dy to y, du to u, and dv to v—each corresponding to an addition of dx to x. So, $y + dy = (u + dn)/(v + dv)$. Multiplying both sides by $v + dv$ and multiplying out the left side, you have a doubly infinitesimal product, $dydv$, which you "throw away" and leave an equation with yv on the left. So, divide through by v and you have: $y + ydv/v + dy = (u + du)/v$. ydv/v has no business with a y in it, or on the left, so substitute $y = u/v$ and put it on the right. Now, you have: $y + dy = u/v + (vdu - udv)/v^2$. Subtract the part that you began with, $y = u/v$, then: $dy = (vdy - udv)/v^2$. Put the dx denominators in to complete the expression.

If $u = f(x)$ and $v = f(x)$ and $y = \dfrac{u}{v}$

Find $\dfrac{dy}{dx}$

$$y = \frac{u}{v}$$

$$y + dy = \frac{u + du}{v + dv}$$

$$(y + dy)(v + dv) = u + du$$

$$yv + ydv + vdy + \cancel{dydv} = u + du$$

$$\boxed{\text{Divide by v}} \quad y + y\frac{dv}{v} + dy = \frac{u + du}{v}$$

$$\boxed{\text{Substitute for y}} \quad y + \frac{udv}{v^2} + dy = \frac{u}{v} + \frac{du}{v}$$

$$y + dy = \frac{u}{v} + \frac{du}{v} - \frac{udv}{v^2}$$

$$= \frac{u}{v} + \frac{vdu - udv}{v^2}$$

$$y \quad = \frac{u}{v}$$

$$dy = \frac{vdu - udv}{v^2}$$

$$\boxed{\frac{d}{dx} \cdot \frac{u}{v} = \frac{v\dfrac{du}{dx} - u\dfrac{dv}{dx}}{v^2}}$$

Checking quotient functions

Here, use that formula on two examples that you already know the answers to. First, turn the powers-of-x thing around: x^4 is the quotient of x^7 divided by x^3. Using this formula, you find the same answer that was obtained directly: $dy/dx = 4x^3$. See the illustration at the top of page 331.

Then, do the other one that was regarded as a product in "Using product formula." $u = \sin 2x$ and $v = x^3$ (where the product formula used: $v = 1/x^3$). $du/dx = 2\cos 2x$, as before. $vdu/dx = 2x^3 \cos 2x$ and $udv/dx = 3x^2 \sin 2x$. Putting them together with v^2, which is x^6, as the denominator, and canceling the x's in numerator and denominator, produces the same answer.

Using the quotient formula

A simple application of the quotient formula is to find the derivative of $\tan x$ because $\tan x$ is $\sin x/\cos x$. Writing $u = \sin x$ and $v = \cos x$, the derivatives are: $du/dx = \cos x$ and $dv/dx = -\sin x$. Substituting into the formula and simplifying, dy/dx is $(\cos^2 x + \sin^2 x)/\cos^2 x$. Dividing through by the denominator, this becomes $dy/dx = 1 + \tan^2 x$. Also, the numerator is 1, by the trig form of Pythagoras, so dy/dx can also be written: $1/\cos^2 x$.

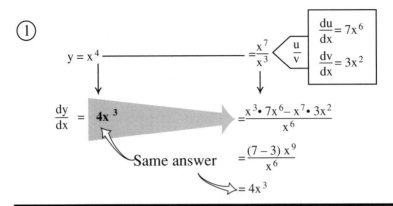

① $y = x^4$

$$= \frac{x^7}{x^3} \quad \frac{u}{v}$$

$$\frac{du}{dx} = 7x^6$$

$$\frac{dv}{dx} = 3x^2$$

$$\frac{dy}{dx} = \mathbf{4x^3}$$

$$= \frac{x^3 \cdot 7x^6 - x^7 \cdot 3x^2}{x^6}$$

$$= \frac{(7-3)x^9}{x^6}$$

Same answer

$$= 4x^3$$

---

② $y = \dfrac{\sin 2x}{x^3}$

$u = \sin 2x$	$\dfrac{du}{dx} = 2\cos 2x$
$v = x^3$	$\dfrac{dv}{dx} = 3x^2$

$$\frac{dy}{dx} = \frac{x^3 \cdot 2\cos 2x - 3x^2 \cdot \sin 2x}{x^6}$$

$$= \frac{2\cos 2x}{x^3} - \frac{3\sin 2x}{x^4}$$

Using Formula

① $y = \tan x = \dfrac{\sin x}{\cos x}$

$u = \sin x$	$\dfrac{du}{dx} = \cos x$
$v = \cos x$	$\dfrac{dv}{dx} = -\sin x$

$$\frac{dy}{dx} = \frac{\cos^2 x - (-\sin^2 x)}{\cos^2 x} = \frac{\cos^2 x + \sin^2 x}{\cos^2 x}$$

$$= 1 + \tan^2 x \quad \boxed{\text{or}} \quad \frac{1}{\cos^2 x}$$

② $y = \dfrac{x^2 + 4}{\sin 2x}$

$u = x^2 + 4$	$\dfrac{du}{dx} = 2x$
$v = \sin 2x$	$\dfrac{dv}{dx} = 2\cos 2x$

$$\frac{dy}{dx} = \frac{2x \sin 2x - (x^2 + 4)\, 2\cos 2x}{\sin^2 2x}$$

$$= \frac{2x}{\sin 2x} - \frac{2(x^2 + 4)}{\tan 2x \cdot \sin 2x} = 2x \operatorname{cosec} 2x - 2(x^2 + 4)\cot 2x \cdot \operatorname{cosec} 2x$$

$\dfrac{1}{\text{sine}} = \text{cosecant}$	$\dfrac{1}{\text{cosine}} = \text{secant}$	$\dfrac{1}{\text{tangent}} = \text{cotangent}$
written *cosec*	written *sec*	written *cot*

For the second example, take $y = (x^2 + 4)/\sin 2x$. Using the formula and making the substitutions, it simplifies to something that still looks rather cumbersome.

You are digging into more frequent uses of functions that involve reciprocals of sine, cosine, and tangent. Other functions enable these equations to be written more simply. The *cosecant* is 1/sine, the *secant* is 1/cosine, and the *cotangent* is 1/tangent. These terms are abbreviated to *sec*, *cosec*, and *cot*, respectively.

Function of a function derivative

Although you might expect this section to be more involved than the ones that covered products or quotients of derivatives, actually it's simpler. The derivation is simple algebra. If $y = f(u)$ and $u = f(x)$, then because dy/dx is too complicated to write directly, du is used as an intermediary. Multiplying together dy/du and du/dx, each of which is relatively simple, the du's cancel and leave dy/dx.

To illustrate and check the method, take $y = \sin^5 x$. To make u the intermediary, it must be $\sin x$. So, $u = \sin x$ and $y = u^5$. Differentiating in stages: $dy/du = 5u^4$, which is $5 \sin^4 x$. Then, du/dx is $\cos x$. So, the product derivative is: $dy/dx = 5 \sin^4 x \cos x$.

$$\text{If } y = f(u) \text{ and } u = f(x)$$
$$\frac{dy}{dx} = \frac{dy}{du} \times \frac{du}{dx}$$

Example $\qquad y = \sin^5 x$

$$\boxed{\begin{array}{l} u = \sin x \\ y = u^5 \end{array}}$$

$$\frac{dy}{dx} = 5u^4 \cdot \cos x = 5 \sin^4 x \cos x$$

Alternate method as check

$$y = \sin^5 x = \frac{1}{16}(10 \sin x - 5 \sin 3x + \sin 5x)$$

$$\frac{dy}{dx} = \frac{1}{16}(10 \cos x - 15 \cos 3x + 5 \cos 5x)$$

$$\boxed{\begin{array}{l} \cos 3x = 4 \cos^3 x - 3 \cos x \\ \cos 5x = 5 \cos x - 20 \cos^3 x + 16 \cos^5 x \end{array}}$$

$$= \frac{1}{16}(10 \cos x + 45 \cos x - 60 \cos^3 x + 25 \cos x - 100 \cos^3 x + 80 \cos^5 x)$$

$$= \frac{1}{16}(80 \cos x - 160 \cos^3 x + 80 \cos^5 x)$$

$$= 5 \cos x (1 - 2 \cos^2 x + \cos^4 x) = 5 \cos x (1 - \cos^2 x)^2$$
$$= 5 \cos x \sin^4 x$$

the same as $5 \sin^4 x \cos x$

To check this equation, convert $\sin^5 x$ to functions of multiples of x (chapter 18). Differentiate this term by term. Now, substitute back for $\cos 3x$ and $\cos 5x$, collect and simplify, to find the same answer in somewhat different form. $\sin^2 x = 1 - \cos^2 x$, so $\sin^4 x = (1 - \cos^2 x)^2$. After this substitution, the result is the same.

Equation of a circle

An interesting use for the function of a function formula is to find the derivative of the equation for a circle. In its simplest form, this equation is $x^2 + y^2 = r^2$. Radius is constant, so r^2 is constant. Turning it around and taking the square root, you have an expression for y. The plus or minus sign in front of the surd means that y can be positive or negative. Actually, a circle has four combinations of the same numeric value of x and y. Both x and y can be either positive or negative and give a point in each quadrant.

The method of finding the derivative is to make u equal to $(r^2 - x^2)^2$ and $y = u^{1/2}$. An interesting way to check the result in all 4 quadrants, is to take the half right angles, where x and y are each r divided by root 2. The slope is unity at each of these points, but of changing sign. By checking, each answer is correct.

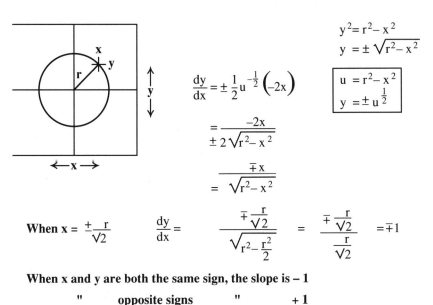

$$y^2 = r^2 - x^2$$
$$y = \pm\sqrt{r^2 - x^2}$$

$$\frac{dy}{dx} = \pm\frac{1}{2}u^{-\frac{1}{2}}(-2x)$$

$$\boxed{\begin{array}{l} u = r^2 - x^2 \\ y = \pm u^{\frac{1}{2}} \end{array}}$$

$$= \frac{-2x}{\pm 2\sqrt{r^2 - x^2}}$$

$$= \frac{\mp x}{\sqrt{r^2 - x^2}}$$

When $x = \dfrac{\pm r}{\sqrt{2}}$ $\qquad \dfrac{dy}{dx} = \dfrac{\mp\dfrac{r}{\sqrt{2}}}{\sqrt{r^2 - \dfrac{r^2}{2}}} = \dfrac{\mp\dfrac{r}{\sqrt{2}}}{\dfrac{r}{\sqrt{2}}} = \mp 1$

When x and y are both the same sign, the slope is − 1

        "   **opposite signs**   **"**   **+ 1**

When x = 0 $\dfrac{dy}{dx} = 0$ $\qquad\qquad\qquad$ **When x = r** $\dfrac{dy}{dx} = \mp\infty$

Successive derivatives of tangent function

If you had difficulty finding the derivative of tan x in question 6 of chapter 17, it will have been cleared up in "Using the quotient formula." Now, apply the function of a function formula to finding successive derivatives of the tangent. Sine and cosine functions run in cycles and come back to the original in the 4th derivative, but not so with the tangent.

The first derivative of $y = \tan x$ was $1 + \tan^2 x$. Make $u = \tan x$ and $y' = 1 + u^2$. This notation is another one used to save space: y' (called y prime) is the 1st derivative of y and it means the same as dy/dx. 2, 3, or more primes are used to indicate the 2nd, 3rd derivative, etc.

Using the function of a function formula, obtain the second derivative. To find the 3rd derivative, substitute u and v for factors in the 2nd, and use the form $y' = 2uv$. In the same way, make similar substitutions in the successive derivatives and use the product formula.

Notice that successive derivatives of tan x have a growing complexity of terms that involve higher powers of tan x: each derivative uses powers up to one higher.

$$y = \tan x$$

$$\frac{dy}{dx} = 1 + \tan^2 x \quad \text{(see p. 3-87)}$$

$$\frac{d^2 y}{dx^2} = 2u \, (1 + \tan^2 x)$$
$$= 2 \tan x \, (1 + \tan^2 x)$$

$\Longleftarrow$
$$\boxed{\begin{array}{l} u = \tan x \\ y' = 1 + u^2 \end{array}}$$

$$\frac{d^3 y}{dx^3} = 2 \left\{ u \, \frac{dv}{dx} + v \, \frac{du}{dx} \right\}$$
$$= 2 \, \{ \tan x \cdot 2\tan x \, (1 + \tan^2 x) + (1 + \tan^2 x) \, (1 + \tan^2 x) \}$$
$$= 2 \, (1 + \tan^2 x) \, (1 + 3\tan^2 x)$$

$\Longleftarrow$
$$\boxed{\begin{array}{c} y'' = 2uv \\ u = \tan x \quad v = 1 + \tan^2 x \end{array}}$$

$$\frac{d^4 y}{dx^4} = 2 \left\{ u \, \frac{dv}{dx} + v \, \frac{du}{dx} \right\}$$

$\Longleftarrow$
$$\boxed{\begin{array}{c} y''' = 2uv \\ u = 1 + \tan^2 x \quad v = 1 + 3w^2 \\ w = \tan x \end{array}}$$

$$= 2 \, \{ (1 + \tan^2 x) \, 6\tan x \, (1 + \tan^2 x) + (1 + 3\tan^2 x) \, 2\tan x \, (1 + \tan^2 x) \}$$
$$= 4\tan x \, (1 + \tan^2 x) \, \{ 3(1 + \tan^2 x) + (1 + 3\tan^2 x) \}$$
$$= 8\tan x \, (1 + \tan^2 x) \, (2 + 3\tan^2 x)$$

y', y'', y''' are first, second, and third **DERIVATIVES** of y

Integration is the reverse of differentiation

Having now gained some familiarity with differentiating a function (also called finding its derivative) it's time to look at the reverse process. What's that? Differentiation plots a growth rate (or the reverse). The reverse takes growth rate and builds whatever grows, called *integration*. You will see this system better as you progress.

Slope, found by differentiation, uses an infinitely small change in *x* to produce infinitely small change in *y*. If you add an infinite number of these infinitely small pieces, you have *x* and *y*, respectively. If you add all the pieces of *y* (called *dy*) together, you have *y*.

If $f'(x)$ is y', two names for the same infinitesimal piece, then the reverse process builds the original *x* and *y*. This process is expressed with the integral sign, which is an old-fashioned long letter *s*. Integration finds what function, when differentiated, will produce the derivative that you began with.

Another sign is also used, with difference explained later: the Greek capital sigma (Σ). From the differentiating you have done, you can start on integration just by taking the reverse process.

If $y = f(x)$	y is a function of x
$dy = f'(x)dx$	dy is a first derivative of the same function times dx

dy is an infinitely small change in y, corresponding to an infinitely small change in x, identified as dx

If the infinitely small changes are added together, the whole is

$$dy = y \qquad\qquad f'(x)dx = f(x)$$

DIFFERENTIATION	**CORRESPONDING INTEGRATION**
$y = ax^n \quad \dfrac{dy}{dx} = anx^{n-1}$	$y = anx^{n-1} \qquad \int y\,dx = ax^n$
	$y = ax^n \qquad \int y\,dx = \dfrac{a}{n+1}x^{n+1}$
$y = a\sin x \quad \dfrac{dy}{dx} = a\cos x$	$y = a\cos x \qquad \int y\,dx = a\sin x$
	$y = a\sin x \qquad \int y\,dx = -a\cos x$

Integration consists of finding what function, when differentiated, will give the one we start with

Patterns in calculations

Reviewing the study of mathematics, always begin with a positive process, then reverse it to produce a negative process. After counting, you got into addition, which was reversed to make subtraction. After shortening multiple addition to make multiplication, it was reversed to parallel multiple subtraction, making division. Then, indices brought powers, then reversed it to find roots.

In each, what began as a negative process, searching for a question to produce an answer, later developed a positive approach to eliminate the search. Integration has a similar relationship to differentiation.

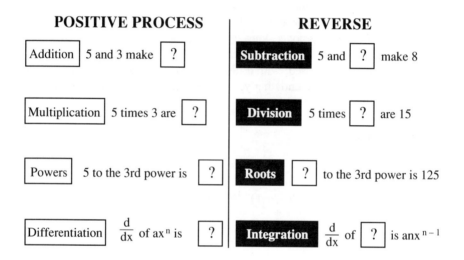

| POSITIVE PROCESS | REVERSE |

Addition | 5 and 3 make | ? | Subtraction | 5 and | ? | make 8

Multiplication | 5 times 3 are | ? | Division | 5 times | ? | are 15

Powers | 5 to the 3rd power is | ? | Roots | ? | to the 3rd power is 125

Differentiation | $\frac{d}{dx}$ of ax^n is | ? | Integration | $\frac{d}{dx}$ of | ? | is anx^{n-1}

The constant of integration

When you differentiate a function, you find its slope at a point or at a whole sequence of points. However, saying that a road has a certain slope (1 in 16, for example) doesn't state how high the road is. It could be at sea level or on top of a mountain.

Looked at mathematically, the three equations here each begin with $y = x^3 - 12x$. Then there is a constant that is shown as $+8$, nothing, or -8. All three equations give the same derivative because derivative of the constant is 0 (e.g., nothing).

By reversing the process, you have no direct means to find the constant. So, you must leave room for an unknown constant, called the *constant of integration*.

If the road began at sea level, then integrating over any distance would find the height of the latest point above sea level.

The Constant of Integration

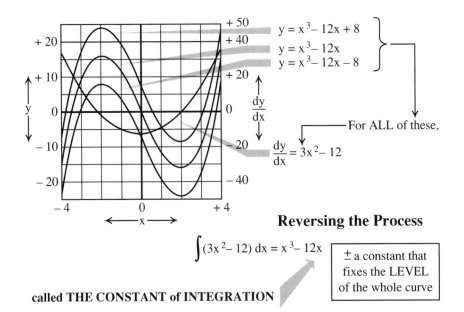

$$y = x^3 - 12x + 8$$
$$y = x^3 - 12x$$
$$y = x^3 - 12x - 8$$

For ALL of these,

$$\frac{dy}{dx} = 3x^2 - 12$$

Reversing the Process

$$\int (3x^2 - 12)\, dx = x^3 - 12x$$

$\pm$ a constant that fixes the LEVEL of the whole curve

called **THE CONSTANT of INTEGRATION**

Definite integrals

This more specific use for integration is practical. The general integral (above) has mainly a theoretical value. A definite integral "adds it up" between definite values. It specifies that the curve starts at a certain point and follows the derivative to a new point.

The integral is written the same way as for the general integral, but limits are put against the long $\int$. The lower one is where it begins, the upper one is where it ends. Next, the integral (the general one, less the constant) is put in square parentheses, with the limits outside at the right. Then, substitute the values of the integral, first at the upper limit, then at the lower limit, which you subtract from the value at the upper limit.

In the graph, the general integral is plotted without the constant of integration. When substituting in the lower value, you could make the starting point by inserting a constant of integration that would make the point zero. However, it is not necessary, because you subtract this value from the upper value. Whatever you make the constant, it disappears when you subtract one value from the other.

By substituting $x = -2$ and $x = +1$, the second produces -11 and the first produces $+16$. By subtracting the first from the second, the change is -27. Substituting values $x = +2$ and $x = +3$, the same process produces the change of y in this range as $+7$.

THE DEFINITE INTEGRAL

$\int_a^b ydx$ Integral ydx between limits of a and b

$[f(x)]_a^b$ Substitute values of x = b and x = a into f(x)

$[f(x = b) - f(x = a)]$ Subtract lower (a) from upper (b)

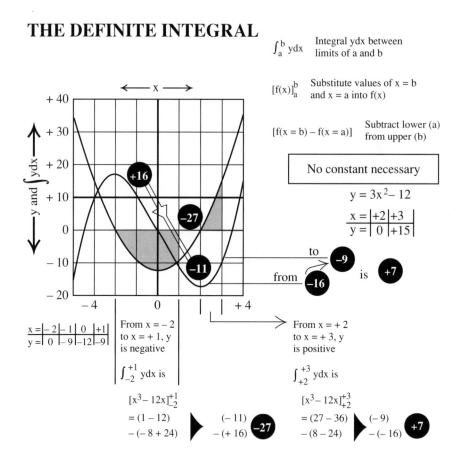

No constant necessary

$y = 3x^2 - 12$

| x = | +2 | +3 |
| y = | 0 | +15 |

| x = | -2 | -1 | 0 | +1 |
| y = | 0 | -9 | -12 | -9 |

From x = -2 to x = +1, y is negative

$\int_{-2}^{+1} ydx$ is

$[x^3 - 12x]_{-2}^{+1}$
$= (1 - 12)$
$-(-8 + 24)$ (-11)
$-(+16)$ **-27**

From x = +2 to x = +3, y is positive

$\int_{+2}^{+3} ydx$ is

$[x^3 - 12x]_{+2}^{+3}$
$= (27 - 36)$
$-(8 - 24)$ (-9)
$-(-16)$ **+7**

Finding area by integration

A most useful application for integration is for finding areas, volumes, and similar things. If *y* is a succession of infinitely small elements in an area or volume, the sum of these elements over a certain range of the curve that is represented by this function will be the area under the curve, which consists of an infinite number of infinitely narrow strips. From the infinite to the finite!

To prove that this method works, take the larger shaded area. The equation of the upper side is: $y = 1/2\,x + 7$. The integral (working the differential formula backwards) is: $1/4\,x^2 + 7x$. By making substitutions for $x = 2$ and $x = 10$ (which were arbitrarily chosen when the equation was written; any other combination could be used) and subtracting the area is 80. Checking it by the geometrical formula proves that you have the right answer.

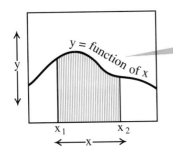

If y represents the height of continuous succession of infinitely small elements of an area,

$$\int_{x_1}^{x_2} y\,dx$$

gives the area under curve y, between points x_1 and x_2

Finding Area by Integration

$$\text{Area} = \int_2^{10} \left(\frac{1}{2}x + 7\right) dx$$

$$= \left[\frac{1}{4}x^2 + 7x\right]_2^{10}$$

$$= \left(\frac{100}{4} + 70\right) - \left(\frac{4}{4} + 14\right)$$

$$= 95 - 15 = \boxed{80}$$

CHECK $\Rightarrow$ Average $y = \frac{1}{2}(8 + 12) = 10$

Area = base $\times$ average height
$$= 8 \times 10 = \boxed{80}$$

Area of a circle

The area does not have to be under a curve. Here, the area of a circle is found by two methods. In the first, the element of area is a wedge from center to circumference, taken at angle (in circular measure) x. The element has an area 1/2 base times height. The height is r, and the base is $r\,dx$. So, the area of the element is $1/2\ r^2\ dx$. Angle x is the variable. Integrating produces $1/2\ r^2\ x$. The lower limit is zero, the upper one is 2π. Substitute and subtract (subtracting zero doesn't alter the upper limit), gives the well-known formula: πr^2.

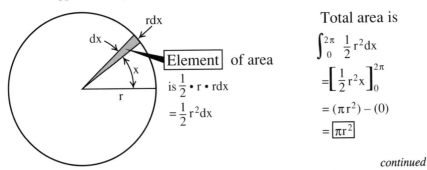

Element of area

is $\frac{1}{2} \cdot r \cdot r\,dx$

$= \frac{1}{2} r^2 dx$

Total area is

$$\int_0^{2\pi} \frac{1}{2} r^2 dx$$

$$= \left[\frac{1}{2} r^2 x\right]_0^{2\pi}$$

$$= (\pi r^2) - (0)$$

$$= \boxed{\pi r^2}$$

continued

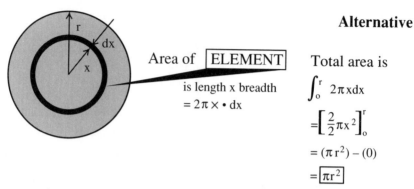

Alternative

Area of ⃞ELEMENT

is length x breadth
$= 2\pi x \cdot dx$

Total area is

$$\int_0^r 2\pi x dx$$

$$= \left[\frac{2}{2}\pi x^2 \right]_0^r$$

$$= (\pi r^2) - (0)$$

$$= \boxed{\pi r^2}$$

The other method uses a thin ring at radius x from the center. Here, the area of the element is the length of the ring, $2\pi x$ times its thickness, dx. Integrating that from zero to r gives the same well-known result. Compare the methods carefully until you understand how each is done.

Curved areas of cylinders and cones

With cylinders and cones, you could find the area two ways, as with the circle. For the cylinder, the length of the element is $2\pi r$ and its width is dx. Integrating from zero to h (the height), produces what, in this case, is fairly obvious, $2\pi rh$. For the cone, instead of being πr^2, it is πrl, and l is the slant height of the cone.

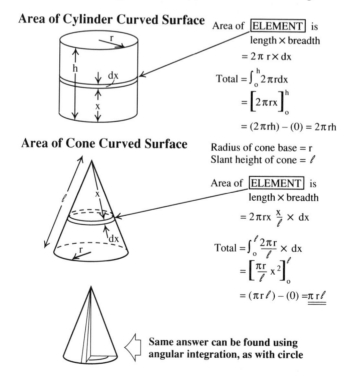

Area of Cylinder Curved Surface

Area of ⃞ELEMENT is
length × breadth
$= 2\pi r \times dx$

$$Total = \int_0^h 2\pi r dx$$

$$= \left[2\pi rx \right]_0^h$$

$$= (2\pi rh) - (0) = 2\pi rh$$

Area of Cone Curved Surface

Radius of cone base = r
Slant height of cone = ℓ

Area of ⃞ELEMENT is
length × breadth
$= 2\pi rx \frac{x}{\ell} \times dx$

$$Total = \int_0^\ell \frac{2\pi r}{\ell} \times dx$$

$$= \left[\frac{\pi r}{\ell} x^2 \right]_0^\ell$$

$$= (\pi r\ell) - (0) = \pi r\ell$$

Same answer can be found using angular integration, as with circle

Surface area of sphere

Move a ring around the sphere from one "end" (if you can imagine a sphere having an "end!") to the other. Measure the position of the ring by angle x. By taking angle x from zero to π radians, the ring will cover the entire area of the sphere's surface.

The circumferential length of the element is $2\pi r \sin x$. Its width is rdx, so area is $2\pi r^2 \sin x \, dx$. Integrating from zero to π: the integral is: $-2\pi r^2 \cos x$. When x is π, $\cos x$ is -1, so the minus times minus makes a plus. When x is zero, $\cos x$ is $+1$. So, $-\cos x$ is -1. Again, minus times minus makes a plus. The answer adds to $4\pi r^2$.

Another way to calculate it would be to use the distance along the axis as the variable, from $-r$ to $+r$. The function produced is more involved and not so easy to integrate, but the result is the same.

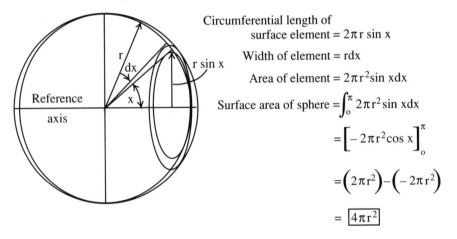

Circumferential length of
surface element = $2\pi r \sin x$

Width of element = rdx

Area of element = $2\pi r^2 \sin x \, dx$

$$\text{Surface area of sphere} = \int_0^\pi 2\pi r^2 \sin x \, dx$$

$$= \left[-2\pi r^2 \cos x \right]_0^\pi$$

$$= \left(2\pi r^2 \right) - \left(-2\pi r^2 \right)$$

$$= \boxed{4\pi r^2}$$

Finding volume by integration

The same method can be used to find volumes. See the illustration at the top of page 342. As the element of an area is a line of width (dx), so the element of a volume is an area of thickness (dx). Taking the volume of this wedge, the area of the element is the area of the base multiplied by x/l: wtx/l. The thickness is dx, so the volume of the element is $wtx \, dx/l$. Integrating, with respect to x, produces $wtx^2/2l$. Substitution again is simple because x starts at zero. The volume is $wtl/2$.

Volume of a pyramid

To make the formula more general, A is used to represent the area of the base and an element of volume at x is taken from the apex. Follow the same method down through. The difference is that the area is proportional to the square of the distance from the apex: $(x/h)^2$. See the example at the bottom of page 342.

By turning the derivative (n. x^{n-1}) around, the integral is $x^{n+1}/n+1$. So, the "active" part of the integral (that is, all except the constants) is $x^{3/3}$).

FINDING VOLUME BY INTEGRATION

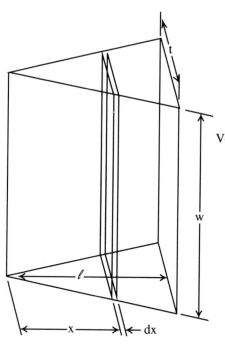

Volume of Wedge

Area of element $= wt\,\dfrac{x}{\ell}$

Thickness of element $= dx$

Volume of element $= \dfrac{wt}{\ell}\,x\,dx$

Volume of whole wedge

$$V = \int_0^\ell \dfrac{wt}{\ell}\,x\,dx$$

$$= \left[\dfrac{wt}{2\ell}\,x^2\right]_0^\ell$$

$$= \left(\dfrac{wt\ell}{2}\right) - \left(0\right)$$

$$= \dfrac{wt\,\ell}{2}$$

VOLUME of PYRAMID

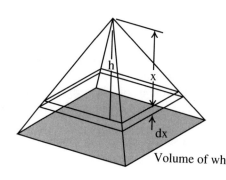

Area of base $= A$

Area of element

$$= A\left(\dfrac{x}{h}\right)^2$$

Thickness $= dx$

Volume of element

$$dV = A\left(\dfrac{x}{h}\right)^2 dx$$

Volume of whole pyramid $V = \int_0^h \dfrac{A}{h^2}\,x^2\,dx$

$$= \left[\dfrac{A}{3h^2}\,x^3\right]_0^h$$

$$= \left(\dfrac{Ah}{3}\right) - \left(0\right)$$

$$= \dfrac{Ah}{3}$$

Volume of cone

The method on the previous page for finding the volume of a pyramid can also be applied directly to the volume of a cone. Its base is a circle whose area (see "Area of a circle") is πr^2. The volume is $1/3\pi r^{2h}$.

Notice that, as compared with the curved surface area, the height is the vertical height, which is measured perpendicular to the case; not the slant height, which is measured up the surface.

Another method to find the volume of a cone uses an element that is a cylindrical shell of radius x. Check it through; the answer should be the same.

Method 1

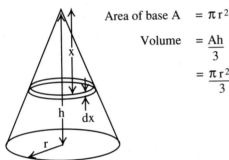

Area of base A $= \pi r^2$

Volume $= \dfrac{Ah}{3}$

$= \dfrac{\pi r^2 h}{3}$

Method 2

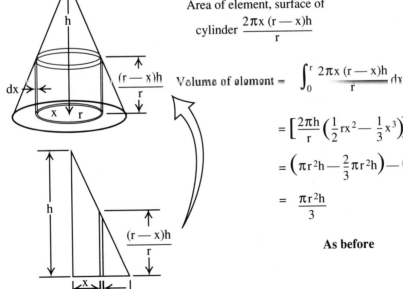

Area of element, surface of

cylinder $\dfrac{2\pi x (r - x)h}{r}$

Volume of element $= \displaystyle\int_0^r \dfrac{2\pi x (r - x)h}{r}\, dx$

$= \left[\dfrac{2\pi h}{r} \left(\dfrac{1}{2} rx^2 - \dfrac{1}{3} x^3 \right) \right]_0^r$

$= \left(\pi r^2 h - \dfrac{2}{3}\pi r^2 h \right) - \left(0 \right)$

$= \dfrac{\pi r^2 h}{3}$

As before

Volume of sphere

Here again, you have a choice of methods to find the volume. The first method takes "slices" of the sphere (very thin, of course) and the second takes cylindrical shells. Taking slices, x goes from $-r$ to $+r$. The cylindrical shell method takes values of x from zero to r. However, the volume of the element has a product function. If you remember the differentiation, it's not difficult. The first method is simpler.

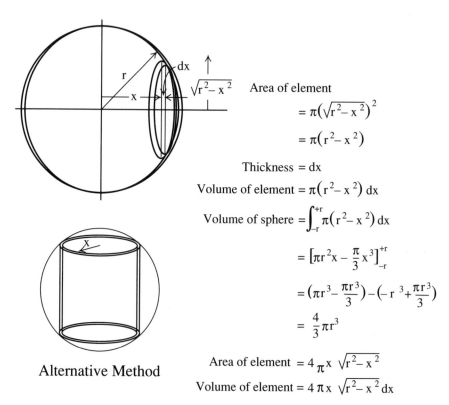

Area of element

$$= \pi\left(\sqrt{r^2-x^2}\right)^2$$

$$= \pi\left(r^2-x^2\right)$$

Thickness $= dx$

Volume of element $= \pi\left(r^2-x^2\right) dx$

Volume of sphere $= \displaystyle\int_{-r}^{+r} \pi\left(r^2-x^2\right) dx$

$$= \left[\pi r^2 x - \frac{\pi}{3}x^3\right]_{-r}^{+r}$$

$$= \left(\pi r^3 - \frac{\pi r^3}{3}\right) - \left(-r^3 + \frac{\pi r^3}{3}\right)$$

$$= \frac{4}{3}\pi r^3$$

Alternative Method

Area of element $= 4\pi x \sqrt{r^2-x^2}$

Volume of element $= 4\pi x \sqrt{r^2-x^2}\, dx$

Questions and problems

1. Using the product formula, differentiate $\sin 3x \cos 2x$.

2. Using the formula for $\sin(A + B)$ and $\sin(A - B)$, find an expression for $\sin 3x \cos 2x$ in terms of $\sin 5x$ and $\sin x$.

3. Differentiate the answer in question 2.

4. Using the formulas for cos($A + B$) and cos ($A - B$), find an expression for the answer to question 1 and thereby show that the result is the same as that found in question 3.

5. Using the quotient formula, find the derivatives for cosecant, and cotangent functions. Check the results by using the function of a function formula.

6. Find the first three derivatives of sec x. Express each in terms of sec and tan.

7. Evaluate the following integrals and check your results by differentiation:

$$\int 4x^3 - dx \qquad \int 6x^5 - dx \qquad \int 9x^8 - dx$$

$$\int x^4 - dx \qquad \int 7x^8 - dx \qquad \int \frac{3}{4}x^{1/4} - dx$$

$$\int 3 - \cos x - dx \quad \int - \sin s - dx \quad \int 3 - \cos 3x - dx$$

$$\int \cos 3x - dx \qquad \int - \sin 3x - dx \quad \int \sin 5x - dx$$

For convenience omit the constants of integration

8. Evaluate the following definite integrals

$$\int_{+3}^{+5} x^2 - dx \qquad \int_{-8}^{-2} \frac{1}{x^2} - dx \qquad \int_{0}^{\pi/2} \cos x - dx$$

$$\int_{0}^{\pi} \cos x - dx \qquad \int_{0}^{\pi/2} \sin x - dx \qquad \int_{0}^{\pi} \sin x - dx$$

9. A four-sided structure is shaped like a pyramid, except that its sides are curved. Each side is formed from a flat piece, cut so that the width at any point is proportional to the square of the distance from the apex. These sides are then curved so that the corners meet all the way up. If the height, measured along the curved surface, is 100 feet, and the base of each side is 30 feet, find the total surface area of the structure.

10. A length of cylindrical pipe is cut off at one end so that the plane of the cut is an angle of 60 degrees (instead of the usual 90 degrees) to the pipe axis. The diameter of the pipe is 2 inches, and the length at the shortest point is 5 inches. Find the surface area of the pipe. HINT: if the pipe is slit at its shortest length and opened out, the slanting cut looks like a sine wave.

11. What would be the area of the pipe surface, if it was cut at 60 degrees at both ends, so that the planes converge at 60 degrees, and the length at the shortest point was still 5 inches?

12. Use a similar element to the one used for the previous two problems to find the surface area of a doughnut-shaped object with the following dimensions: the hole through the middle has radius r; the cross section through the body of the doughnut is circular also with radius r.

13. A pyramid-like structure, similar to question 9, has its curve determined by the dimension from the structure's vertical axis to its surface, which is proportional to the square of the distance from the apex. If it was 100 feet high (actual height this time) with a 10-foot square base, what would be its volume?

14. A regular tetrahedron is a pyramid that has an equilateral triangle for its base and three more equilateral triangles for its sides. Find (a) the distance from the center of the base to the center of each base side when the base side is 5 inches; (b) the tetrahedron's height; and (c) its volume.

15. The first derivative of a function of x has its zero values: $x = -6$ and $x = +8$, and it equals -135 at $x = +3$. The original function has one zero value at $x = +3$. Find the original function, its derivative, and the other values for which the original function is zero.

16. For a certain function of x, the value of the original function and of its second derivative are both -6 when x is zero; the second derivative is 0 when x is $+1$; the first derivative is 0 when $x = -1$ or $+3$. Find the function.

17. For a function of x whose highest index is 4, the 1st and 3rd derivatives are each zero when $x = 0$. Show this function with no terms in x or x^3. If the function itself has a zero value when x is -4, -2, $+2$, and $+4$, and a value of $+64$ when $x = 0$, find the function.

18. All the facts in question 17 are the same, except the value of the function when $x = 0$, which is $+16$. Find the function and carefully check your deductions.

19. An equation has the form: $y = a \sin x + b \cos 3x$. When $x = \pi/4$, the value of y is $3/\sqrt{2}$ when x has the same value. dy/dx is known to be $1/\sqrt{2}$. Find the values of a and b.

20. $y = a \sin x + b \cos 4x$, when $x = \pi 4$, the value of y is 2.5, and the value of dy/dx is 1. Find a and b.

20
CHAPTER

Combining calculus with other tools

Maxima and minima

Finding maximum and minimum points of a function, by means of differentials, is easy to do and can be a big help in many ways. When slope is zero, a function momentarily does not vary. If elsewhere it varies, the value identified by the zero derivative is either a maximum or a minimum in the function itself.

To plot a curve for $y = x^4 - x^3 - 18x^2 + 27x$, tabulate values of x and corresponding values of y, then mark them out on paper to plot the curve. The table here lists nine value sets. For some curves, nine points would be enough to draw quite an accurate curve, but not when such high powers are included.

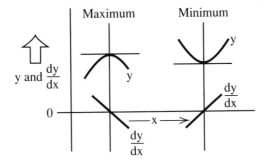

Find the maximum and minimum points on
$$y = x^4 - x^3 - 18x^2 + 27x$$

Plot a graph ▷

Values of x	x^4	$-x^3$	$-18x^2$	$+27x$	Values of y
-4	$+256$	$+64$	-288	-108	-76
-3	$+81$	$+27$	-162	-81	-135
-2	$+16$	$+8$	-72	-54	-102
-1	$+1$	$+1$	-18	-27	-43
0	0	0	0	0	0
$+1$	$+1$	-1	-18	$+27$	$+9$
$+2$	$+16$	-8	-72	$+54$	-10
$+3$	$+81$	-27	-162	$+81$	-27
$+4$	$+256$	-64	-288	$+108$	$+12$

Maximum and minimum points

On the graph paper at the left, the nine points are marked. Although they give an idea where the curve goes, they are not precise enough to be sure that it is accurate. A greater number of points might help, but the difficult spots are the maximum and minimum points. Apparently, a minimum is at or near $x = -3$, a maximum is at or near $x = +1$, and another minimum is at or near $x = +3$. The derivative $dy/dx = 4x^3 - 3x^2 - 36x + 27$ is a cubic equation in x, which can have 3 roots.

Try $x + 3$ as a factor to represent the root at $x = -3$. It factors to: $(x + 3)$ $(4x^2 - 15x + 9)$. Finding the roots of the second factor can now be solved as a quadratic. They are $x = +3$ (the factor is $x - 3$) and $x = +3/4$ (the factor is $4x - 3$). Now, you know that the two minima are precisely at $x = -3$ and $x = +3$, and the maximum is at $x = 3/4$. This information is a big help when plotting the curve accurately at these points.

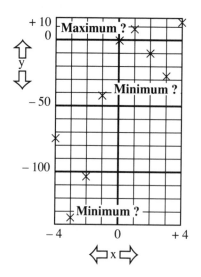

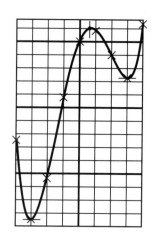

$$\frac{dy}{dx} = 4x^3 - 3x^2 - 36x + 27 = (4x - 3)(x^2 - 9) = (4x - 3)(x - 3)(x + 3)$$

Maximum or minimum when $4x - 3 = 0$ $x = \dfrac{3}{4}$

$$x - 3 = 0 \quad x = 3$$

$$x + 3 = 0 \quad x = -3$$

$x = \dfrac{3}{4}$ $x^4 = \dfrac{81}{256}$ $x^3 = \dfrac{27}{64}$ $18x^2 = \dfrac{81}{8}$ $27x = \dfrac{81}{4}$

$$y = 10\frac{5}{256}$$

Point of inflection

What might be even more difficult by simply plotting points, without help from the derivative, is a curve of the type represented by: $y = x^4 - 6x^2 + 8x$. Six points are tabulated, from $x = -3$ to $x = +2$. Without better knowledge, you might look for a mistake in the region between $x = 0$ and $x = -1$. It doesn't look like a smooth curve.

Values of x	x^4	$-6x^2$	$+8x$	Values of y
-3	8 1	-54	-24	$+3$
-2	16	-24	-16	-24
-1	1	-6	-8	-13
0	0	0	0	0
$+1$	1	-6	$+8$	$+3$
$+2$	16	-24	$+16$	$+8$

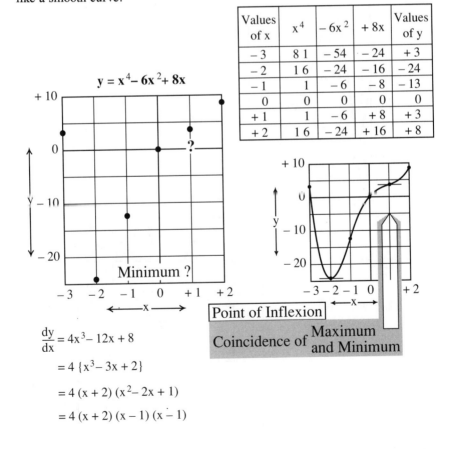

$y = x^4 - 6x^2 + 8x$

Minimum ?

Point of Inflexion

Coincidence of Maximum and Minimum

$$\frac{dy}{dx} = 4x^3 - 12x + 8$$

$$= 4\{x^3 - 3x + 2\}$$

$$= 4(x + 2)(x^2 - 2x + 1)$$

$$= 4(x + 2)(x - 1)(x - 1)$$

The derivative is: $dy/dx = 4x^3 - 12x + 8$. This equation can be factored to: $4(x + 2)(x - 1)(x - 1)$. Notice that the two identical roots are $x = +1$. Mathematically, both a maximum and minimum occur at $x = +1$. Such a point is called a *point of inflection*. It means that, right at the point of the root ($x = +1$), the curve is momentarily level.

Second derivative gives more information

Go back to the function considered in "Maximum and minimum points." The second derivative is: $d^2y/dx^2 = 12x^2 - 6x - 38$. Equating this to zero finds two more special values of x. These roots are $x = -1.5$ and $x = +2$. The second derivative has points where the first derivative has maxima or minima. A maximum in the first derivative is a point of maximum slope in the original function.

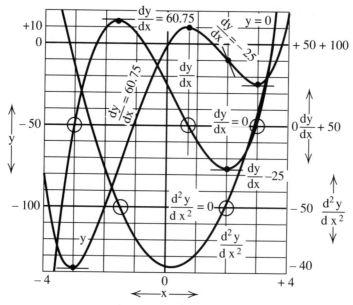

$y = x^4 - x^3 - 18x^2 + 27x$

$\dfrac{dy}{dx} = 4x^3 - 3x^2 - 36x + 27$

$\dfrac{d^2y}{dx^2} = 12x^2 - 6x - 36$

x	y	$\dfrac{dy}{dx}$	$\dfrac{d^2y}{dx^2}$	
− 4	− 76	− 133	+ 180	
− 3	− 135 ←	− 0	+ 90 ←	Minimum
− 2	− 102	+ 55	+ 24	
− 1.5	− 72.5625	+60.75 ←	− 0 ←	Maximum
− 1	− 43	+ 56	− 18	Slope +
0	0	+ 27	− 36	Maximum
+ .75	+ 10.01953125 ←	− 0	−33.75	
+ 1	+ 9	− 8	− 30	
+ 2	− 10	− 25 ←	− 0 ←	Maximum
+ 3	− 27 ←	− 0	+ 54	Slope −
+ 4	+ 12	+ 91	+ 132	Minimum

Most often, the second derivative finds points of maximum slope in the original function. Whichever it finds, it provides more information that enables the original function curve to be plotted more accurately.

More help from second derivatives

Now, it's time to take two more functions of x. The second derivative has two roots, $x = -1$ and $x = +1$. The original curve has a maximum slope at $x = -1$. The other point is the one that was already identified as the point of inflection, which is a minimum slope—zero!

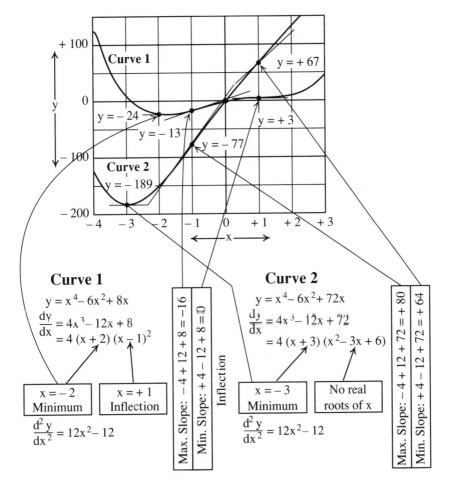

Curve 1

$$y = x^4 - 6x^2 + 8x$$

$$\frac{dy}{dx} = 4x^3 - 12x + 8$$
$$= 4(x + 2)(x - 1)^2$$

$x = -2$	$x = +1$
Minimum	Inflection

$$\frac{d^2y}{dx^2} = 12x^2 - 12$$

Max. Slope: $-4 + 12 + 8 = -16$

Min. Slope: $+4 - 12 + 8 = 0$

Inflection

Curve 2

$$y = x^4 - 6x^2 + 72x$$

$$\frac{dy}{dx} = 4x^3 - 12x + 72$$
$$= 4(x + 3)(x^2 - 3x + 6)$$

$x = -3$	No real
Minimum	roots of x

$$\frac{d^2y}{dx^2} = 12x^2 - 12$$

Max. Slope: $-4 + 12 + 72 = +80$

Min. Slope: $+4 - 12 + 72 = +64$

Maximum and Minimum Slopes
at x = – 1   x = + 1

The other curve merely changes the coefficient of the x term. Because it doesn't reach the second derivative, it will have the same roots. Here, you see what happens when the curve is tilted—what changing the x coefficient does. Instead of a point of inflection, it has a point of minimum slope.

If the x coefficient was changed the opposite way, the coincident maximum and minimum at the point of inflection would be separated, and that point would become one of maximum negative slope. The first derivative would have three different roots, instead of having two of them coincident.

Maximum area with constant perimeter

One practical use of differentiation is to find such a maximum. Derive the formula for the area of this oblong with constant perimeter. It is a quadratic: area = $x(p - x)$ (when the perimeter is $2p$). The derivative is: $dA/dx = p - 2x$. The derivative is zero when $2x = p$ or $x = 1/2\,p$. So, the maximum area is when the oblong is a square with equal dimensions both ways.

Had the area been kept constant, the perimeter varied, and the minimum perimeter solved for, the result would have been the same.

$$\textbf{OBLONG} \begin{cases} \text{Constant Perimeter} \\ \text{Maximum Area} \end{cases}$$

Perimeter = 2p

Area = x (p – x) = px – x^2

$\dfrac{dA}{dx} = p - 2x$ This is zero when $p = 2x$ *or* $x = \dfrac{1}{2}\,p$

$p - x = p - \dfrac{1}{2}\,p$ *or*   also $= \dfrac{1}{2}\,p$ $\Big\}$ square

Boxes with minimum surface area

The previous section shows that the best shape of an area for minimum perimeter is square. So, how high should a box be for minimum surface area?

Assume that the volume is V and the base is x square; the height must be V/x^2. You have a top and bottom that each have an area of x^2, and four sides that each have an area of x times V/x^2, which reduces to V/x. So, the total surface area is $4V/x + 2x^2$. Differentiating, $dA/dx = -4V/x^2 + 4x$, which will be zero when $V = x^3$, making the height equal to the sides of the square base. In other words, the figure is a cube.

To make this exercise more interesting, suppose that the box only needs the square base—no top. Now, the total area is $4V/x + x^2$. Differentiating, $dA/dx = -4V/x^2 + 2x$, which will be zero when $V = 1/2\,x^3$, making the height half the length of the square sides.

BOX  Constant Volume
Minimum Surface Area

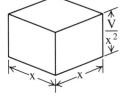

Surface area: Top and bottom $2x^2$

$$\text{Four sides}\;\; 4x \cdot \frac{V}{x^2} = \frac{4V}{x}$$

$$\text{Total surface area} = \frac{4V}{x} + 2x^2$$

$$\frac{d}{dx}\left(\frac{4V}{x} + 2x^2\right) = -\frac{4V}{x^2} + 4x \qquad \text{This is zero when } x^3 = V$$

 Cube

$$\frac{V}{x^2} = x$$

OPEN-TOPPED BOX

Surface area: Bottom x^2

$$\text{Four sides}\;\; 4x \cdot \frac{V}{x^2} = \frac{4V}{x}$$

$$\text{Total surface area} = \frac{4V}{x} + x^2$$

$$\frac{d}{dx}\left(\frac{4V}{x} + x^2\right) = -\frac{4V}{x^2} + 2x \qquad \text{This is zero when } x^3 = 2V$$

$$V = \frac{x^3}{2} \qquad \text{Height} = \frac{V}{x^2} = \frac{x^3}{2x^2} = \frac{x}{2}$$

Height is half side length

Cylindrical container with minimum surface area

The volume of a cylinder is πr^h. The surface area is πr^2 for both top and bottom and $2\pi rh$ for the curved surface. The total surface area is: $2\pi r(r + h)$. To obtain only one variable, write h in terms of the constant V and the variable r, which is: $h = V/\pi r^2$. Now, the equation for total area becomes: $A = 2\pi r^2 + 2V/r$. The derivative is: $dA/dr = 4\pi r - 2V/r^2$. The derivative is zero when: $r^3 = V/2\pi$. To find h, substitute in the expression for h already found, $h = 2r$. Twice the radius is diameter, so the minimum area is when height equals diameter.

Suppose you want a cylindrical container with no top. Following the same method for the open box, the height needs to be equal to the radius—half the diameter. In short, if a closed container is cut in two, two open containers of minimum surface area are formed.

CYLINDRICAL CONTAINER
Minimum Surface Area for Constant Volume

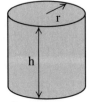

VOLUME $\boxed{V = \pi r^2 h}$

SURFACE AREA $A = 2\pi r^2$ ◁ (top and bottom)
$+ 2\pi r h$ ◁ (curve surface)
$+ 2\pi r \, (r + h)$

To get only one variable   substitute $h = \dfrac{V}{\pi r^2}$

$$A = 2\pi r \left(r + \frac{V}{\pi r^2} \right) = 2\pi r^2 + \frac{2V}{r}$$

To find minimum $\dfrac{dA}{dr} = 4\pi r - \dfrac{2V}{r^2}$ This is zero when $\boxed{2\pi r^3 = V}$ OR $r^3 = \dfrac{V}{2\pi}$

Substitue to find h: $h = \dfrac{V}{\pi r^2}$

$$= \frac{2\pi r^3}{\pi r^2} = 2r$$

Diameter equals height

Same Container Without Top

$$A = \pi r^2 + 2\pi r h$$
$$= \pi r^2 + \frac{2V}{r}$$

$\dfrac{dA}{dr} = 2\pi r - \dfrac{2V}{r^2}$ This is zero when $\pi r^3 = V$ OR $r^3 = \dfrac{V}{\pi}$

$$h = \frac{V}{\pi r^2} = \frac{\pi r^3}{\pi r^2} = r$$

Height equals radius, or half diameter

Conical container

What is the best shape for an ice cream cone? You want one in which the cone will accommodate the same volume with minimum surface area.

Volume is: $V = 1/3\pi r^2 h$. Surface area: $A = \pi r \ell$ which, to put h "in the picture," is: $A = r(r^2 + h^2)^{1/2}$ Substituting to get h in terms of V and r, then substituting that into the one for A, differentiating and equating to zero, you find that h, for minimum surface area, is equal to root 2 times r (diameter divided by root 2).

Do you not see many ice cream cones this shape? Maybe the manufacturer doesn't want to give you so much ice cream for the cost of the cone!

CONICAL CONTAINER
Constant Volume Minimum Surface (Curved)

VOLUME $V = \dfrac{1}{3}\pi r^2 h$

SURFACE AREA $A = \pi r \ell$
$$= \pi r \sqrt{r^2 + h^2}$$

Substitute $h = \dfrac{3V}{\pi r^2}$: $A = \pi r \sqrt{r^2 + \dfrac{9V^2}{\pi^2 r^4}} = \sqrt{\pi^2 r^4 + \dfrac{9V^2}{r^2}}$

$$\dfrac{dA}{dr} = \dfrac{1}{2\sqrt{\pi^2 r^4 + \dfrac{9V^2}{r^2}}}\left(4\pi^2 r^3 - \dfrac{18V^2}{r^3}\right)$$ This is zero when $2\pi^2 r^6 = 9V^2$
$$\sqrt{2}\,\pi r^3 = 3V$$

Substitute to find h: $h = \dfrac{3V}{\pi r^2} = \dfrac{\sqrt{2}\,\pi r^3}{\pi r^2} = \sqrt{2}\,r$

Equations for circles, ellipses, and parabolas

At left is the equation (in its simplest form) for each of these curves. Using those equations, both the circle and ellipse are centered at the origin (where both x and y are zero). The parabola is different.

At right is a more general equation for the same curve, which is not so centered. Deriving these general forms is a simple matter, merely substitute $(x - a)$ for x and $(y - b)$ for y, in the original simple form, then multiply out. Notice that the relationship between the second-order terms (those that involved x or y squared) is unchanged by this shift. This fact lets you recognize curves that are circles, ellipses, or parabolas from their respective equations, when you might otherwise not be sure what they are.

In the circle, the *r* represents radius. In the ellipse, two constants replace it, designated *q* and *s*. They are half the principal axes of the ellipse. These principal axes can be regarded as maximum (major) and minimum (minor) diameters of the ellipse.

SECOND ORDER CURVES

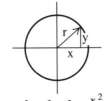

1. Circle

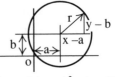

General $(x - a)^2 + (y - b)^2 = r^2$

Special $x^2 + y^2 = r^2$ or $\dfrac{x^2}{r^2} + \dfrac{y^2}{r^2} = 1$

$\boxed{x^2 + y^2} - 2ax - 2by + a^2 + b^2 = r^2$

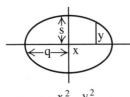

2. Ellipse

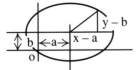

General $\dfrac{(x - a)^2}{q^2} + \dfrac{(y - b)^2}{s^2} = 1$

Special $\dfrac{x^2}{q^2} + \dfrac{y^2}{s^2} = 1$

$\boxed{\dfrac{x}{q^2} + \dfrac{y^2}{s^2}} - \dfrac{2ax}{q^2} - \dfrac{2by}{s^2} + \dfrac{a^2}{q^2} + \dfrac{b^2}{s^2} = 1$

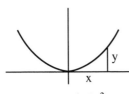

3. Parabola

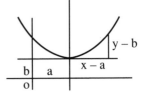

Special $y = fx^2$

General $y - b = f(x - a)^2$

$\boxed{fx^2} - 2afx - y + fa^2 + b = 0$

Directrix, focus, and eccentricity

On the previous page, the parabola was the "odd man out," as compared with the circle and the ellipse. A circle is generated with a compass that has its point at the center. An ellipse is generated with a little more complicated device that uses two centers to elongate it. But a parabola?

An alternative way to generate curves describes these curves more effectively. If you drove around a circular track, you would not have a compass attaching you to the center of the track! You would direct your course with the steering wheel. Paralleling that idea, visualize going along a parabolic course, positioning yourself by two things; a focus and a directrix.

For the parabola, you keep the two distances, from the focus and the directrix, equal. That distance, at a given point, is designated u. Where the focus is opposite the directrix, those two distances are each f (called focal length). Drawing a line through the focus parallel to the directrix, you can divide the u, measuring your distance from the directrix, into $2f$ and y.

Fact 1 relates to the u that measures your distance from the focus. Fact 2 relates to the other u. Combine them, rearrange, and you have an equation that is of the same form as a parabola. The condition described here is, in fact, another way to generate a parabola. Because the two u's are equal, this curve has an *eccentricity* of 1 (unity).

DIRECTRIX and FOCUS

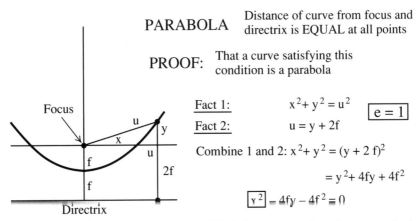

PARABOLA   Distance of curve from focus and directrix is EQUAL at all points

PROOF: That a curve satisfying this condition is a parabola

Fact 1: $x^2 + y^2 = u^2$ $\boxed{e = 1}$

Fact 2: $u = y + 2f$

Combine 1 and 2: $x^2 + y^2 = (y + 2f)^2$

$$= y^2 + 4fy + 4f^2$$

$$\boxed{x^2} - 4fy - 4f^2 = 0$$

Which is an equation for a parabola

The ellipse and the circle

Make eccentricity less than 1. Use the same method, but the distance from the focus is eu instead of u. Apply the same two facts and combine them, as you did for the parabola. Finish with an equation for the curve that takes the same form as an ellipse.

As eccentricity (e) decreases, the distance from the directrix increases, and eu gets to be closer to f in value. If you make eccentricity zero, the same equation simplifies so that it represents a circle, and f is then r.

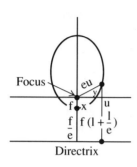

ELLIPSE   Distance of curve from focus and directrix is constant ratio at all points

Distance from focus on axis = f

Distance from directrix on axis $= \dfrac{f}{e}$ $\boxed{e < 1}$

PROOF: That a curve satisfying this condition is an ellipse

Fact 1: $x^2 + y^2 = e^2 u^2$ Fact 2: $y + f\left(1 + \dfrac{1}{e}\right) = u$

Combine 1 and 2: $x^2 + y^2 = e^2 \left\{ y + f\left(1 + \dfrac{1}{e}\right) \right\}^2$

$$= e^2 y^2 + 2ef(1 + e)\, y + f^2(1 + e)^2$$

Rearrange $\boxed{x^2 + (1 - e^2)\, y^2} - 2ef(1 + e)\, y = f^2(1 + e)^2$

Which is an equation for an ellipse

As e becomes smaller, distance from directrix becomes greater; focus is nearer center of ellipse; ellipse becomes more like a circle

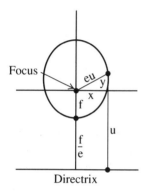

A CIRCLE is a second-order curve with an ECCENTRICITY $\boxed{e = 0}$

$$x^2 + y^2 = f^2 \quad (f \text{ is then } r)$$

Relationships between focus, directrix, and eccentricity

Look at these three curves in terms of these new parameters. The circle has a single focus, which is the *center*. The directrix removes itself to infinity. In fact, because it's removed to infinity, you could regard it as being at infinity in every direction. A circle of infinite radius would be a straight line. Viewed as an algebraic equation for a circle the second order terms, x^2 and y^2 are equal.

The ellipse has two foci at finite distance from each other. The ellipse is symmetrical about those two foci. An ellipse is eccentric. Mathematically or numerically, that eccentricity is greater than zero (the circle's value) but less than 1. Like the circle, its algebraic equation has two second order terms, x^2 and y^2, but their coefficients are not equal. With two symmetrical foci, it has two directrices at finite distances.

The parabola can be viewed (more about this later) as having two foci: one finite and one infinite. Its single directrix is at a distance equal to the focal length, thus eccentricity is 1 (unity). Its algebraic equation has only one second-order term.

S U M M A R Y	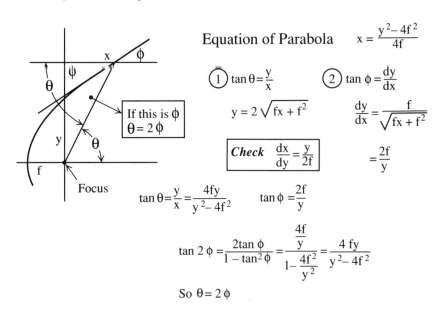		
Focus	Single	2 Finite	1 Finite 1 Infinite
Directrix	At Infinity	2 at finite distance	Single
Eccentricity	0	< 1	= 1
2nd-Order Terms	Equal	Unequal	Only one

Focus property of parabolas

Rearranging the equation for a parabola forms an equation for x in terms of the focal length and y. Now, look at the angles at a specific point on the curve, θ between a line from the focus and the parabola's axis, and ϕ, the slope of the curve at the same specific point. Tan θ is equal to y/x. Tan ϕ is equal to dy/dx. Using these facts, you can deduce the relationship $\theta = 2\phi$. This relationship about the parabola is important.

Equation of Parabola $\quad x = \dfrac{y^2 - 4f^2}{4f}$

① $\tan\theta = \dfrac{y}{x}$ ② $\tan\phi = \dfrac{dy}{dx}$

$y = 2\sqrt{fx + f^2}$ $\dfrac{dy}{dx} = \dfrac{f}{\sqrt{fx + f^2}}$

If this is ϕ
$\theta = 2\phi$

$\boxed{Check \quad \dfrac{dx}{dy} = \dfrac{y}{2f}}$ $= \dfrac{2f}{y}$

$\tan\theta = \dfrac{y}{x} = \dfrac{4fy}{y^2 - 4f^2}$ $\tan\phi = \dfrac{2f}{y}$

$\tan 2\phi = \dfrac{2\tan\phi}{1 - \tan^2\phi} = \dfrac{\dfrac{4f}{y}}{1 - \dfrac{4f^2}{y^2}} = \dfrac{4fy}{y^2 - 4f^2}$

So $\theta = 2\phi$

Focus property of ellipses

Applying the same idea to an ellipse, write equations for the minor and major axes. For convenience, a and b are half of each. Relating these to the coefficients of x^2 and y^2, finish with conversion factors, write the ratio of a/b, in terms of eccentricity e, and a value for e, in terms of the ratio a/b (or more precisely, b/a, because it yields the square root of a number less than 1).

RELATION BETWEEN MAJOR AXES and ECCENTRICITY of ELLIPSE

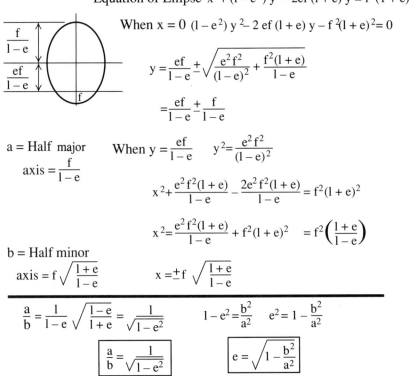

Equation of Ellipse $x^2 + (1 - e^2)\,y^2 - 2ef\,(1 + e)\,y = f^2(1 + e)^2$

When $x = 0$ $(1 - e^2)\,y^2 - 2\,ef\,(1 + e)\,y - f^2(1 + e)^2 = 0$

$$y = \frac{ef}{1 - e} \pm \sqrt{\frac{e^2 f^2}{(1 - e)^2} + \frac{f^2(1 + e)}{1 - e}}$$

$$= \frac{ef}{1 - e} \pm \frac{f}{1 - e}$$

a = Half major

$\text{axis} = \dfrac{f}{1 - e}$

When $y = \dfrac{ef}{1 - e}$ $\quad y^2 = \dfrac{e^2 f^2}{(1 - e)^2}$

$$x^2 + \frac{e^2 f^2 (1 + e)}{1 - e} - \frac{2e^2 f^2 (1 + e)}{1 - e} = f^2(1 + e)^2$$

$$x^2 = \frac{e^2 f^2 (1 + e)}{1 - e} + f^2(1 + e)^2 \quad = f^2\left(\frac{1 + e}{1 - e}\right)$$

b = Half minor

$\text{axis} = f\sqrt{\dfrac{1 + e}{1 - e}}$ $\qquad x = \pm f\sqrt{\dfrac{1 + e}{1 - e}}$

$\dfrac{a}{b} = \dfrac{1}{1 - e}\sqrt{\dfrac{1 - e}{1 + e}} = \dfrac{1}{\sqrt{1 - e^2}}$ $\qquad 1 - e^2 = \dfrac{b^2}{a^2} \quad e^2 = 1 - \dfrac{b^2}{a^2}$

$\boxed{\dfrac{a}{b} = \dfrac{1}{\sqrt{1 - e^2}}}$ $\qquad \boxed{e = \sqrt{1 - \dfrac{b^2}{a^2}}}$

Reflection properties of ellipses and parabolas

The angular relationship of the parabola explains why a light source at the focus reflects from all points along the surface of the parabola into a parallel beam. An optical law of reflection is that angle of incidence equals the angle of reflection. The fact that angle ϕ is half of angle θ means that the angle of the line from the surface point to the focus is equal to the angle between the surface point and the line parallel to the axis.

Extending this to the ellipse, instead of focusing into a parallel beam as the parabola does, an ellipse focuses from one focal point to the other.

REFLECTION PROPERTIES

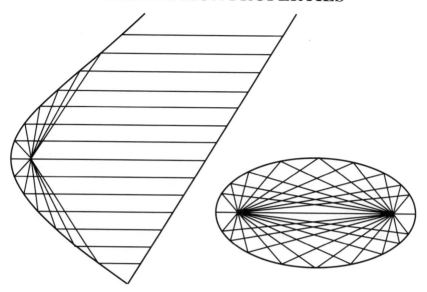

Hyperbolas: eccentricity greater than unity

The circle has zero eccentricity; the ellipse is less than unity; the parabola has an eccentricity of 1. So, what happens if the eccentricity is greater than 1? The result is a curve, called a *hyperbola*. Look at the equations: for the circle, the coefficients of x^2 and y^2 are equal; for the ellipse, they are unequal; for the parabola, one is zero; for the hyperbola, one reverses its sign.

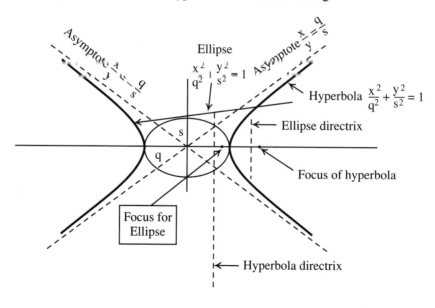

Another parameter helps define hyperbolas. To derive it, write the algebraic equations in yet another form. For the circle, the equation: $x^2 + y^2 = r^2$ is divided through by r^2, so the left-hand side consists of two ratios squared, which together make 1. For the ellipse, the half axes are labeled q and s. The sum of x^2/q^2 and y^2/s^2 is 1. The hyperbola reverses one of these signs, so the difference is equal to 1. For the parabola, one of these second-order terms disappears.

Asymptotes

The new parameter that helps construct these curves is called an *asymptote*. It is a straight line through the origin (the point where $x = 0$ and $y = 0$) whose equation is: $x/y = q/s$. The curve approaches this line more and more closely as it moves further from the origin. Notice the various lines identified in the previous section for the complementary ellipse and hyperbolas.

On this page, notice one particular hyperbola, the right hyperbola. An ellipse has two pairs of complementary hyperbolas, both of which use the same pair of asymptotes. A right hyperbola has four identically shaped curves; all touch a central circle in different quadrants.

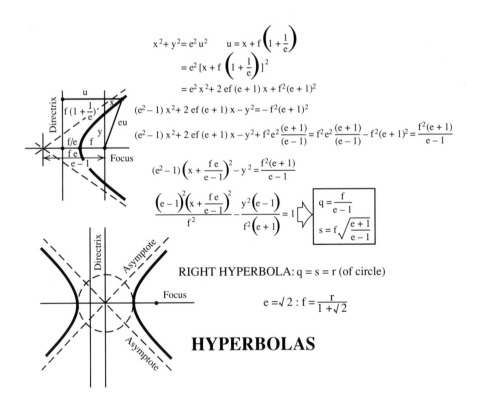

$$x^2 + y^2 = e^2 u^2 \qquad u = x + f\left(1 + \frac{1}{e}\right)$$

$$= e^2\left[x + f\left(1 + \frac{1}{e}\right)\right]^2$$

$$= e^2 x^2 + 2\,ef\,(e + 1)\,x + f^2(e + 1)^2$$

$$(e^2 - 1)\,x^2 + 2\,ef\,(e + 1)\,x - y^2 = -f^2(e + 1)^2$$

$$(e^2 - 1)\,x^2 + 2\,ef\,(e + 1)\,x - y^2 + f^2 e^2 \frac{(e + 1)}{(e - 1)} = f^2 e^2 \frac{(e + 1)}{(e - 1)} - f^2(e + 1)^2 = \frac{f^2(e + 1)}{e - 1}$$

$$(e^2 - 1)\left(x + \frac{f\,e}{e - 1}\right)^2 - y^2 = \frac{f^2(e + 1)}{e - 1}$$

$$\frac{(e - 1)^2\left(x + \frac{f\,e}{e - 1}\right)^2}{f^2} - \frac{y^2(e - 1)}{f^2(e + 1)} = 1$$

$$q = \frac{f}{e - 1}$$

$$s = f\sqrt{\frac{e + 1}{e - 1}}$$

RIGHT HYPERBOLA: $q = s = r$ (of circle)

$$e = \sqrt{2} : f = \frac{r}{1 + \sqrt{2}}$$

HYPERBOLAS

Second-order curves

To prepare for the next step, which relates all of these curves to conic sections, notice the different characteristics of these curves. A circle has a single focus and only 1 curve. Viewed as a special ellipse, you could say that it has a single focus because the two elliptical foci coincide. It's not difficult to understand zero eccentricity—that's what "round" means to some people! Both second-order terms have equal coefficients. Its directrix is at infinity in any direction.

An ellipse begins eccentricity which, by the definition given in mathematics, is less than 1 for any completed curve (circles and ellipses). An ellipse has two directrices that are parallel with the minor axis and closer to the ellipse as the ellipse itself elongates. Its equation has two second-order terms that are unequal but of the same sign.

A parabola is a marginal curve—the first one not to complete itself within finite dimensions. Theoretically, it has 2 foci, one finite, the other infinitely removed. Its eccentricity is 1 and it has only one second-order term.

A hyperbola goes a step further than the parabola by having 2 finite foci and 2 directrices. The hyperbola's eccentricity is greater than unity, second-order terms are of opposite sign, and it is characterized by asymptotes.

SUMMARY	CIRCLE	ELLIPSE	PARABOLA	**HYPERBOLA**
Focus	Single	2 Finite	1 Finite 1 Infinite	**2 Finite**
Curves	1	1	1	**2**
Directrix	At Infinity	2 Finite	1 Finite	**2 Finite**
Eccentricity	0	< 1	$= 1$	> 1
Second-Order Terms	Equal Same Sign	Unequal Same Sign	Only one	**Opposite Signs**
				2 Asymptotes

Conic sections produce second-order curves

Here is another way to relate this succession of second-order curves. A circle or an ellipse are seen as a section through a cylinder at right angles for the circle and obliquely for the ellipse. Also, slightly less obviously, they can be seen as sections through a cone (4). Making a cut parallel to the opposite face of the cone (5) yields a parabola. Cutting the cylinder more obliquely results in hyperbolas because the cutting plane intersects both cone extensions. You can visualize the extended cone by picturing the slanting side that protrudes beyond the apex and turning it "inside out," or giving that part a negative volume. If the plane cuts parallel to the cone's axis (8), the sloping sides of the cone coincide with its asymptotes. If it cuts obliquely (11), the parabola still has asymptotes, but they do not coincide with the sides of the cone. The dashed lines (12) indicate the asymptotes in that case.

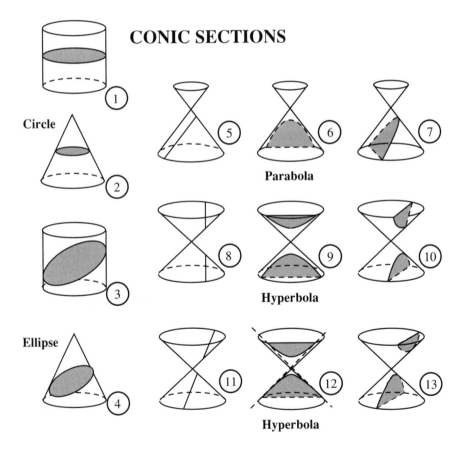

CONIC SECTIONS

Questions and problems

1. By differentiation, find the maximum and minimum points on the curve for $y = \sin x + 3 \sin 2x$. Find the values of y at which they occur and the approximate values of y at $x = 0, 1, 2,$ and 3.

2. Find maximum and minimum points through the first cycle of the curve: $y = \sin x + 1/3 \sin 3x$.

3. Find the approximate points and values at the points where slope of the curves in questions 1 and 2 are a maximum.

4. A curve that involves the first five exponents of x, has the maxima and minima at $x = -5, x = -3, x = +1,$ and $x = +5$. At $x = 0$, the curve value is 0 and at $x = 1$, the curve value is 24.1. Find the equation for the curve, the values at the maxima and minima named, and the points at which maximum slope occurs and values at these points.

5. By resolving its initial velocity into two components (vertical and horizontal), find the angle at which to fire a trajectile to achieve maximum distance for a fixed initial velocity, assuming no friction. Assume that the target is level with the firing location and that vertical deceleration and acceleration is uniform.

6. A container consists of a cylindrical section finished at the bottom with a hemisphere (half sphere); the top is open. Find what proportions will enable a given volume to be contained in minimum container surface (a) with the hemisphere extending outside the cylindrical part and (b) with it inverted inside the cylinder's end.

7. Find the equation for a parabola whose axis coincides with the x axis. The focus is the origin and focal length (f) is 5.

8. An ellipse has principal axes equal to 10 and 20. Find values for e and f.

9. A hyperbola equation is: $x^{2/a} - y^{2/b} = 1$. What is the effect of changing the sign on the right-hand side, to: $x^2 - y^2 = -1$? If a is greater than b, how will the eccentricities of the two hyperbolas be related?

10. Find the volumes of 2 sections of a sphere that are intersected by a plane at a distance of one half its radius from the center. Check your answer by adding the volume of both pieces, divided by the cut, and verifying this answer against the sphere's total volume.

21
CHAPTER

Introduction to coordinate systems

Two-dimensional systems of coordinates

Systems of coordinates can be used to locate points in a plane or in space, and to describe lines, areas, volumes, and shapes within the same systems. Most of the time, shapes were considered entities. The rectangular system of coordinates, by using x and y as coordinates for graphs, was touched on. However, they were used more as plots of quantities than as dimensions in a system of measurement. This chapter begins to cover how to specify the precise position of a point in a plane of two dimensions, or in three-dimensional space.

Graphs that use rectangular coordinates are usually in the form of a "box," with values of x plotted horizontally between specific values and values of y plotted vertically between other specific values. The *origin* is a concept drawn from systems of coordinates.

Mark the origin with an "O." In plane rectangular coordinates, the X and Y axes both pass through O; the X axis horizontally and the Y axis vertically. A position in those coordinates is measured by values of x and y; x is measured horizontally and y is measured vertically. Convention makes the values of x to the right of the origin positive and values to the left negative. Values of y above the X axis are positive and those below are negative.

Plane polar coordinates also use an origin. Instead of having two axes that the rectangular kind use, it has just one reference axis. The position of a point is measured by the direct distance from the origin and by the angle the line measuring that distance makes to the reference axis.

Before moving on, notice the equivalence between the two systems. First, assume that you know the location of a point in rectangular coordinates, as x, y.

366

To convert them to polar coordinates, the radius r is the square root of the sum of the squares of x and y. Angle θ has a tangent of y/x. At one time, the English used a negative index after the ratio symbol. This would be written $\tan^{-1} y/x$. The method commonly used today is arctan y/x, which represents exactly the same.

The other conversion is simpler. If the polar coordinates are r and θ, the equivalent rectangular coordinates are $x = r \cos \theta$, $y = r \sin \theta$.

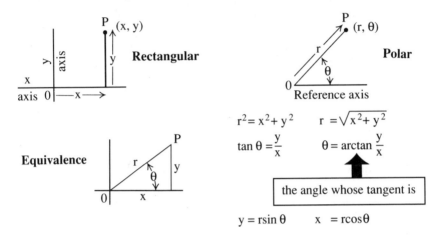

$$r^2 = x^2 + y^2 \qquad r = \sqrt{x^2 + y^2}$$

$$\tan \theta = \frac{y}{x} \qquad \theta = \arctan \frac{y}{x}$$

the angle whose tangent is

$$y = r\sin \theta \qquad x = r\cos\theta$$

Equation of a straight line

You have written many equations that represent straight lines (also curves). A general form for a straight line, in rectangular (sometimes called *cartesian* for Rene Descartes, a 16th-century French philosopher) coordinates is: $y = ax + b$ (see figures at the top of page 368). Here, a is the ratio between rates of change of x and y (or dy/dx) and b is the point where the line crosses the y axis. At the y axis, x is 0, so at that point: $y = b$.

Making direct substitutions for y and x, the equation in polar coordinates becomes $r \sin \theta = ar \cos \theta + b$. That, too, is the general form. Both forms are simplified considerably if the line passes through the origin. Then, $b = 0$, $y = ax$, and $\theta = \arctan a$, where θ is constant, not a variable angle, as in the general equation.

Equation for a circle

In general terms, we write an expression, using rectangular coordinates, x and y, the circle's radius, R (using a capital R here, to avoid confusion with the r polar coordinate), and the coordinates of its center, a and b. If the circle is centered on the origin all the terms with a or b in disappear, leaving the more familiar form: $x^2 + y^2 = R^2$ (see figure at the bottom of page 368).

By substituting, $x = r \cos \theta$ and $y = r \sin \theta$, we derive an expression for the same circle in polar coordinates. The two terms $r^2 \cos^2 \theta + r^2 \sin^2\theta$ add together to make r^2, and all the other terms are transferred to the other side.

EQUATION of a STRAIGHT LINE

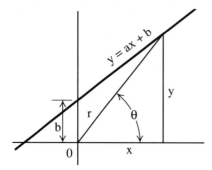

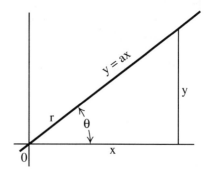

$y = ax + b \quad \dfrac{y}{x} = a + \dfrac{b}{x}$

$\theta = \text{arc tan} \left(a + \dfrac{b}{x}\right)$

$r = \sqrt{x^2 + (ax + b)^2}$

$$a = \frac{dy}{dx}$$

b is intercept
on y axis

General $\quad r \sin\theta = ar \cos\theta + b$

$y = ax \quad \dfrac{y}{x} = a$

$\theta = \text{arc tan } a$

$r = \sqrt{x^2 + a^2 x^2}$

Special $\quad b = 0$

EQUATION of a CIRCLE

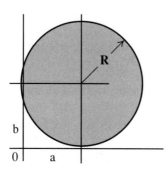

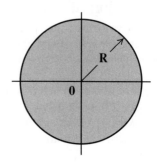

$(x - a)^2 + (y - b)^2 = R^2$

$x^2 - 2ax + a^2 + y^2 - 2by + b^2 = R^2$

$r^2 = x^2 + y^2$

$\quad = R^2 + 2ax + 2by - a^2 - b^2$

$x = r \cos\theta \qquad y = r \sin\theta$

$r^2 = R^2 + 2ar \cos\theta + 2br \sin\theta - a^2 - b^2$

$r^2 - 2\,[a \cos\theta + b \sin\theta]\,r + (a^2 + b^2) - R^2 = 0$

General

$x^2 + y^2 = R^2$

$r^2 = R^2 \qquad r = R$

Special

$a = 0 \qquad b = 0$

Three-dimensional systems of coordinates

Both these systems can be extended to the third dimension. Just as two coordinates in either will specify a point in a plane, three coordinates will specify a point in three-dimensional space.

In the rectangular system, add a third direction mutually at right angles to the other two, usually called the Z axis. Coordinates x, y, and z completely locate a point P in three-dimensional space, with respect to origin O and axes X, Y, and Z.

The polar system can use an additional angle in several ways. It still uses an origin and a reference axis as the starting point. The simple method is to use a reference plane in addition to the reference axis. A plane that contains the point P and the reference axis will have a specific angle to the reference plane. That angle has the symbol ϕ, the Greek letter phi. Then, starting from the reference line, within the plane already defined by ϕ, measure angle θ, the Greek letter theta, to the radius r, which measures the distance from the origin. Thus, the point is completely defined by r, ϕ, and θ. As in rectangular coordinates, it is specified by x, y, and z.

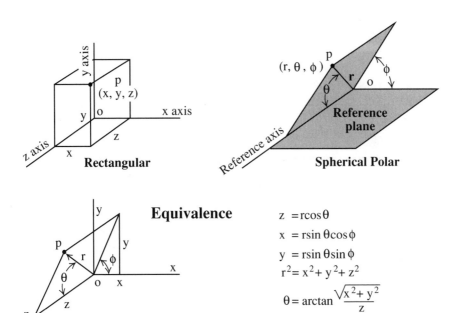

$$z = r\cos\theta$$
$$x = r\sin\theta\cos\phi$$
$$y = r\sin\theta\sin\phi$$
$$r^2 = x^2 + y^2 + z^2$$
$$\theta = \arctan\frac{\sqrt{x^2+y^2}}{z}$$
$$\phi = \arctan\frac{y}{x}$$

Equations of line and plane in rectangular coordinates

In a plane, the only linear form is a line. Three-dimensional space has two linear forms: a line and a plane. The equation for a line in three-dimensional space is really two equations. Using the intercept designation from the linear equation in a plane, *a* is the slope of the line, with reference to the XY plane: $a = dy/dx$. This equation denotes the slope of the plane that contains the line in the XY plane. *b* is the intercept of the plane that contains the line on the YZ plane. Next, $c = dy/dz$ is in the other direction and *d* is its intercept on the XY plane. This data enables the equation to be written: $y = ax + b = cz + d$.

The other linear form is the equation for a plane. Just as a simple equation represents a line in a plane, so a simple equation represents a plane in three-dimensional space. Using the XZ plane as the reference, *c* can be the intercept of the plane on the y axis, *a* the slope dy/dx, and *b* the slope dy/dz. To avoid confusion with the coordinate angles in polar coordinates, use the angles α and β (the Greek alpha and beta) to correspond with their tangents, *a* and *b*. This identification helps to identify angles, as well as the slopes with which they correspond.

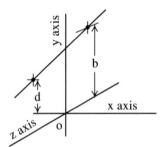

Equation of Line in Three Dimensions

$$y = ax + b = cz + d$$

$$a = \frac{dy}{dx} \quad \text{b is intercept on yz plane}$$

$$c = \frac{dy}{dz} \quad \text{d is intercept on xy plane}$$

Equation of Plane in Three Dimensions

$$y = ax + bz + c$$

$$a = \tan\alpha = \frac{dy}{dx} \qquad b = \tan\beta = \frac{dy}{dz}$$

c is intercept on y axis

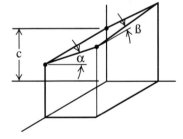

Equations in spherical polar coordinates

Using the equivalent from "Three-dimensional systems of coordinates," convert each set of values on the previous page to spherical polar coordinates, and make a double equation with two equals signs, as for rectangular coordinates. Regarding this as two equations, you can derive separate equations in polar coordinates. Both coordinates are necessary to define a line. By itself, each defines a plane. The line that the two equations put together define the intersection between the two planes. The substitutions convert the rectangular coordinate equation for a plane to one that uses spherical polar coordinates.

Any of these expressions, like virtually all equations that include trig functions, can take a variety of forms. Use a conversion that changes slope factor *a* of the rectangular system to angle ψ (Greek psi) so that tan = *a*. You could write arctan *a* instead of ψ. Use whichever is most appropriate in a specific problem.

EQUATION of LINE in
SPHERICAL POLAR COORDINATES

$$x = r \sin\theta\cos\phi \qquad y = r \sin\theta\sin\phi \qquad z = r\cos\theta$$

$$r \sin\theta\sin\phi = ar \sin\theta\cos\phi + b = cr\cos\theta + d$$

$$\sin\phi - a\cos\phi = \frac{b}{r\sin\theta} \qquad a\sin\theta\cos\phi + \frac{b}{r} = c\cos\theta + \frac{d}{r}$$

Make a = tan ψ $\qquad \sin\psi = \dfrac{a}{\sqrt{1 + a^2}}$

$$\cos\psi = \frac{1}{\sqrt{1 + a^2}} \qquad \boxed{\frac{b - d}{r} = c\cos\theta - a\sin\theta\cos\phi}$$

$$\sin\phi - a\cos\phi = \sqrt{1 + a^2}\sin(\phi - \psi)$$

$$\boxed{\sqrt{1 + a^2}\sin(\phi - \arctan a) = \frac{h}{r\sin\theta}}$$

EQUATION of PLANE in
SPHERICAL POLAR COORDINATES

$$x = r \sin\theta\cos\phi \qquad y = r \sin\theta\sin\phi \qquad z = r\cos\theta$$

$$r \sin\theta\sin\phi = ar \sin\theta\cos\phi + b\cos\theta + c$$

$$\sin\theta\sin\phi - a\sin\theta\cos\phi - b\cos\theta = \frac{c}{r}$$

$$\boxed{\sqrt{1 + a^2}\sin\theta\sin(\phi - \arctan a) - b\cos\theta = \frac{c}{r}}$$

Three-dimensional second-order curves

Equations can represent any three-dimensional shape. Equations for cones and other simple shapes are equally simple. Here is one that is not so easy. It can be viewed as a parabolic curve rotated around the Y axis, touching the origin. In rectangular coordinates, it is $y = a(x^2 + z^2)$.

Suppose you need to know the volume that is enclosed under this three-dimensional shape, within a square area defined in the XZ plane as 2s each way. To tie down the dimensions of the parabola, the height of the solid at its corners (where it is a maximum), is h, measured parallel to the Y axis.

The simplest method to solve this problem uses multiple integration in two directions. This is the same as simple integration, done one after the other. It makes no difference which direction is taken first. In this case, it is obvious because the x and z dimensions are interchangeable, however, even if they weren't it makes no difference.

MULTIPLE INTEGRATION

Example

Rectangular parabolic section:

$$y = a (x^2 + z^2)$$

Element $= ydxdz$

$\qquad = a(x^2 + z^2)\,dxdz$

Maximum value of $y = a(s^2 + s^2)$

$\quad a = \dfrac{h}{2s^2} \qquad = 2as^2$

$\qquad\qquad\qquad = h$

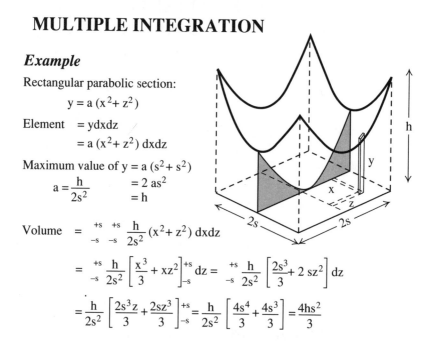

Volume $= \displaystyle\int_{-s}^{+s}\int_{-s}^{+s} \dfrac{h}{2s^2}(x^2 + z^2)\,dxdz$

$= \displaystyle\int_{-s}^{+s} \dfrac{h}{2s^2}\left[\dfrac{x^3}{3} + xz^2\right]_{-s}^{+s} dz = \displaystyle\int_{-s}^{+s} \dfrac{h}{2s^2}\left[\dfrac{2s^3}{3} + 2sz^2\right] dz$

$= \dfrac{h}{2s^2}\left[\dfrac{2s^3 z}{3} + \dfrac{2sz^3}{3}\right]_{-s}^{+s} = \dfrac{h}{2s^2}\left[\dfrac{4s^4}{3} + \dfrac{4s^3}{3}\right] = \dfrac{4hs^2}{3}$

Using Numbers $S = 5, h = 12$ Volume $= 400$

Questions and problems

Some of the following questions are to test you on your work in this course before this chapter.

1. The equation for a straight line is $y = 1.5x + 3$. Find the equation for a line at right angles to this one, which intersects it at point (6, 12).

2. Find the equation, in both rectangular and polar systems of coordinates for a circle whose center is on the Y axis and whose circumference passes through the origin; the radius is *R*. Find the coordinates of a point 45 degrees above and to the right of its center on the circumference. Also, find the equation for a line that makes a tangent with the circle at this point.

3. In the spherical polar coordinate equations for a line, substitute $b - d = 0$. What does the resulting equation show? Verify your conclusion from the rectangular coordinate form.

4. If in the same equations, both *b* and *d* are zero, what do the equations reduce to and what do the reductions signify?

5. In the equation for a plane, make $c = 0$. Show that this equation represents a plane through the origin. By using α for arctan *a* and β for arctan *b*, find expressions for θ in terms of α, β, and ϕ, and for ϕ in terms of α, β, and θ. Verify your expressions by considering cases where the plane contains various axes of the rectangular system, but is not coincident with the planes of that system.

6. A pyramid is *h* high at the apex, has a triangular base with a front side *b* wide, and a back corner distance *d* perpendicularly behind the front. Write a formula for its volume.

7. Assume that the pyramid has all its faces curved so that the linear dimension of each edge is proportional to the square of the vertical distance of the horizontal section from the apex. Now what will the volume be?

8. In a system of notation, the following facts are noted: to check if a number is divisible by 7, add the digits. If the sum is divisible by 7, the number itself is. To check if a number is divisible by 4, the last digit must be a 4 or 0. What is the number base?

9. Find the 10th root of 1000, correct to five decimal places by using a binomial series. NOTE: the 10th root of 1024 is 2.

10. A number in a certain system is written 3576. Its binary equivalent is 11101111110. What number base is it?

11. An ancient weight system if found in a remote village that uses weights that appear to be in 1, 3, 9, and 27 units. How could this system of weights be used to weigh articles whose weights, in the system used, are: 2, 5, 7, and 18 units? Show that those four weights can be used to weigh articles at intervals from 1 to 40 units. What is the basic unit in this system of notation?

12. Using the fact that dy/dx is the reciprocal of dy/dx, and rewriting the original expression to give x in terms of y, find dy/dx when y is: (a) arcsin x; (b) arccos x; (c) arctan x.

13. The cost of a long electrical transmission line depends on the insulation and wire size that must be used. The cost caused by insulation is \$$av$ per mile; by wire size, it is \$$b/v$ per mile (where a and b are constants based on the cost of materials and v is transmission voltage used). Find the transmission voltage that would allow the line to have a minimum cost, in terms of a and b.

14. Differentiate $y = 3x^2$ arcsin x with the product function formula.

15. A curve has inflection points at $x = -2$ and $x = +2$, and a maximum slope of $+20$ at $x = 0$. Write the equation for this curve.

16. A waveform follows the equation: $y = \sin x - a \sin 3x$. Find the maximum value a can have before the cycle has two maxima and minima.

17. Derive an equation for a cone that has an apex at the origin, an axis that is the Y axis, and an angle between the surface and axis that is arctan a.

18. Derive an equation for a plane that is parallel to the Z axis at an angle of arctan b to the Y axis (or arctan $1/b$ to the X axis) and with intercept c on the Y axis. Substitute in the equation written for question 17 to find the equation for the conic section produced (a) as a projection on the XZ plane and (b) in the plane of the section. Use u at the coordinate that corresponds to the x direction in this plane.

Part 4

Developing algebra,
geometry,
trigonometry,
and calculus
as analytical methods
in mathematics

22
CHAPTER

Complex quantities

Imaginary quantities

Imaginary numbers were introduced in Part 2 of this book to show how you could find extra roots that aren't in the ordinary numbering system. Once mankind didn't accept fractions—only whole or integral numbers could be counted. Then, people could only count positive quantities, negative numbers were impossible. Finally, the square root of a negative number was allowed. What else can be had?

It's a help to think of numbers as being measured in various directions. Negative numbers are measured in a 180-degree reversal from positive. Imaginary numbers allowed another pair of directions: 90 and 270 degrees. *Complex* angles combine positive or negative real numbers with positive or negative imaginary numbers.

To get this concept clear, a geometric illustration in rectangular coordinates visualizes the relationship. Multiplying by a negative produces rotation through 180 degrees. Multiplying by a negative again produces another rotation through 180 degrees, which brings us back to the original positive. With simple positive and negative numbers, you don't need to decide whether the rotation is clockwise or counterclockwise. When you multiplied by imaginary operator i, you thought of it algebraically as root minus 1. Geometrically, think of it as rotating through 90 degrees, and use the counterclockwise rotation because dependent variable y uses upwards for positive values.

Now, every time you multiply by i, as well as performing the numerical multiplication, the *vector* (as it is called) is rotated counterclockwise through 90 degrees. Multiply by i again and this time a second 90 degrees completes 180. Because i was defined algebraically as root minus 1, when you square it, it becomes -1. Multiplying by i four times results in a positive real quantity again.

IMAGINARY QUANTITIES

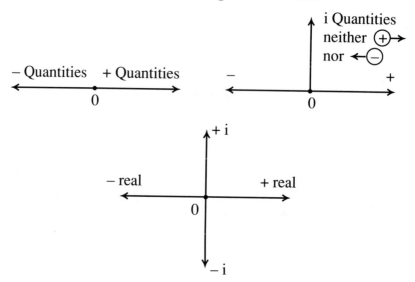

IMAGINARY OPERATOR i

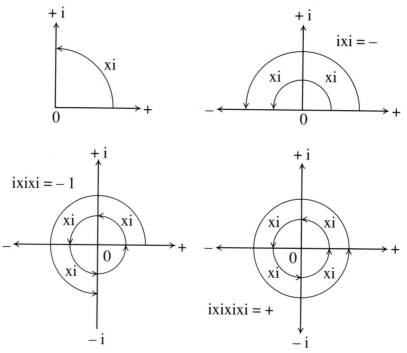

The complex plane

These rectangular coordinates can now be regarded as a complex plane, in which you can plot quantities that are part real and part imaginary. The real part is measured right for positive and left for negative. An imaginary part is measured up for positive and down for negative. The complex quantity can be written algebraically as: $a + ib$.

If both parts are positive, the quantity is in the first quadrant. If the real part is negative and the imaginary part is positive, the quantity is in the second quadrant. If both parts are negative, the quantity is in the third quadrant. Finally, if the real part is positive, but the imaginary part is negative, the quantity is in the fourth quadrant.

Now look at the cube root of -1 (see Part 2, where it was the cube root of $+1$). The root in the first quadrant is at what would be called 60 degrees, and it could be called the first cube root of -1. Now squaring that puts the product in the second quadrant, at what would be called 120 degrees. Cubing it verifies it as a cube root, by turning 180 degrees, on the negative end of the real axis.

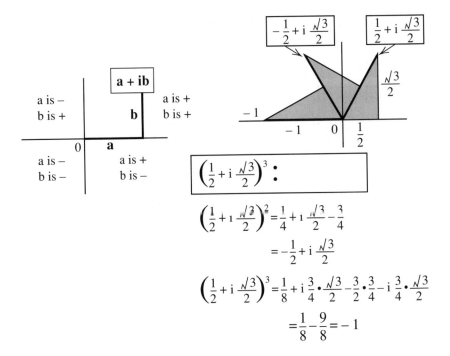

Complex quantities

The Pythagorean theorem, whether you think in terms of geometry, trigonometry, or algebra, shows that the magnitude of each root, which is the length of the vector from the origin, is 1. You have taken three steps with the cube root of -1.

Suppose that you start with the same quantity in the second quadrant (120 degrees), and write in the polar coordinates as 1, with the angle at 120 degrees. Multiply this first cube root of +1 by itself, to get the square at 240 degrees, and again to get the cube when it is 360 degrees, which proves that the quantity you began with is the cube root of +1.

Check your work with this technique to gain confidence in the use of complex quantities.

See the complex quantity interpretation of the three cube roots of +1. Cubing the first root at 120 degrees multiplies that angle by 3, making 360 degrees, which is the same as 0, thus it is positive. Cubing the second at 240 degrees multiplies that by 3, making 720 degrees, where it becomes positive after two revolutions. Cubing the third at 360 degrees (a.k.a. 0 degrees) makes either no revolutions, keeping it positive or 3 revolutions, where it arrives at positive again.

In the last two sections, everything was kept simple by taking only quantities that had a magnitude of 1. How are quantities with magnitudes other than 1 represented? Look at the cube root of 8. Using the method already used a few times now, the first cube root of +8 is −1 plus i root 3, which is 120 degrees. Now, multiply that by itself and simplify: −2 −i 2 root 3. Multiplying by the root again returns the answer to +8; the imaginary part disappears to prove that the cube root is correct.

$$\sqrt{\left(\frac{1}{2}\right)^2 + \left(\frac{\sqrt{3}}{2}\right)^2} = \sqrt{\frac{1}{4} + \frac{3}{4}} = 1$$

$$\frac{1}{2} + i\,\frac{\sqrt{3}}{2} = 1\underline{/60°}$$

$$\left(\frac{1}{2} + i\,\frac{\sqrt{3}}{2}\right)^2$$

$$= -\frac{1}{2} + i\,\frac{\sqrt{3}}{2} = 1\underline{/120°}$$

$$\left(\frac{1}{2} + i\,\frac{\sqrt{3}}{2}\right)^3$$

$$= -1 = 1\underline{/180°}$$

$$-\frac{1}{2} + i\,\frac{\sqrt{3}}{2} = 1\underline{/120°}$$

$$\left(-\frac{1}{2} + i\,\frac{\sqrt{3}}{2}\right)^2 = \frac{1}{4} - i\,\frac{\sqrt{3}}{2} - \frac{3}{4}$$

$$= -\frac{1}{2} - i\,\frac{\sqrt{3}}{2} = 1\underline{/240°}$$

COMPLEX QUANTITIES

$$\left(-\frac{1}{2} + i\,\frac{\sqrt{3}}{2}\right)^3 = -\frac{1}{8} + i\,\frac{3}{4}\cdot\frac{\sqrt{3}}{2} + \frac{3}{2}\cdot\frac{3}{4} - i\,\frac{3}{4}\cdot\frac{\sqrt{3}}{2}$$

$$= -\frac{1}{8} + \frac{9}{8} = 1 \quad 1\underline{/360°}$$

Three Cube Roots of 1

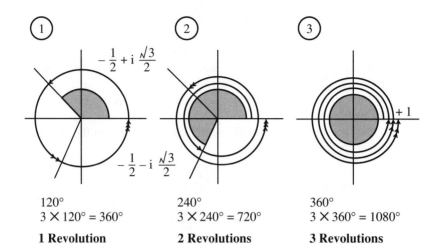

① $-\frac{1}{2} + i\frac{\sqrt{3}}{2}$

$-\frac{1}{2} - i\frac{\sqrt{3}}{2}$

120°
3 × 120° = 360°

1 Revolution

240°
3 × 240° = 720°

2 Revolutions

360°
3 × 360° = 1080°

3 Revolutions

$$-1 + i\sqrt{3} = 2\underline{/120°}$$

$$(-1 + i\sqrt{3})^2 = 1 - i2\sqrt{3} - 3 = -2 - i2\sqrt{3} = 4\underline{/240°}$$

$$(-1 + i\sqrt{3})^3 = -1 + i3\sqrt{3} + 9 = i3\sqrt{3} - 9 - 1 = 8\underline{/360°}$$

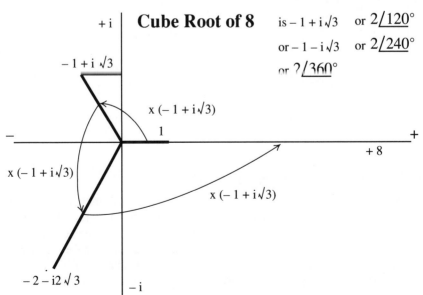

Cube Root of 8

is $-1 + i\sqrt{3}$ or $2\underline{/120°}$

or $-1 - i\sqrt{3}$ or $2\underline{/240°}$

or $2\underline{/360°}$

Multiplying complex quantities

Now confirm the method by multiplying any two complex quantities. You have two quantities, each written in two different ways: in rectangular coordinates as real and imaginary parts or in polar coordinates as a magnitude and an angle. In the picture, two quantities set out, each beginning at the positive real (X) axis. Beginning at the magnitude of the first quantity and multiplying each part of the second by the magnitude of the first, erect the third shaded triangle. This triangle brings you to the product, in magnitude and angle, or it can be read in rectangular coordinates as real and imaginary parts.

Study the diagram to see how the quantities that appear in the algebra are reproduced in coordinate geometry.

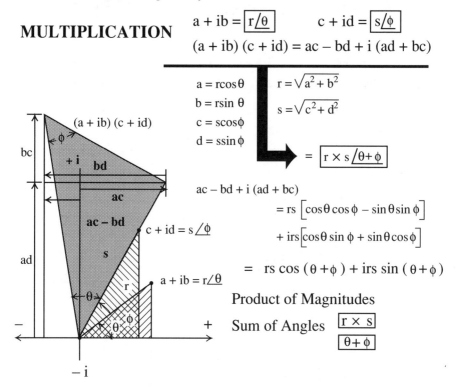

MULTIPLICATION

$a + ib = \boxed{r\underline{/\theta}}$ $c + id = \boxed{s\underline{/\phi}}$

$(a + ib)(c + id) = ac - bd + i(ad + bc)$

$a = r\cos\theta$

$b = r\sin\theta$

$c = s\cos\phi$

$d = s\sin\phi$

$r = \sqrt{a^2 + b^2}$

$s = \sqrt{c^2 + d^2}$

$= \boxed{r \times s\ \underline{/\theta + \phi}}$

$ac - bd + i(ad + bc)$

$= rs\left[\cos\theta\cos\phi - \sin\theta\sin\phi\right]$

$+ irs\left[\cos\theta\sin\phi + \sin\theta\cos\phi\right]$

$= rs\cos(\theta + \phi) + irs\sin(\theta + \phi)$

Product of Magnitudes

$+$ Sum of Angles $\boxed{r \times s}$

$\boxed{\theta + \phi}$

Reciprocal of complex quantities

If the quantity $a + ib$ has a magnitude r that is greater than unity, its reciprocal will have the magnitude $1/r$, which will be less than unity. The larger shaded area uses unit magnitude on the positive real axis for its base, and the magnitude r of the quantity $a + ib$ for its top side. Scaling this area down to make the longest side fit unit magnitude on the positive real axis, the side that was 1 in the bigger triangle is now the reciprocal of the original complex quantity, in both magnitude and polar angle.

The algebra shows how to calculate these values. Having a complex quantity in the denominator is not easy to handle. However, multiplying top and bottom by $a - ib$, the denominator becomes the sum of two squares, which reverses the sign of the imaginary part in the numerator.

In the study and use of complex quantities, a quantity such as $a - ib$ is called the *conjugate* of $a + ib$, or vice versa.

RECIPROCAL EXPRESSIONS

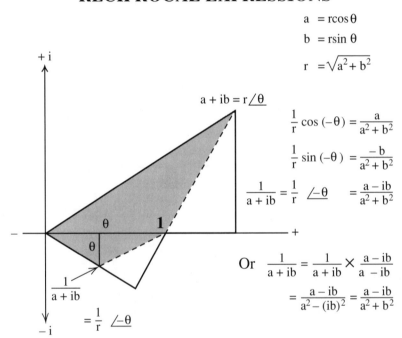

$$a = r\cos\theta$$
$$b = r\sin\theta$$
$$r = \sqrt{a^2 + b^2}$$

$$\frac{1}{r}\cos(-\theta) = \frac{a}{a^2 + b^2}$$

$$\frac{1}{r}\sin(-\theta) = \frac{-b}{a^2 + b^2}$$

$$\frac{1}{a + ib} = \frac{1}{r}\angle{-\theta} = \frac{a - ib}{a^2 + b^2}$$

Or $\quad \dfrac{1}{a + ib} = \dfrac{1}{a + ib} \times \dfrac{a - ib}{a - ib}$

$$= \frac{a - ib}{a^2 - (ib)^2} = \frac{a - ib}{a^2 + b^2}$$

$a + ib = r\angle\theta$

Division of complex quantities

The next logical step is division. Take two complex quantities and divide one by the other. From the actual size of the divisor (shaded area in left part of sketch), change the magnitude of the longest side to fit the longest side of the dividend, maintaining its shape or proportion. The quotient is then the side of the proportionate area of the divisor that was unit positive on the real axis before it was reduced.

As the right part of the sketch shows, the angle of the quotient is found by subtracting the angle of the divisor from the angle of the dividend. Magnitudes are simply divided, represented by r/s. Check the consistency of the methods that represent the operation, shown here.

DIVISION

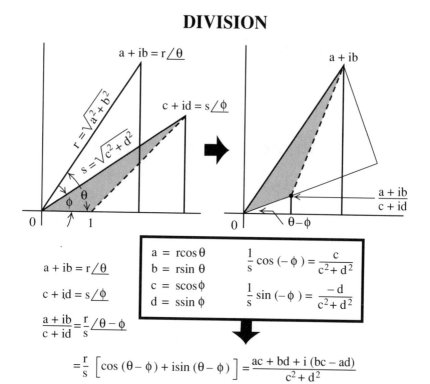

$$= \frac{r}{s} \left[\cos(\theta - \phi) + i\sin(\theta - \phi) \right] = \frac{ac + bd + i(bc - ad)}{c^2 + d^2}$$

Rationalization

You have already performed rationalization without naming it as such. In complex quantities, rationalization is what simplification is in fractions and similar subjects. A complex quantity is a simple combination of a real part and an imaginary part. Complex quantities can, as numerators, share the same denominator, as a matter of convenience or simplicity, but the denominator should be entirely real. Otherwise, the real and imaginary parts of the numerator are not truly real and imaginary, but each are complex.

Take a simplification that consists of two complex quantities multiplied together in the numerator and two more multiplied together in the denominator. When these quantities are multiplied, the numerator and denominator can each be simplified to single real and imaginary parts. Now, to rationalize, the numerator and denominator are each multiplied by the conjugate of the denominator, so only the numerator contains both real and imaginary parts. If desired, the whole quantity can be written separately: as a real part and as an imaginary part.

Rationalization

Evaluate $\dfrac{(a + ib)\,(c + id)}{(e + if)\,(g + ih)}$

$(a + ib)(c + id) = ac - bd + i(bc + ad) \Longrightarrow$ | Write $k = ac - bd$; $\ell = bc + ad$
$= k + i\,\ell \longleftarrow$

$(e + if)(g + ih) = eg - fh + i(fg + eh) \Longrightarrow$ Write $m = eg - fh$; $n = fg + eh$
$= m + in \longleftarrow$

$$\frac{(a + ib)\,(c + id)}{(e + if)\,(g + ih)} = \frac{k + i\,\ell}{m + in} = \frac{(k + i\,\ell)(m - in)}{(m + in)(m - in)} = \frac{km + \ell n + i(\ell m - kn)}{m^2 + n^2}$$

Example $\dfrac{(3 + i4)(5 - i6)}{(4 + i3)(1 - i2)}$ $= \dfrac{15 + 24 + i(20 - 18)}{4 + 6 + i(3 - 8)}$

$$= \frac{39 + i2}{10 - i5} = \frac{39 + i2}{5(2 - i1)}$$

$\times$ $\boxed{\dfrac{2 + i1}{2 + i1}}$ $= \dfrac{(39 + i2)(2 + i1)}{5(2^2 + 1^2)}.$

$$= \frac{78 - 2 + i(4 + 39)}{5 \times 5'} = \frac{76 + i43}{25}$$

Checking results and summarizing

It is easy to make mistakes when handling many numbers—even if you use a calculator! Often, the numbers happen to be convenient for making some relatively simple checks. Here, the first factor in the numerator has the same magnitude as the first factor in the denominator. One factor is $3 + i4$ and the other is $4 + i3$. The magnitude of both is 5. So, the whole expression will have the same magnitude if these two factors are removed; only the angle is changed. You could check to see that the magnitude is the same with these two factors removed.

Study the summary about complex quantities. First, study which quadrant a quantity falls in, according to the sign combination of its real and imaginary parts. Next, study the significance of addition, subtraction, multiplication, division, powers, and roots, in terms of magnitude, angle, and real and imaginary parts.

You cannot add or subtract magnitude and/or the angle of complex quantities; you have to work on the real and imaginary parts. In multiplication and division, magnitudes multiply and divide, but angles add and subtract. For powers and roots, the magnitudes take powers and roots, but angles multiply and divide by the indices. Any of these operations can be performed with real and imaginary parts by using the operator, i.

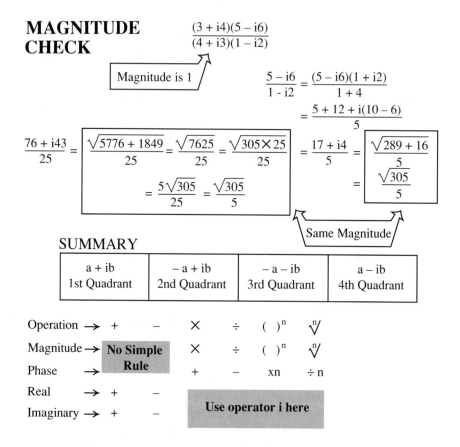

MAGNITUDE CHECK

$$\frac{(3 + i4)(5 - i6)}{(4 + i3)(1 - i2)}$$

Magnitude is 1

$$\frac{5 - i6}{1 - i2} = \frac{(5 - i6)(1 + i2)}{1 + 4}$$

$$= \frac{5 + 12 + i(10 - 6)}{5}$$

$$\frac{76 + i43}{25} = \frac{\sqrt{5776 + 1849}}{25} = \frac{\sqrt{7625}}{25} = \frac{\sqrt{305 \times 25}}{25}$$

$$= \frac{17 + i4}{5} = \frac{\sqrt{289 + 16}}{5}$$

$$= \frac{5\sqrt{305}}{25} = \frac{\sqrt{305}}{5}$$

$$= \frac{\sqrt{305}}{5}$$

Same Magnitude

SUMMARY

$a + ib$ 1st Quadrant	$-a + ib$ 2nd Quadrant	$-a - ib$ 3rd Quadrant	$a - ib$ 4th Quadrant

Operation →	+	−	×	÷	( )^n	$\sqrt[n]{}$
Magnitude →	No Simple Rule		×	÷	( )^n	$\sqrt[n]{}$
Phase →			+	−	xn	$\div n$
Real →	+	−				
Imaginary →	+	−	Use operator i here			

Use of a complex plane

This understanding of complex quantities led to the use of a complex plane in various ways, which we sample later. In the earlier graphic representation of quantities, the quantity measured horizontally was x and vertically was y. Sometimes t was horizontal. The independent variable was horizontal and the dependent was vertical.

The independent variable has a plane, so it can be complex. The real part is measured left or right, but the imaginary part is measured at right angles to it. The direction for the dependent variable is vertical.

Here, a relatively simple example uses a complex plane that shows two properties of this kind of presentation. At all points, magnitude of y is measured vertically. Quantities x and iz are in the complex horizontal plane. To make the working easier to follow, y is expressed as a complex fraction—real and imaginary parts in both numerator and denominator.

At a point where the denominator is zero (1), the value of y goes to infinity, called a *pole*. Where the numerator goes to zero (2), the value of y is zero, called a *zero*.

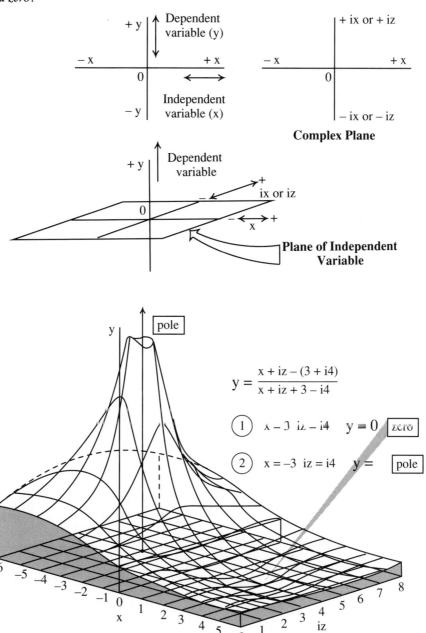

Quadratic roots with complex quantities

Earlier quadratics problems were craftily avoided where the quantity under the surd, in the formula method, would be negative because such a problem would have "no real roots." You probably thought the problem could not be solved. With imaginary numbers, such a problem can have an imaginary or complex solution with very definite mathematical meaning—no figment of the imagination!

Here are two examples. In the first, the quadratic uses real quantities that result in conjugate complex roots. In the second, the quadratic itself is complex and has complex roots that are not conjugate. So, quadratic equations "work" with complex quantities, as well as with real ones.

① $x^2 + x + 1 = 0$ By formula $x = -\dfrac{1}{2} \pm \sqrt{\dfrac{1}{4} - 1}$

$$= -\dfrac{1}{2} \pm \dfrac{i\sqrt{3}}{2}$$

Check $x = -\dfrac{1}{2} + \dfrac{i\sqrt{3}}{2}$ $x^2 = -\dfrac{1}{2} - \dfrac{i\sqrt{2}}{2}$ $x^2 + x + 1 = 0$ ✔

$x = -\dfrac{1}{2} - \dfrac{i\sqrt{3}}{2}$ $x^2 = -\dfrac{1}{2} + \dfrac{i\sqrt{2}}{2}$ $x^2 + x + 1 = 0$ ✔

② $x^2 + ix - 1 = 0$ By formula $x = -i\dfrac{1}{2} \pm \sqrt{-\dfrac{1}{4} + 1}$

$$= -i\dfrac{1}{2} \pm \dfrac{\sqrt{3}}{2}$$

$$\text{or} \pm \dfrac{3}{2} - i\dfrac{1}{2}$$

Check $x = \dfrac{\sqrt{3}}{2} - i\dfrac{1}{2}$ $ix = \dfrac{1}{2} + i\dfrac{\sqrt{3}}{2}$ $x^2 = \dfrac{1}{2} - i\dfrac{\sqrt{3}}{2}$ $x^2 + ix - 1 = 0$ ✔

$x = -\dfrac{\sqrt{3}}{2} - i\dfrac{1}{2}$ $ix = \dfrac{1}{2} - i\dfrac{\sqrt{3}}{2}$ $x^2 = \dfrac{1}{2} + i\dfrac{\sqrt{3}}{2}$ $x^2 + ix - 1 = 0$ ✔

Roots by complex quantities

From the pattern summarized in "Checking results and summarizing," you can simplify finding roots by using complex quantities. For example, you know that 32 has a fifth root that is 2. But your advancing knowledge of mathematics suggests that 32 has 4 more roots. What are they? You could develop a set of simultaneous equations, but an easier method is more visual. Rather obviously, these roots are of magnitude 2, with angles that divide the four quadrants into five parts. Using values of sine and cosine of these angles, we can derive the complex roots.

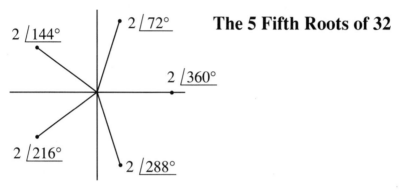

$2\,\underline{/144°}$ $2\,\underline{/72°}$ **The 5 Fifth Roots of 32**

$2\,\underline{/360°}$

$2\,\underline{/216°}$

$2\,\underline{/288°}$

1   $2\left[\cos72° + i\sin72°\right]$

2   $2\left[\cos144° + i\sin144°\right] \;=\; 2\left[-\cos36° + i\sin36°\right]$

3   $2\left[\cos216° + i\sin216°\right] \;=\; 2\left[-\cos36° - i\sin36°\right]$

4   $2\left[\cos288° + i\sin288°\right] \;=\; 2\left[+\cos72° - i\sin72°\right]$

5   $2\left[\cos360° + i\sin360°\right] \;=\; 2$

Roots $0.618 \pm i1.9022$
$-1.618 \pm i1.1756$
2

Question and problems

1. Multiply $0.6 + i0.8$ by $0.8 + i0.6$ and verify that the product is wholly imaginary and equal to 1.

2. Square each of the quantities in question 1, then multiply the two squares together and verify that the product is wholly real and of value 1.

3. Raise $0.6 + i0.8$ and $0.8 + i0.6$ each to the third power (cube) and fourth power. Multiply the cubes and fourth powers together, and verify that the product of the cubes is wholly imaginary (negative) while the product of the fourth powers is positive real, and that both have value 1.

4. By geometrical construction in the complex plane, deduce the two square roots of $i8$. Verify by squaring each complex root obtained so that its square is simply $i8$.

5. Using the rotational vector principle for powers and roots, find the 7th roots of 128. 7 roots should be possible.

6. Find the eight 8th roots of 256.

7. Find the ten 10th roots of 1024.

8. Find the nine 9th roots of 512.

9. Find the values of x or x^2 for which the following expression is (a) wholly imaginary, (b) wholly real:

$$1 - 10x^2 + 5x^4 + i(5x - 10x^3 + x^5)$$

10. A complex expression can be written in the form:

$$(1 + iax)(1 + ibx)(1 - icx)(1 - idx)$$

Deduce two conditions for this product to be real for all values of x, and from these conditions show that the only satisfying condition is that the pair of constants (a and b) must be the same as the pair (c and d).

11. Rationalize the expression: $(1 + iax)(1 + 1/ibx)$. From the result, deduce that the condition for the expression being wholly real is $x^2 = 1/ab$.

12. Show that the real part of the expression in question 11 has a constant value: $1 + a/b$.

13. Rationalize the following expression and verify its magnitude by canceling factors of equal magnitude in numerator and denominator

$$\frac{(12 + i5)(24 - i7)}{(3 + i4)(12 - i5)}$$

14. Rationalize and find the magnitude of the following expression:

$$\frac{15 + i8}{84 + i13}$$

15. Solve the following quadratic equations:

$$x^2 - 2x + 2 = 0 \qquad x^2 - 2x + 10 = 0 \qquad 13x^2 - 4x + 1 = 0$$
$$x^2 - i2x - 10 = 0 \qquad x^2 - i2x + 8 = 0 \qquad 3x^2 - i8x + 3 = 0$$
$$3x^4 - 8x^2 - 3 = 0 \qquad 9x^4 - 24x^2 + 25 = 0 \qquad 9x^4 - i24x^2 - 25 = 0$$

16. Find locations for a pole and zero in the complex plane for:

$$y = \frac{x - 3 + i4}{x - 5 + i12}$$

17. Find the complex solutions of the quadratic:

$$x^2 - (8 + i16)x - 33 + i56 = 0$$

18. The simplest nonrationalized form of an expression is:

$$\frac{x^2 - (8 + i10)x - 9 + i38}{x^2 + (8 + i10)x - 9 + i38}$$

By factoring, find the poles and zeros in the complex plane.

19. Find the values of x or x^2 for which:

$$\frac{1 - x^2 + i2x}{1 - 3x^2 + i(3x - x^3)}$$

(a) wholly real, (b) wholly imaginary.

20. Rationalize and thus show the condition for

$$\frac{1 + iax}{1 + ibx}$$

to be wholly real or wholly imaginary.

23

Making series do what you want

A pattern to a series

Returning to the study of series expansions, you can make the formation of a series for any function easier by looking for the patterns by which they can be made. Most series of a given type conform to a certain pattern. Assume here that the series takes the form of rising powers of x as the basic variable. It might begin with a constant, which is x^0, after which follow terms in x, x^2, x^3, x^4, and so on. Three samples of this type are: *binomial expansion* of $(a+x)^n$, $\sin x$, and $\cos x$.

Binomial

$$(a+x)^n = a^n + na^{n-1}x + \frac{n(n-1)}{2!}a^{n-2}x^2 + \frac{n(n-1)(n-2)}{3!}a^{n-3}x^3 + \frac{n(n-1)(n-2)(n-3)}{4!}a^{n-4}x^4 + \frac{n(n-1)(n-2)(n-3)(n-4)}{5!}a^{n-5}x^5$$

$$\sin x = x - \frac{x^3}{3!} + \frac{x^5}{5!}$$

$$\cos x = 1 - \frac{x^2}{2!} + \frac{x^4}{4!}$$

$$f(x) = f(0) + f_1(0)x + f_2(0)\frac{x^2}{2!} + f_3(0)\frac{x^3}{3!} + f_4(0)\frac{x^4}{4!} + f_5(0)\frac{x^5}{5!}$$

n	= 0	1	2	3	4	5
$f_n(x)$	$(a+x)^n$	$n(a+x)^{n-1}$	$n(n-1)(a+x)^{n-2}$	$n(n-1)(n-2)(a+x)^{n-3}$	$n(n-1)(n-2)(n-3)(a+x)^{n-4}$	$n(n-1)(n-2)(n-3)(n-4)(a+x)^{n-5}$
$f_n(0)$	a^n	na^{n-1}	$n(n-1)a^{n-2}$	$n(n-1)(n-2)a^{n-3}$	$n(n-1)(n-2)(n-3)a^{n-4}$	$n(n-1)(n-2)(n-3)(n-4)a^{n-5}$
$f(x)$	a^n	$+na^{n-1}x$	$+n(n-1)a^{n-2}\frac{x^2}{2!}$	$+n(n-1)(n-2)a^{n-3}\frac{x^3}{3!}$	$+n(n-1)(n-2)(n-3)a^{n-4}\frac{x^4}{4!}$	$+n(n-1)(n-2)(n-3)(n-4)a^{n-5}\frac{x^5}{5!}$

$f_n(x)$	$\sin x$	$\cos x$	$-\sin x$	$-\cos x$	$\sin x$	$\cos x$
$f_n(0)$	0	+1	0	-1	0	+1
$f(x)$		x		$-\frac{x^3}{3!}$		$+\frac{x^5}{5!}$

$f_n(x)$	$\cos x$	$-\sin x$	$-\cos x$	$\sin x$	$\cos x$	$-\sin x$
$f_n(0)$	1	0	-1	0	+1	0
$f(x)$	1		$-\frac{x^2}{2!}$		$+\frac{x^4}{4!}$	

The panel shows the basic pattern, called the *Maclaurin series*. If $f(x)$ is the function that you want to expand into a series, find the successive derivatives of the function, as in the second line under the panel (below the n line). Then, find the value of this derivative when $x = 0$, which you write in the next line. This value, divided by factorial n, becomes the coefficient of x^n in the expansion. This number is then written in the last line, and it agrees with the original series (as derived earlier). When no term exists for a given power of x, the value of that derivative when x is zero is also 0.

Pursuing the pattern

Now you have a tool to let you expand a function series more quickly. Try the expansion of $(a + 1/x)^n$. You run into trouble right away! When x is zero, the function's value is infinity. You need another rule: you cannot take just any function of x. You must choose it so that all values of the function and its derivatives, when x is zero, are finite.

Now, look for a function that fits a need you have found. The derivative is either equal to or is proportional to its value at each point. It forms the basis for natural rates of growth or decay. If the slope is upwards, it's growth; if downward, it's decay. In both, the rate of change is proportional to the value at the instant. This function does not yet have a name.

$$\left(a + \frac{1}{x}\right)^n = \ ? \qquad f(0) = (a + \infty)$$

Must use x so that all values of $f_n(0)$ are zero or finite

To find f(x) so that $f_1(x) = f(x)$ $\quad$ or $f_1(x) = -f(x)$

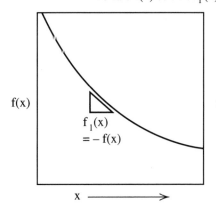

f(x)

$f_1(x)$
$= -f(x)$

x ⟶

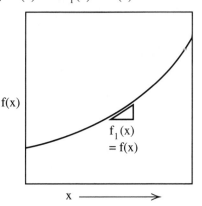

f(x)

$f_1(x)$
$= f(x)$

x ⟶

Natural growth and decay functions

The easiest way to find a series for such a function is to use the method used for the sine and cosine series in Part 3. Only alternative derivatives appeared in that final series, here every one does. Equate each successive term in the derived expansion: first, $b = a$; then, $c = b/2$; $d = c/3$, $e = c/4$; and so on. By multiplying the successive coefficients (the function is given by a, the constant), outside of the parentheses, which contain a series that begins $1 + x$, subsequent terms take the general form $x^{n/n!}$; the nth power of x divided by n factorial.

Try another approach. Expand $(a + 1/n)^n$. Show the expansion, while n is still finite, in two forms. However, make n infinite and it changes, because $n-1$ is substantially equal to n. All those infinite n's cancel, leaving just the reciprocal factorial series, which is the same as $f(x)$, if you make $x = 1$ and $a = 1$.

If you use the binomial expansion of $(1 + 1/n)^{nx}$, and again make n infinity, the expansion becomes the same as $f(x)$ for any value of x, not just $x = 1$. What does this expansion show you?

It demonstrates that the expansion of $(1 + 1/n)^n$, when n is infinity, represents a "number" to which is given a universal symbol, ϵ. If this number is raised to the power x, which makes ϵ^x, you have the function you are looking for. If $f(x) = \epsilon^x$ then $f'(x) = \epsilon^x$ also. In the way of writing used earlier, if $y = \epsilon^x$, then also $dy/dx = \epsilon^x$.

$$f(x) = a + bx + cx^2 + dx^3 + ex^4 + fx^5 +$$

$$f_1(x) = b + 2cx + 3dx^2 + 4ex^3 + 5fx^4 +$$

$$b = a \quad c = \frac{b}{2} \quad d = \frac{c}{3} \quad e = \frac{d}{4} \quad f = \frac{e}{5}$$

$$f(x) = a\left[1 + x + \frac{x^2}{2!} + \frac{x^3}{3!} + \frac{x^4}{4!} + \frac{x^5}{5!} + \ldots\right]$$

BINOMIAL APPROACH

$$\left(1 + \frac{1}{n}\right)^n = 1 + n\frac{1}{n} + \frac{n(n-1)}{2!}\frac{1}{n^2} + \frac{n(n-1)(n-2)}{3!}\frac{1}{n^3} + \frac{n(n-1)(n-2)(n-3)}{4!}\frac{1}{n^4}$$

$$= 1 + 1 + \frac{1 - \frac{1}{n}}{2!} + \frac{\left(1-\frac{1}{n}\right)\left(1-\frac{2}{n}\right)}{3!} + \frac{\left(1-\frac{1}{n}\right)\left(1-\frac{2}{n}\right)\left(1-\frac{3}{n}\right)}{4!}$$

If $n = \infty$

$$\left(1 + \frac{1}{n}\right)^n = 1 + 1 + \frac{1}{2!} + \frac{1}{3!} + \frac{1}{4!} + \frac{1}{5!} + \ldots$$

Same as $f(x)$ when $x = 1$ and $a = 1$

$$\left(1+\frac{1}{n}\right)^{nx} = 1 + nx\,\frac{1}{n} + \frac{nx(nx-1)}{2!}\,\frac{1}{n^2} + \frac{nx(nx-1)(nx-2)}{3!}\,\frac{1}{n^3} + \frac{nx(nx-1)(nx-2)(nx-3)}{4!}\,\frac{1}{n^4} +$$

$$= 1 + x + \frac{\left(1-\frac{1}{nx}\right)}{2!}\,x^2 + \frac{\left(1-\frac{1}{nx}\right)\left(1-\frac{2}{nx}\right)}{3!}\,x^3 + \frac{\left(1-\frac{1}{nx}\right)\left(1-\frac{2}{nx}\right)\left(1-\frac{3}{nx}\right)}{4!}\,x^4 +$$

If $n = \infty$

$$\left(1+\frac{1}{n}\right)^{nx} = 1 + x + \frac{x^2}{2!} + \frac{x^3}{3!} + \frac{x^4}{4!} + \frac{x^5}{5!} + \dots$$

Same as $f(x)$

If $\left(1 + \frac{1}{\infty}\right)^{\infty} = \varepsilon = 1 + 1 + \frac{1}{2!} + \frac{1}{3!} + \frac{1}{4!} + \frac{1}{5!} + \dots$

Then $\left(1 + \frac{1}{\infty}\right)^{\infty x} = \varepsilon^x = 1 + x + \frac{x^2}{2!} + \frac{x^3}{3!} + \frac{x^4}{4!} + \frac{x^5}{5!} + \dots$

Value of ϵ

The value of ϵ is always expressed by a form of the letter *e*. It might be the lower-case English letter (e), the Greek lowercase epsilon (ϵ), or a script letter ($\mathscr{E}$). Whichever form you see, it will be reserved for this special "number." You will quickly grasp its meaning once you understand its special significance.

Calculate its numerical value, which you can do on your calculator, to more decimal places, if you wish. Enter 1 in memory. Dividing 1 by 1 is still 1. Add that into memory. Dividing by 2 provides 0.5. Add that into memory. Dividing by 3 produces 0.16666. Add that into memory. Keep dividing by the next higher integer till you have enough decimal places.

$$\varepsilon = 1 + 1 + \frac{1}{2!} + \frac{1}{3!} + \frac{1}{4!} + \frac{1}{5!} + \dots$$

$\div 3$	.1 6 6 6 6 6 6	7
$\div 4$	.0 4 1 6 6 6 6	7
$\div 5$	.0 0 8 3 3 3 3	3
$\div 6$	.0 0 1 3 8 8 8	9
$\div 7$	.0 0 0 1 9 8 4	1
$\div 8$	.0 0 0 0 2 4 8	0
$\div 9$	.0 0 0 0 0 2 7	6
$\div 10$	.0 0 0 0 0 0 2	8
$\div 11$	.0 0 0 0 0 0 0	2
	2 .7 1 8 2 8 1 8	3

continued

$$f(x) = \varepsilon^{ax}$$

$$= 1 + ax + \frac{a^2 x^2}{2!} + \frac{a^3 x^3}{3!} + \frac{a^4 x^4}{4!} + \frac{a^5 x^5}{5!} + \dots$$

$$f_1(x) = \quad a + a^2 x + \frac{a^3 x^2}{2!} + \frac{a^4 x^3}{3!} + \frac{a^5 x^4}{4!} + \dots$$

$$= \quad a \left[1 + ax + \frac{a^2 x^2}{2!} + \frac{a^3 x^3}{3!} + \frac{a^4 x^4}{4!} + \dots \right]$$

$$= \quad a \varepsilon^{ax}$$

Series for arctan x

As an exercise in using the series from "A pattern to a series," find a series for arctan x. First, find the series of functions: $f'(x), f''(x)$ and $f'''(x)$, when $f(x) =$ arctan x.

If $y =$ arctan x, then $x = \tan y$. You already know that $dx/dy = 1 + \tan^2 y = 1 + x^2$. So, $dy/dx = 1/(1 + x^2)$. That is $f'(x)$. From there on it's easy, using the quotient formula. Having done a few, you substitute in $x = 0$. Arctan $0 = 0$. $f'(x) = 1$. $f''(x)$ is 0 again. $f'''(x) = -2$, and so on. Writing the series in the form from "A pattern to a series"and simplifying, arctan $x = x - x^{3/3} + x^{5/5} - x^{7/7} + \dots$

$$f(x) = \text{arctan } x \quad f_1(x) = \frac{1}{1 + x^2} \quad f_2(x) = \frac{-2x}{(1 + x^2)^2} \quad f_3(x) = \frac{2(3x^2 - 1)}{(1 + x^2)^3}$$

$$f(0) = 0 \qquad f_2(0) = 1 \qquad f_2(0) = 0 \qquad f_3(0) = -2$$

$$f_4(x) = \frac{24x(1 - x^2)}{(1 + x^2)^4} \qquad f_5(x) = \frac{24(1 - 10x^2 + 5x^4)}{(1 + x^2)^5}$$

$$f_4(0) = 0 \qquad\qquad f_5(0) = 24$$

$$f(x) = x - \frac{2}{3!} x^3 + \frac{24}{5!} x^5 \quad \dots \quad = x - \frac{x^3}{3} + \frac{x^5}{5} \dots$$

Concept of logarithms

The inverse function of $y = \varepsilon^x$ has some unexpectedly useful properties. To keep up with the convention of making x the independent variable and y the dependent, start by writing the equation the other way: $x = \varepsilon^y$. The inverse of this equation is read, "*y* equals log to the base ϵ of *x*." "Log" is short for the word *logarithm*, a word that is derived from the Greek that means "words about numbers," the study of numbers. The "arithm" part is also used in "arithmetic."

What the statement means is that to obtain *x*, ϵ has to be raised to the power *y*. Thus it is the exact inverse of the statement $x = \epsilon^y$: the same fact, stated the opposite way, starting at *x* to find *y*, instead of vice versa.

No simpler way exists to define a logarithm. A better understanding of what logarithms are comes from using them or applying them in mathematics. The graph shows values of *x* corresponding to values of *y* at 1, 2, and 3. Logarithms allow you to calculate values of *y* for values of *x* between 1, 2.7183, 7.3891, 20.085, etc. Having calculated such inbetween logarithms to any required accuracy, multiplication and division can be replaced with simple addition and subtraction, and indices can be replaced with simple multiplication and division.

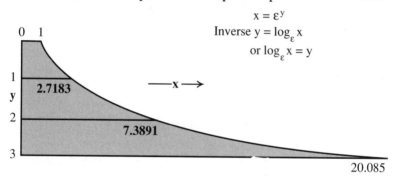

$$x = \epsilon^y$$
$$\text{Inverse } y = \log_\epsilon x$$
$$\text{or } \log_\epsilon x = y$$

If $\log_\epsilon x = m$ $\epsilon^m = x$

$\log_\epsilon y = n$ $\epsilon^n = y$

$$xy = \epsilon^{m+n}$$

$\log_\epsilon xy = m + n$

Similarly $\log_\epsilon \dfrac{x}{y} = m - n$

A gap in the series of derivatives

At the top of the next page, the functions of *x* are listed in the form $f(x) = ax^n$ with their derivatives in the form $f'(x) = anx^{n-1}$. The table is plain sailing for positive values of *n* and for negative values of *n*. You can even go down to fractional values of *n*, positive and negative, but a gap shows up in the sequence of derivatives at $n = 0$. As the derivative's coefficient gets near zero, its index approaches -1, but it never gets there!

Here is a derivative with no function for it to belong to: $f'(x) = 1/x$. But what is $f(x)$?

n	$f(x) = ax^n$	$f_1(x) = anx^{n-1}$
1	ax	a
2	ax^2	$2ax$
3	ax^3	$3ax^2$
-1	a/x	$-a/x^2$
-2	a/x^2	$-2a/x^3$
-3	a/x^3	$-3a/x^4$
0.5	$a/x^{1/2}$	$1/2ax^{-1/2}$
-0.5	$ax^{-1/2}$	$-1/2ax^{-3/2}$
0.1	$ax^{0.1}$	$0.1ax^{-0.9}$

Another Approach Needed

$$\left(1 + \frac{1}{\infty}\right)^\infty = \varepsilon$$

or

$$(1 + 0)^\infty = \varepsilon$$

$$f(x) = \infty \cdot x^0 \qquad f_1(x) = 1/x$$

?

Coefficient approaches zero

What is ?

Logarithmic function in calculus

Start with the inverse function: $x = \epsilon^y$. You can differentiate this function, giving $dx/dy = \epsilon^y = x$: that is how ϵ was defined. The reciprocal of that is what you were looking for: $dy/dx = 1/x$. So, if $x = \epsilon^y$, then $\log_\epsilon x = y$. That is the missing function in the series.

However, applying Maclaurin's series to this one sets a problem because $f(0)$ is negative infinity. The logarithm of a number is the power to which ϵ must be "raised" to produce that number. Zero is achieved only by "raising" any number (including ϵ) to negative infinity. The first derivative $f'(0)$ has a value of positive infinity. None of the numbers is finite. You need some other way to find this series.

If $\quad x = \varepsilon^y \quad \dfrac{dx}{dy} = \varepsilon^y = x$

Another Approach Needed

$$\frac{dy}{dx} = \frac{1}{x}$$

$\text{Log}_\varepsilon x = y$

$f(x) = \log_\varepsilon x \qquad\qquad f_1(x) = \dfrac{1}{x}$

$f(x) = \log_\varepsilon x$	$f_1(x) = \dfrac{1}{x}$	$f_2(x) = -\dfrac{1}{x^2}$
$f(0) = -\infty$	$f_1(0) = \infty$	$f_2(0) = -\infty^2$

Functions of ϵ

You calculated a series for ϵ and ϵ^x in the section, "Natural growth and decay functions." It represents natural, or exponential growth (or decay). As x increases in linear increments, ϵ^x increases in exponential increments: its rate of growth is proportional to itself, ϵ^x. Fairly obviously $\epsilon-^x$, which is the same as $1/\epsilon^x$, yields the same series of terms, but with signs alternately changed to plus and minus. You can find this by substituting $(-x)$ for x in the series.

You can also find the series for ϵ^{ix} and $\epsilon-^{ix}$ by making a similar substitution in the original series. Here, you have evaluated the terms for the $-x$ series for a short distance.

$$\epsilon^x = 1 + x + \frac{x^2}{2!} + \frac{x^3}{3!} + \frac{x^4}{4!} + \frac{x^5}{5!} + \dots$$

$$\epsilon^{-x} = 1 - x + \frac{x^2}{2!} - \frac{x^3}{3!} + \frac{x^4}{4!} - \frac{x^5}{5!} + \dots$$

$$\epsilon^{ix} = 1 + ix - \frac{x^2}{2!} - \frac{ix^3}{3!} + \frac{x^4}{4!} + \frac{ix^5}{5!} - \dots$$

$$\epsilon^{-ix} = 1 - ix - \frac{x^2}{2!} + \frac{ix^3}{3!} + \frac{x^4}{4!} - \frac{ix^5}{5!} - \dots$$

$$i^2 = -1$$
$$i^3 = -i$$
$$i^4 = 1$$

ϵ^{-1}: $+$ terms	$-$ terms
1.0	1.0
0.5	0.16666667
0.04166667	0.00833333
0.00138889	0.00019841
0.00002480	0.00000276
0.00000028	0.00000002
1.54308064	1.17520119
1.17520119	

$$\epsilon^{-1} = 0.36787945$$

Relationship between exponential and trigonometric series

Did you notice the similarity between the series for sine and cosine, and the exponential series in the previous section? If you multiply $\sin x$ by i, every term in the series is multiplied by i. Now, you can put these together so that ϵ^{ix} consists of $\cos x$ and $i \sin x$ added together, and that $\epsilon-^{ix}$ consists of $\cos x - i \sin x$.

You can turn that equivalence around by simply adding and subtracting those equations and dividing them by 2. So, you have a family of exponential functions. The real series is representative of exponential growth and decay. Imaginary series combine to make the series for sine and cosine.

$$\sin x = x - \frac{x^3}{3!} + \frac{x^5}{5!} - .... \qquad \cos x = 1 - \frac{x^2}{2!} + \frac{x^4}{4!} - ....$$

$$i\sin x = ix - \frac{ix^3}{3!} + \frac{ix^5}{5!} - ....$$

$\varepsilon^{ix} = \cos x + i\sin x$	
$\varepsilon^{-ix} = \cos x - i\sin x$	
$\cos x = \dfrac{\varepsilon^{ix} + \varepsilon^{-ix}}{2}$	
$\sin x = \dfrac{\varepsilon^{ix} - \varepsilon^{-ix}}{i2}$	

ε^{x}	Exponential growth
ε^{-x}	Exponential decay
ε^{ix} and ε^{-ix}	Both combinations of real and imaginary trigonometry (sin and cos) functions

Convergence of exponential and trigonometric series

The terms in these four series are all the same: only the pattern of signs changes. So, from term to term, each will have the same rate of convergence. Here, to show how rate of convergence changes with values of x, successive terms are tabulated. The bigger x is, the further into the series expansion before the terms turn to converge. Notice that, whatever finite value x has, the series eventually begins to converge.

$\begin{array}{c}\varepsilon^{\pm x} \\ \varepsilon^{\pm ix}\end{array}$	1	x	$\dfrac{x^2}{2!}$	$\dfrac{x^3}{3!}$	$\dfrac{x^4}{4!}$	$\dfrac{x^5}{5!}$	$\dfrac{x^6}{6!}$
x = 0	1	0	0	0	0	0	0
x = 1	1	1	0.5	0.1$\dot{6}$	0.041$\dot{6}$	0.008$\dot{3}$	0.00138
x = 2	1	2	2	1.$\dot{3}$	0.$\dot{6}$	0.2$\dot{6}$	0.04
x = 3	1	3	4.5	4.5	3.375	2.025	1.0125
x = 4	1	4	8	10.$\dot{6}$	10.$\dot{6}$	8.5$\dot{3}$	5.6$\dot{8}$

Significance of exponential series

To see what the series do, the example at the top of the facing page shows a succession of plotted terms, both for ϵ^x and $\epsilon - ^x$. In the first, adding successive terms sweeps the curve upward, and each term brings in further curvature when the preceding terms are inadequate to reach the exponential.

In the negative series, the effect is more easily seen. Alternate terms, by themselves, pull the curve alternately up and down, but the overall effect is an exponential decay curve. Study these curves and pursue the series further, if you wish.

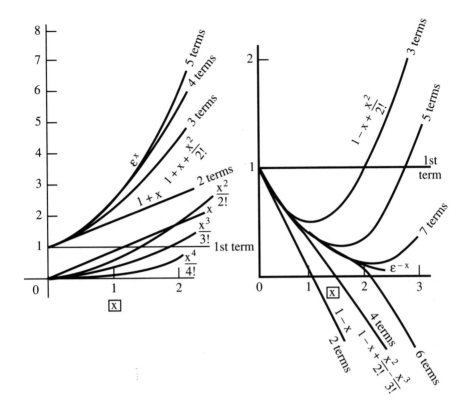

Significance of ϵ^{ix}

Following the same method for ϵ^{ix}, real and imaginary terms are plotted separately (see figure on top of page 402). The first real term is a constant. Successive terms pull the curve up and down, approaching something that looks like an expanding sine wave.

The first imaginary term (second term of the series) is a linear upward slope, which represents the beginning of a sine wave. Beyond this slope, the combination is similar to the real terms.

Remember that the real and imaginary terms are quite separate entities in the complex quantity. If you could visualize them in the complex plane, the effect would be rotational—especially if you could combine positive and negative imaginary series, according to the formula in, "Relationship between exponential and trigonometric series." However, the imaginary terms can be isolated as a series by adding the conjugate series, which is what the formula does.

Complex exponential functions

Functions, such as ϵ^{a+ib} or ϵ^{a-ib}, are complex exponential functions in the sense that the exponential base (ϵ) is raised to a power or uses an exponent that is part

Significance of ϵ^{ix}

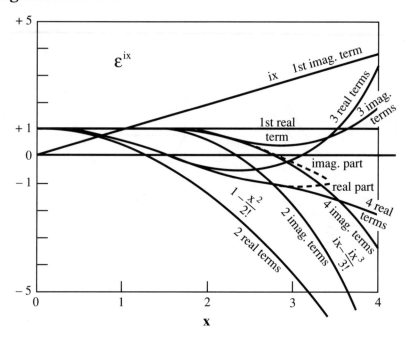

Complex exponential functions

$$\epsilon^{a+ib} = \epsilon^a \, \epsilon^{ib} = \epsilon^a \, (\cos b + i \sin b)$$

$$\epsilon^{a-ib} = \epsilon^a \, \epsilon^{-ib} = \epsilon^a \, (\cos b - i \sin b)$$

$$\frac{\epsilon^{a+ib} + \epsilon^{a-ib}}{2} = \epsilon^a \cos b$$

Real part of ϵ^{a+ib} or ϵ^{a-ib} is $\epsilon^a \cos b$

Real part of → $\epsilon^{(a+i\omega)t}$	=	ϵ^{at}	$\cos \omega t$
		Exponential Coefficient	Sine-wave Curve

real, part imaginary (see figure at bottom of page 402). These numbers should not be confused with imaginary functions of ϵ, such as ϵ^{ix}, which are complex quantities or complex functions of x, but are not complex exponential functions of x.

Notice that the conjugate pairs of complex exponential functions can be combined to produce a real function. The writing can be simplified by assuming the conjugate to be used to obtain the real function, but instead using only one of the pair with the words, "the real part of."

Whichever method of expression you use, the complex exponential function . (either a conjugate pair of them or the real part of one) is a convenient way to represent a sine-wave curve with an exponential coefficient.

Complex p plane

The complex p plane is the significance of the real part of the complex exponential function or of the conjugate pair. The envelope lines at top and bottom follow equations: ϵ^{at} and $-\epsilon^{at}$. Between these amplitude boundaries, as amplitude markers, is the expanding cosine wave of stated angular frequency. Angular frequency, using the symbol lowercase Greek omega ω, is 2π times periodic frequency.

At the right is the concept of a complex p plane. With axes $-a$ and $+a$, horizontally, and positive and negative i times angular frequency vertically, the plane accommodates the complex quantity $p = a + i\omega$. The quantity pertinent to the problem under study is not p itself, but ϵ^p. Used this way, the plane represents exponents of ϵ, and is called the complex p plane.

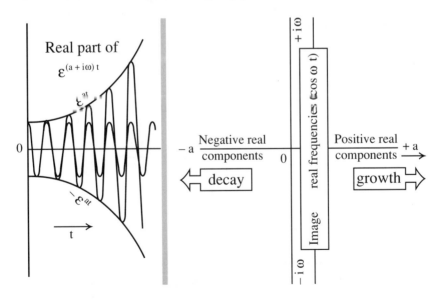

For ϵ^p where $p = a + i\omega$

Points on the horizontal axis represent real exponential functions: to the left for decay, to the right for growth or expansion. Points on the vertical axis are imaginary exponential functions and represent (in pairs, at equal distance from the origin, above and below) real frequency of unchanging amplitude. Conjugate pairs of points, located in upper and lower quadrants to the same side of the vertical axis, represent exponentially decaying (left side) or growing (right side) sinusoids, similar to the one at the left, which would be to the right.

Complex frequency plane

To engineers, the complex p plane is not very "real." On it, growth and decay are real elements, and frequency is represented by conjugate imaginary elements. If performance is analyzed in terms of frequency, it might be more convenient to regard frequency as a real quantity. To do this, rotate the plane through 90 degrees by writing $p = a + i\omega = iq$. Now, $q = -ia$. The plane so formed can be called a *q plane*, from this derivation, or because of its purpose, the complex frequency plane.

Real frequencies always appear in pairs (positive and negative) of equal value. Formulas for resonance always involve a solution in the form f^2 or ω^2, which always, mathematically, has equal positive and negative roots. In the q plane, simple exponential growth or decay is shown by single imaginary values on the vertical axis, above or below the horizontal axis. Growing or decaying sine waves are presented by conjugate pairs, to right and left of the vertical axis, in which, for this presentation, *conjugate* has a slightly different significance. The usual conjugate pair has a real pair of the same sign, with the imaginary pair of opposite sign. In this use, the imaginary pair have the same sign and the real pair have the opposite sign.

$$\text{If } p = a + i\omega = iq \qquad q = \omega - ia$$

Quantity q is in complex frequency plane

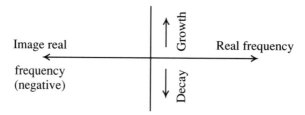

Hyperbolic functions

The trigonometric functions, such as cos x and sin x, represent an angle in radians, stated in circular measure. In this measure, 2π radians represents a complete rotation (360 degrees). This relationship is easy to visualize.

To complete the pattern, just as cos x is equal to a complementary pair of imaginary exponential functions (so is sin x), you need functions that consist of complementary pairs of real exponential functions. To correspond with normal circular measure, these functions are called hyperbolic cosine and hyperbolic sine. Following the trig ratio pattern, a hyperbolic tangent also exists. These use the symbols or abbreviations, *cosh*, *sinh*, and *tanh*, adding the *h* to represent hyperbolic. In reading, they are pronounced "cosh," "shine," and "tank," which are easier to pronounce.

These functions are hyperbolic because the ratio is measured along the curve of a hyperbola, instead of around a circle. Notice that however large a hyperbolic "angle" becomes, it remains in the same quadrant, unlike the circular measured angle, which rotates through four quadrants. NOTE: in this diagram, rotation is drawn the opposite direction from the usual convention for space reasons only.

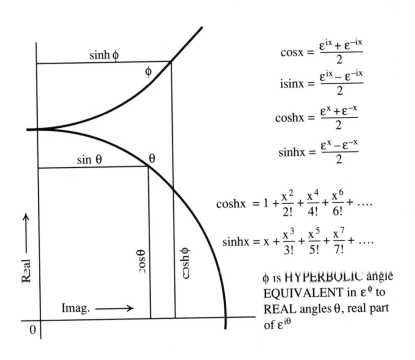

$$\cos x = \frac{\varepsilon^{ix} + \varepsilon^{-ix}}{2}$$

$$i\sin x = \frac{\varepsilon^{ix} - \varepsilon^{-ix}}{2}$$

$$\cosh x = \frac{\varepsilon^{x} + \varepsilon^{-x}}{2}$$

$$\sinh x = \frac{\varepsilon^{x} - \varepsilon^{-x}}{2}$$

$$\cosh x = 1 + \frac{x^2}{2!} + \frac{x^4}{4!} + \frac{x^6}{6!} + \ldots$$

$$\sinh x = x + \frac{x^3}{3!} + \frac{x^5}{5!} + \frac{x^7}{7!} + \ldots$$

ϕ is HYPERBOLIC angle EQUIVALENT in ε^{ϕ} to REAL angles θ, real part of $\varepsilon^{i\theta}$

Questions and problems

1. Using Maclaurin's series, find a series for tan x.

2. Evaluate the expression $(1 + 1/n)^n$ for values of n from $n = 1$ to $n = 5$. Compare the result of successively higher values of n with the expansion when n is infinity, for the same number of terms.

3. Evaluate $\epsilon^{0.5}$ and $\epsilon^{1.5}$ each to four places of decimals.

4. By approximate multiplication, check that the product of $\epsilon^{0.5}$ and $\epsilon^{1.5}$ is the same as the expansion of ϵ^2.

5. Convert the following exponential quantities to the equivalent trigonometrical and hyperbolic functions:

ϵ^{a+ib} $\qquad\qquad$ $\epsilon^{a+ib} + \epsilon^{a-ib}$ $\qquad$ $\epsilon^{a+ib} + \epsilon^{-a-ib}$

$\epsilon^{a+ib} - \epsilon^{a-ib}$ $\qquad$ $\epsilon^{a+ib} - \epsilon^{-a-ib}$ $\qquad$ $\epsilon^{a+ib} + \epsilon^{-a+ib}$

$\epsilon^{a+ib} - \epsilon^{-a+ib}$ $\qquad$ $\epsilon^{a-ib} + \epsilon^{-a-ib}$ $\qquad$ $\epsilon^{a-ib} - \epsilon^{-a-ib}$

$\epsilon^{-a+ib} + \epsilon^{-a-ib}$ $\qquad$ $\epsilon^{-a+ib} - \epsilon^{-a+ib}$

6. Convert the expressions derived in answer to question 5 back to their original form to verify your work.

7. By converting to trigonometrical functions, multiplying out and converting back to exponential form, verify that:

$$\epsilon^{ix} \times \epsilon^{-ix} = \epsilon^0.$$

8. Verify the sum and difference formulas for $\sin(A + B)$, $\sin(A - B)$, $\cos(A + B)$, and $\cos(A - B)$ by converting them to complex exponential functions.

9. By the same method as question 8, deduce formulas for $\sinh(A + B)$, $\sinh(A - B)$, $\cosh(A + B)$, and $\cosh(A - B)$.

10. From the formulas found in answer to question 9, find formulas for $\tanh(A + B)$, and $\tanh(A - B)$ to correspond with those usually given for $\tan(A + B)$ and $\tan(A - B)$.

11. Tabulate functions of $\cosh nA$ and $\sinh nA$ up to $n = 5$, in the form of powers of $\cosh A$ and $\sinh A$.

12. Tabulate functions of $\cosh A$ and $\sinh A$ in terms of multiple functions of $\cosh nA$ and $\sinh nA$.

13. Find derivatives (d/dx) of sinh x, cosh x, and tanh x.

14. Using derivatives of $f(x)$, find series for tanh x and arctanh x (using Maclaurin's series).

15. Using the appropriate series, evaluate the following:

$$\begin{array}{ccc} \cosh 0.1 & \sinh 0.1 & \cosh 0.2 \\ \sinh 0.2 & \cosh 0.5 & \sinh 0.5 \end{array}$$

16. Show that the following expressions are true:

$$\begin{array}{cc} \sinh ix = i \sin x & \cosh ix = \cos x \\ \sin ix = i \sinh x & \cos ix = \cosh x \end{array}$$

17. Evaluate, correct to 4 places of decimals, the complex quantities: $\epsilon^{i\,0.5}$; $\epsilon^{-0.5}$. By approximate multiplicationn, verify the results: the result should approximate unity.

18. Evaluate, to four places of decimals sinh 1 + sin 1 and cos 1 + cosh 1.

19. A curve has the properties shown here: When $x = 1$, $y = 1$, $dy/dx = 3$; when $x = 2$, $y = 3$, $dy/dx = 2$, (shown to scale here). Find (a) the constants for an equation of the type: $y = a + bx + cx^2 + dx^3$ to fit these values; (b) the intercept on the Y axis (when $x = 0$); and (c) the point and value of minimum slope.

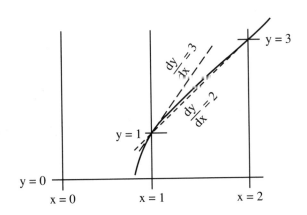

20. A decay curve, shown here, might have an equation of the form $e = E\epsilon^{-at}$ cos ωt. The wave is a decaying sinusoid of frequency given by: $f = \omega/2\pi$, which

has a period of $1/f$. On the drawing, is the time $1/f$ shown by distance A or distance B? A is the distance between points where the sinusoid touches the envelope decay curve (dashed line), and B is the distance between successive peaks in the same direction. This is an exercise in analysis: prove your conclusion by appropriate differentiation.

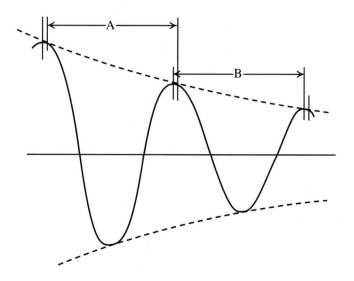

24
CHAPTER

The world of logarithms

Logarithmic series

Maclaurin's series cannot be used to find a series for log x, so another method must be found. The first step, yielding a basic logarithmic series, changes the variable, a step which is very useful as you proceed into more involved mathematics. Instead of using log x as the variable, use log $(1 + x)$, which produces finite values for successive derivatives when $x = 0$. So, you return to Maclaurin's series.

$$f(x) = \log_\varepsilon x \qquad f_1(x) = \frac{1}{x} \qquad f_1(0) = \infty$$

Use $f(x) = \log_\varepsilon(1 + x)$ $\qquad f_1(x) = \frac{1}{1 + x} \qquad f_1(0) = 1$

$$f_0(0) = \log_\varepsilon 1 = 0 \qquad f_2(x) = -\frac{1}{(1 + x)^2} \qquad f_2(0) = -1$$

$$\varepsilon^0 = 1 \qquad\qquad f_3(x) = \frac{2}{(1 + x)^3} \qquad f_3(0) = 2$$

$$f_4(x) = \frac{-6}{(1 + x)^4} \qquad f_4(0) = -6$$

$$\log_\varepsilon(1 + x) = x - \frac{x^2}{2!} + \frac{2x^3}{3!} - \frac{6x^4}{4!} \ldots\ldots$$

$$= x - \frac{x^2}{2} + \frac{x^3}{3} - \frac{x^4}{4} + \frac{x^5}{5} \ldots\ldots$$

409

When the coefficients are simplified by dividing by the factorial number series, they are a sort of harmonic series that doesn't converge very rapidly. Numerators are successive powers of x, and denominators are the simple number series, not the factorial.

You'll want logs of numbers bigger than 2. Here, the rate of convergence is shown in finding log 2 by this method. Two things slow its convergence: the only diminishing factor is essentially a harmonic series of the integral number reciprocals. It straddles the ultimate value, which means it has to converge much further to reach its ultimate value.

Find Log$_\varepsilon$ 2

$$\text{Log}_\varepsilon 2 = 1 - \frac{1}{2} + \frac{1}{3} - \frac{1}{4} + \frac{1}{5} - \frac{1}{6} + \frac{1}{7} - \frac{1}{8}$$

1	→0.7833
− 0.5	− 0.1667
0.5	0.6166
+ 0.3333	+ 0.1429
0.8333	0.7595
− 0.25	− 0.125
0.5833	0.6345
+ 0.2	+ 0.1111
0.7833	0.7456

There must be a Quicker Way

Logarithmic series: modified

Here is a trick that logarithms are made for. If you modify the variable again, using $(1 + x)/(1 - x)$, by the principles of logarithms, the log of this variable will be the log of $(1 + x)$ minus the log of $(1 - x)$.

First, the series of log $(1 - x)$ was a succession of powers of x divided by the harmonic succession of integral numbers, alternating in sign. The series for log $(1 - x)$ uses the same numerical terms, but all the signs are minus. Remember, you're going to subtract it from log $(1 + x)$, which turns all those minus signs positive in the final value.

This method does two things: it knocks out the even powers of x and combines them. The series is contained in a big parentheses, multiplied by 2.

To show how much more quickly this series converges, use it to calculate log 2, which by the first method would take forever. Make $(x + 1)/(x - 1) = 2$. That's another variable change. Solving that equation, the variable in the series is not 1, but 1/3. Since every other term has dropped out, successive terms diminish by x^2 (or 1/9). This ratio results in much quicker convergence. It converges so rapidly that only 4 terms are now needed to obtain log 2 to 4 places of decimals.

$$f(x) = \log_\varepsilon (1 - x) \qquad f_1(x) = \frac{-1}{1-x} \qquad\qquad f_1(0) = -1$$

$$f_2(x) = \frac{-1}{(1-x)^2} \qquad\qquad f_2(0) = -1$$

$$f_3(x) = \frac{-2}{(1-x)^3} \qquad\qquad f_3(0) = -2$$

$$f_4(x) = \frac{-6}{(1-x)^4} \qquad\qquad f_4(0) = -6$$

$$\log_\varepsilon(1 - x) = -x - \frac{x^2}{2!} - \frac{2x^3}{3!} - \frac{6x^4}{4!} - \ldots$$

$$= -x - \frac{x^2}{2} - \frac{x^3}{3} - \frac{x^4}{4} - \ldots$$

$$\log_\varepsilon \frac{1+x}{1-x} = \log_\varepsilon(1 + x) - \log_\varepsilon(1 - x)$$

$$= x - \frac{x^2}{2} + \frac{x^3}{3} - \frac{x^4}{4} + \frac{x^5}{5} - \ldots$$

$$+ x + \frac{x^2}{2} + \frac{x^3}{3} + \frac{x^4}{4} + \frac{x^5}{5} + \ldots$$

$$= 2\left[x + \frac{x^3}{3} + \frac{x^5}{5} + \ldots\right]$$

For $\log_\varepsilon 2$ $\qquad \dfrac{1+x}{1-x} = 2 \qquad\qquad$ $1 + x = 2(1 - x)$

$$3x = 1$$

$$x = \frac{1}{3}$$

$$\text{Log}_\varepsilon 2 \;=\; 2\left[\frac{1}{3} + \frac{1}{3}\left(\frac{1}{3}\right)^3 + \frac{1}{5}\left(\frac{1}{3}\right)^5 + \frac{1}{7}\left(\frac{1}{3}\right)^7 + \ldots\right]$$

.6666 | 667
.0246 | 914
.6913 | 581
.0016 | 461
.6930 | 042
.0001 | 306
.6931 | 348

Much Quicker Convergence

Calculating logarithms

Here you calculate two logarithms to find a comparison in convergence rate. To calculate log 1.1, make $x = 1/21$. Successive terms now converge by more than 400:1. Three terms of the series produce the log correct to six decimal places.

As you already saw, to calculate log 2, $x = 1/3$, so convergence is about one decimal place for each extra term. For accuracy to six places, seven terms are required.

Now try log 3; $x = 1/2$. The series converges much more slowly, but try another way. You've already "done" log 2. Log 3 = log 2 + log 1.5, because 3 = 2 × 1.5. So, find log 1.5 and add it to log 2. Log 1.5 uses $x = 1/5$ and it converges faster than log 2 did. Now you have a quicker reliable 6-figure value for log 3.

$$\text{Log}_\varepsilon 1.1: \quad \frac{1+x}{1-x} = 1.1 \qquad x = \frac{1}{21}$$

$$\text{Log}_\varepsilon 1.1 = 2 \left[\frac{1}{21} + \frac{1}{3}\left(\frac{1}{21}\right)^3 + \frac{1}{5}\left(\frac{1}{21}\right)^5 + \ldots \right]$$

$$
\begin{array}{l}
.095238 \,|\, 1 \\
.000072 \,|\, 0 \\
.000000 \,|\, 0 \\
\hline
.095310 \,|\, 1
\end{array}
$$

$$
\text{Log}_\varepsilon 2 = 2 \left[
\begin{array}{ll}
\dfrac{1}{3} & .666666 \,|\, 7 \\[2ex]
+\dfrac{1}{3}\left(\dfrac{1}{3}\right)^3 & .024691 \,|\, 4 \\[2ex]
+\dfrac{1}{5}\left(\dfrac{1}{3}\right)^5 & .001646 \,|\, 1 \\[2ex]
+\dfrac{1}{7}\left(\dfrac{1}{3}\right)^7 & .000130 \,|\, 6 \\[2ex]
+\dfrac{1}{9}\left(\dfrac{1}{3}\right)^9 & .000011 \,|\, 3 \\[2ex]
+\dfrac{1}{11}\left(\dfrac{1}{3}\right)^{11} & .000001 \,|\, 0 \\[2ex]
+\dfrac{1}{13}\left(\dfrac{1}{3}\right)^{13} & .000000 \,|\, 1 \\[1ex]
\hline
& .693147 \,|\, 2
\end{array}
\right.
$$

$Log_\varepsilon 3$: $\dfrac{1+x}{1-x} = 3$ $x = \dfrac{1}{2}$ $Log_\varepsilon 3 = 2\left[\dfrac{1}{2} + \dfrac{1}{3}\left(\dfrac{1}{2}\right)^3 + \dfrac{1}{5}\left(\dfrac{1}{2}\right)^5 + ...\right]$

$$2\left[\begin{array}{l} \dfrac{1}{2} \\[2mm] +\dfrac{1}{3}\left(\dfrac{1}{2}\right)^3 \\[2mm] +\dfrac{1}{5}\left(\dfrac{1}{2}\right)^5 \\[2mm] +\dfrac{1}{7}\left(\dfrac{1}{2}\right)^7 \\[2mm] +\dfrac{1}{9}\left(\dfrac{1}{2}\right)^9 \\[2mm] +\dfrac{1}{11}\left(\dfrac{1}{2}\right)^{11} \\[2mm] +\dfrac{1}{13}\left(\dfrac{1}{2}\right)^{13} \end{array}\right.$$

	=	
	= 1.000000	0
	= .083333	3
	= .012500	0
	= .002232	1
	= .000434	0
	= .000088	8
	= .000018	8

$\longrightarrow Log_\varepsilon 2$ .693147 | 1

Much slower convergence than $\log_\varepsilon 2$

$Log_\varepsilon 3 = Log_\varepsilon 2 + Log_\varepsilon 1.5$

$Log_\varepsilon 1.5$: $x = \dfrac{1}{5}$

$Log_\varepsilon 1.5 = 2\left[\dfrac{1}{5} + \dfrac{1}{3}\left(\dfrac{1}{5}\right)^3 + \dfrac{1}{5}\left(\dfrac{1}{5}\right)^5 + ...\right]$

$$+2\left[\begin{array}{l} \dfrac{1}{5} \\[2mm] +\dfrac{1}{3}\left(\dfrac{1}{5}\right)^3 \\[2mm] +\dfrac{1}{5}\left(\dfrac{1}{5}\right)^5 \\[2mm] +\dfrac{1}{7}\left(\dfrac{1}{5}\right)^7 \\[2mm] +\dfrac{1}{9}\left(\dfrac{1}{5}\right)^9 \end{array}\right.$$

.400000	0
.005333	3
.000128	0
.000003	7
.000000	1

$Log_\varepsilon 3$ =   1.098612 | 2

In the example at the top page 413, you tackle finding all the logs for integers up to 10. Notice the short cuts you can take. Log 4 is twice log 2. You can derive it either from $4 = 2 \times 2$ or from $4 = 2^2$. Log 5 is log 4 + log 1.25. Log 6 is log 2 + log 3. Log 7 is log 4 + log 1.75. Log 8 is 3 times log 2 because 8 is 2^3. Log 9 is twice log 3 because $9 = 3^2$. Finally, log 10 is log 2 + log 5.

$\text{Log}_\varepsilon 4 \;=\; 2\log_\varepsilon 2$		1.386294
$\text{Log}_\varepsilon 5 \;=\; \log_\varepsilon 4 + \log_\varepsilon 1.25 \quad x = \dfrac{1}{9}$		1.609438
$\text{Log}_\varepsilon 6 \;=\; \log_\varepsilon 2 + \log_\varepsilon 3$		1.791759
$\text{Log}_\varepsilon 7 \;=\; \log_\varepsilon 4 + \log_\varepsilon 1.75 \quad x = \dfrac{3}{11}$		1.945910
$\text{Log}_\varepsilon 8 \;=\; 3\log_\varepsilon 2$		2.079441
$\text{Log}_\varepsilon 9 \;=\; 2\log_\varepsilon 3$		2.197224
$\text{Log}_\varepsilon 10 \;=\; \log_\varepsilon 2 + \log_\varepsilon 5$		2.302585

Common logarithms

Although all logarithms must be calculated in their basic form to the base ε, sometimes called *hyperbolic* or *Naperian logarithms* (from the name of the discoverer of logarithms), they are generally called either *natural logarithms* or *log base ε*.

$$\text{Log}_{10} x = y \quad x = 10^y \qquad \text{Log}_\varepsilon 10 = t \qquad \varepsilon^t = 10$$

$$\text{So} \quad x = (\varepsilon^t)^y = \varepsilon^{ty} \qquad \text{Log}_\varepsilon x = ty$$

$$\boxed{\text{So} \quad \text{Log}_{10} x = \frac{\text{Log}_\varepsilon x}{\text{Log}_\varepsilon 10}}$$

$\text{Log}_{10} 2$:

```
                 .301030
       2.302585 |.6931471
                 .6907755
                  23716
                  23026
                    690
                    691
```

$\text{Log}_{10} 3$:

```
                  .477121
       2.302585 |1.0986122
                  .9210340
                  1775782
                  1611810
                   163972
                   161181
                     2791
                     2303
                      488
                      461
                       27
                       23
                        4
```

$$\text{Log}_{10} 10 = 1 \qquad \text{Log}_{10} 100 = 2 \quad \text{Etc.}$$

A hindrance to using natural logs for everyday use lies in the fact that 10 log = 2.302585093. If logs used base 10, then log 10 to base 10 is 1. You change the base by dividing the natural logarithm by log 10. The resulting figure is the same number's logarithm in base 10.

Using logarithms: multiplication and division

Most students of this edition will use logarithms from their pocket calculator— it's so much easier than using tables. A calculator that provides logs has both kinds, natural and common. The key for common logs is marked *log* and the one for natural logs is marked *ln*. Both are useful.

The examples on this page were prepared from 4-figure log tables. Your calculator probably lists more figures than the tables did. On my calculator, I enter log 32 and get 1.505149978; 256 produces 2.408239965. Adding them is 3.9133889944. Using the *shift*, which reverses the action, the answer is 8192 exactly!

The last example shows another difference with tables. The table only gave the mantissa—the decimal part. You had to insert the *characteristic*—the whole number to the left of the decimal point that tells where the point is in the number itself. 0.0969 is the mantissa (in 4-figure tables) for the digits 125. The bar over the 1 indicates that the characteristic is negative. So, the log is $-1 + 0.0969$. My calculator reads -0.903089987. However, if I enter 1.25 instead of 0.125, it reads 0.096910013. If the number is larger than 1, the mantissa doesn't change; only the characteristic changes as the decimal point shifts.

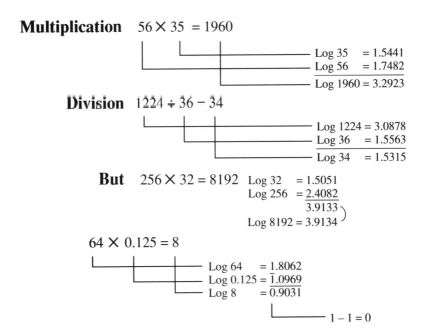

Multiplication $56 \times 35 = 1960$

Log 35   = 1.5441
Log 56   = 1.7482
Log 1960 = 3.2923

Division $1224 \div 36 - 34$

Log 1224 = 3.0878
Log 36   = 1.5563
Log 34   = 1.5315

But $256 \times 32 = 8192$

Log 32   = 1.5051
Log 256   = 2.4082
          3.9133
Log 8192 = 3.9134

$64 \times 0.125 = 8$

Log 64   = 1.8062
Log 0.125 = $\overline{1}$.0969
Log 8   = 0.9031

$1 - 1 = 0$

Using logarithms: indices

Here again, the examples on these two pages were prepared from 4-figure tables. A pocket calculator can find the answers far more accurately. In fact, most calculators have one key, x^y, which saves the use of the log key altogether. However, just look at this page and "work it over" with a calculator.

Log 12 reads 1.079181246. x 3 makes it 3.237643738. Using the *shift* and *log* produces 1728 exactly. Enter 12 again. Press x^y, then *3*, and =. The calculator reads 1728 again.

In the next example, log 2 is 0.301029995, the correct answer again. However, entering log 1024 lists 3.010299957 one more place.

The previous pages used logs or the x^y key where the indices were fairly obvious. Sometimes the answer isn't that simple. Here's one: $35^{4/5}$. Doing it by calculator logs: Log 35 = 1.544068044. Times 0.8 = 1.235254435. Shift log = 17.18915135. Using the x^y key produces the same answer.

You could also calculate it by binomial expansion and take it further, if your calculator is equipped with adequate memory and paren features. You needn't recalculate each term. After the second term, you can multiply/divide by the additional factors. For example, to get the third term from the second, multiply by 3 and divide by 320, and so on. This series converges very rapidly.

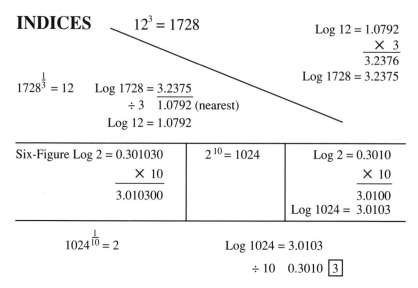

INDICES $12^3 = 1728$

Log 12 = 1.0792
   $\times$ 3
   3.2376
Log 1728 = 3.2375

$1728^{\frac{1}{3}} = 12$ Log 1728 = 3.2375
   $\div$ 3   1.0792 (nearest)
   Log 12 = 1.0792

Six-Figure Log 2 = 0.301030	$2^{10} = 1024$	Log 2 = 0.3010
$\times$ 10		$\times$ 10
3.010300		3.0100
		Log 1024 = 3.0103

$1024^{\frac{1}{10}} = 2$ Log 1024 = 3.0103
   $\div$ 10   0.3010 $\boxed{3}$

Roots are more accurate than powers

$0.0625^{\frac{1}{4}} = 0.5$ Log 0.0625 = $\overline{2}$.7959
   $\div$ 4 = $\overline{1}$.6990 ◀ Borrow $\overset{+}{2}$
Log 0.5

35 $^{4/5}$ By Logs

Log 35 = 1.5441

$\times$ 0.8 = 1.2353

Antilog 1.2353 = 17.19

By Binomial Expansion

$$(32+3)^{4/5} = 16 + \frac{4}{5}\cdot\frac{1}{2}\cdot 3 + \frac{\frac{4}{5}\cdot -\frac{1}{5}}{2}\cdot\frac{1}{64}\cdot 9 + \frac{\frac{4}{5}\cdot -\frac{1}{5}\cdot -\frac{6}{5}}{3\cdot 2}\cdot\frac{1}{2048}\cdot 27$$

```
    16  ←
  + 1.2  ←
    17.2
  - 0.01125  ←
    17.18875
  + 0.00042  ←
    17.18917
```

50 $^{1.5}$ By Logs

Log 50 = 1.6990

$\times$ 1.5 = 2.5485

Antilog 2.5485 = 353.6

By Binomial Expansion

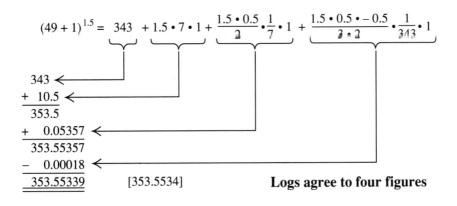

$$(49+1)^{1.5} = 343 + 1.5\cdot 7\cdot 1 + \frac{1.5\cdot 0.5}{2}\cdot\frac{1}{7}\cdot 1 + \frac{1.5\cdot 0.5\cdot -0.5}{3\cdot 2}\cdot\frac{1}{343}\cdot 1$$

```
  343  ←
+ 10.5  ←
  353.5
+   0.05357  ←
  353.55357
-   0.00018  ←
  353.55339       [353.5534]       Logs agree to four figures
```

In this one, 4-figure logs are rather limited. Using the same calculator, either with logs or the x^y key, the result is 353.5533906. The choice of binomial expression produced all except the last two digits with only 4 terms.

Of course, your calculator won't pick up a binomial series for you. That's just an exercise to show that the binomial works to check your calculating. How does the calculator do it? It has built-in programs that run the log series—so fast it will read an answer in a fraction of a second. Remember, it works in binary, even though it keeps track of decimal digits. It's fast, but you have the brains!

Using logarithms with a formula

The formula here relates pressure and volume in the physical expansion and compression of a true gas. It is typical of many application formulas. The quantities p and v are variable. k and index n are both constants. In this tabulation, $k = 1000$ and $n = 1.4$.

Tabulate values of v from 10 to 30 (assuming this covers the range needed in our specific problem) and use logarithms to calculate the corresponding value of p (in the last column). The 3rd column lists values of $0.4 \log v$ as an aid in finding $1.4 \log v$. Tabulating with this method made the work easier before the advent of calculators.

$$\boxed{pv^n = k} \quad n = 1.4 \qquad k = 1000 \qquad\qquad p = k/v^n$$

v	Log v	0.4Log v	1.4Log v	3–1.4Log v	Antilog
10	1.0000	0.4000	1.4000	1.6000	39.81
12	1.0792	0.4317	1.5109	1.4891	30.84
14	1.1461	0.4584	1.6045	1.3955	24.86
16	1.2041	0.4816	1.6857	1.3143	20.62
18	1.2553	0.5021	1.7574	1.2426	17.48
20	1.3010	0.5204	1.8214	1.1786	15.09
22	1.3424	0.5370	1.8794	1.1206	13.20
24	1.3802	0.5521	1.9323	1.0677	11.69
26	1.4150	0.5660	1.9810	1.0190	10.45
28	1.4472	0.5789	2.0261	0.9739	9.417
30	1.4771	0.5908	2.0679	0.9321	8.553

⇧ v   p ⇧

200	2.3010	0.9204	3.2214	$\bar{1}.7786$	0.6006

The 4th column subtracts from 3, which is log 1000. To do this on a calculator, you have a choice: use logs or the x^y key. Either way, you have a twist to get k into it. If k was other than a power of 10, it would complicate matters a little. One method is to use the $1/x$ (reciprocal) key, then multiply by 1000 (or whatever k is).

Finding the law by logarithms

Here, the process of the previous section is reversed. You know that a few pairs of values for v and p relate, according to a law of the type: $pv^n = k$. This shows how it was done by logs, again before the advent of calculators. You can use your calculator here, but the option of using the x^y key is not so easy; using the log key is easier.

Take the logs of the p values: 1.361727836 and 1.176091259. Subtract, getting 0.185636579. Take the logs of the v values: 1.176091259 and 1.301029996. Again subtract: 0.124938736. Divide the first subtraction by the second: $0.185636579/0.124938736 = 1.485820827$, the value for n. This calculation involves use of your calculator's memory or a scratch pad. Realize that all those digits are unnecessarily "accurate." The numbers you began with are probably only accurate to 2 significant figures.

P	23	15
V	15	20

Find n and k in $pv^n = k$

$\log p + n \log v = \log k$

(1) $1.3617 + 1.1761\,n = \log k$
(2) $1.1761 + 1.3010\,n = \log k$

Subtract $0.1856 - 0.1249\,n = 0$

$$n = \frac{0.1856}{1.1249} = \underline{1.486}$$

Log 0.1856 = $\bar{1}$.2686
Log 0.1249 = $\bar{1}$.0966
   0.1720
Antilog = 1.486

Log 1.3010 = 0.1142
Log n   = 0.1720
    0.2862
Antilog = 1.933

$1.1761 + \lfloor 1.3010\,n \rfloor$
= $1.1761 + 1.933 = 3.1091 = \log k$

Antilog: $k = \underline{1285}$

Questions and problems

1. Calculate the natural logarithms of the following numbers, correct to six decimal places: 1.1, 1.2, 1.25, 1.3, 1.4, 1.5, 1.6, 1.7, 1.8, 1.9, 2.

2. From the results of question 1, verify the following:

$$\log 1.2 + \log 1.25 = \log 1.5$$
$$\log 1.2 + \log 1.5 \;\; = \log 1.8$$
$$\log 1.6 + \log 1.25 = \log 2$$

3. Using the logs already calculated, and natural log 10, calculate logs for 10 times the same numbers.

4. Calculate the natural logs for numbers from 1 to 20, then convert them to common logs and verify your results.

5. Use logarithms to calculate the musical interval of $2^{1/12}$. As a check, evaluate $0.5^{1/12}$ and verify that the result is the reciprocal of $2^{1/12}$.

6. Evaluate the remaining musical intervals: $2^{1/12}$, $2^{1/6}$, $2^{1/4}$, $2^{1/3}$, $2^{5/12}$, $2^{1/2}$, $2^{7/12}$, $2^{2/3}$, $2^{3/4}$, $2^{5/6}$, and $2^{11/12}$ by logarithms.

7. In acoustics and electronics, the decibel scale (db) logarithmically represents amplitude or magnitude. 20 db represents a pressure ratio of 10:1, 40 db a ratio of 100:1. Use whatever source of logs you have to find the ratios for db units from 1 to 19.

8. Assuming you found yourself in a country where the numbering system was duodecimal, find the value of ϵ in that system, correct to four places of duodecimals.

9. Using the natural log of 12, convert logs of numerals from 1 to 12 (in the duodecimal system) to logs to the base 12. (Use t for ten and e for eleven.)

10. Using the graph at the top of page 421, find the law for the straight line drawn as an angle across this log-log paper, in the form: $y = Ax^b$.

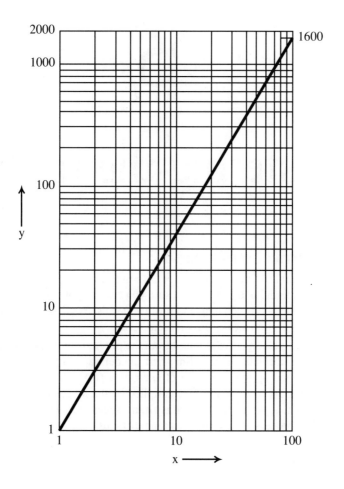

25

CHAPTER

Mastering the tricks

Trigonometrical series: tan x

Were you successful in finding a series for tan x in answer to question 1 in chapter 23? Perhaps you found one and weren't sure you had it right because the successive derivatives included such long and involved terms. Well, now you can check your result. This section works the first four terms.

Each successive derivative is found from the previous one by using the function of a function formula. In each line, first the derivative is found as f (tan x), then it is multiplied by the derivative of tan x as a $f(x)$, which is always $(1 + \tan^2 x)$. Thus, the last factor is always multiplied by the longer factor as a simple algebraic product, again differentiated as f(tan x) and the factor $(1 + \tan^2 x)$ is appended. This process is more tedious than difficult.

$f(x) = \tan x$

$f(x) = \tan x$ $f(0) = 0$

$f_1(x) = 1 + \tan^2 x$ $f_1(0) = 1$

$f_2(x) = 2\tan x\,(1 + \tan^2 x)$ $f_2(0) = 0$

$f_3(x) = 2(1 + 3\tan^2 x)\,(1 + \tan^2 x)$ $f_3(0) = 2$

$f_4(x) = 8(2\tan x + 3\tan^3 x)(1 + \tan^2 x)$ $f_4(0) = 0$

$f_5(x) = 8(2 + 15\tan^2 x + 15\tan^4 x)\,(1 + \tan^2 x)$ $f_5(0) = 16$

$f_6(x) = 16(17\tan x + 60\tan^3 x + 45\tan^5 x)(1 + \tan^2 x)$ $f_6(0) = 0$

$f_7(x) = 16(17 + 231\tan^2 x + 525\tan^4 x + 315\tan^6 x)(1 + \tan^2 x)$ $f_7(0) = 16 \cdot 17$

$$\tan x = x + \frac{x^3}{3} + \frac{2x^5}{15} + \frac{17x^7}{315} + \cdots$$

Series for sec *x*

The series for sec *x* method is similar to the working that you did for tan *x*. However, the zero and finite terms, when equating the derivative to 0, alternate with those for tan *x*. Because the first derivative of sec *x* can take the product form, sec *x* tan *x*, sec *x* is a factor of every derivative. Still, it is simpler to write the rest of the factors in terms of tan *x* and its powers.

The second derivative is found from the product formula, where *u* is sec *x* and *v* is tan *x*. Following derivatives must be treated first by the function of a function formula and then completed with the product formula.

f(x) = secx

$f(x) = \sec x$ $f(0) = 1$

$f_1(x) = \sec x \tan x$ $f_1(0) = 0$

$f_2(x) = \sec x \,(1 + 2\tan^2 x)$ $f_2(0) = 1$

$f_3(x) = \sec x \tan x \,(5 + 6\tan^2 x)$ $f_3(0) = 0$

$f_4(x) = \sec x \,(5 + 28\tan^2 x + 24\tan^4 x)$ $f_4(0) = 5$

$f_5(x) = \sec x \tan x \,(61 + 180\tan^2 x + 120 \tan^4 x)$ $f_5(0) = 0$

$f_6(x) = \sec x \,(61 + 662\tan^2 x + 1320\tan^4 x + 720\tan^6 x)$ $f_6(0) = 61$

$$\sec x = 1 + \frac{x^2}{2!} + \frac{5x^4}{4!} + \frac{61x^6}{6!} + \cdots$$

Series for arcsin *x*, arccos *x*

You already found a series for arctan *x* in chapter 23. The series for arcsin *x* and arccos *x* are found in the same way (as shown at the top of page 423), but they are not quite so simple as the one for arctan *x* because the derivative is not quite so simple. The denominator always contains $(1 - x^2)$ with an exponent that has an odd half, but the method is exactly the same. When the sine is 0, the angle is 0, and no constant term is in the arcsin series. As with other trig series, alternate terms have zero coefficient.

For the arccos series, the angle whose cosine is zero is $\pi/2$ (commonly called 90 degrees). So, the series has a constant term of precisely this value. After that, the only difference from the arcsin series is that the following terms all have negative signs. The arccos is simply $\pi/2$ minus the arcsin. You should have known this, but it's interesting to have the math prove it for us!

f (x) = arcsinx

If $y = \arcsin x$

$\quad x = \sin y$

$\dfrac{dx}{dy} = \cos y = \sqrt{1 - \sin^2 y} = \sqrt{1 - x^2}$ $\dfrac{dy}{dx} = \dfrac{1}{\sqrt{1 - x^2}}$

$f(x) = \arcsin x$ $f(0) = 0$

$f_1(x) = \dfrac{1}{\sqrt{1 - x^2}}$ $f_1(0) = 1$

$f_2(x) = \dfrac{x}{(1 - x^2)^{3/2}}$ $f_2(0) = 0$

$f_3(x) = \dfrac{1 + 2x^2}{(1 - x^2)^{5/2}}$ $f_3(0) = 1$

$f_4(x) = \dfrac{9x + 6x^3}{(1 - x^2)^{7/2}}$ $f_4(0) = 0$

$f_5(x) = \dfrac{9 + 72x^2 + 24x^4}{(1 - x^2)^{9/2}}$ $f_5(0) = 9$

$f_6(x) = \dfrac{225x + 600x^3 + 120x^5}{(1 - x^2)^{11/2}}$ $f_6(0) = 0$

$f_7(x) = \dfrac{225 + 4050x^2 + 5400x^4 + 720x^6}{(1 - x^2)^{13/2}}$ $f_7(0) = 225$

$\arcsin x = x + \dfrac{x}{3!} + \dfrac{9x^5}{5!} + \dfrac{225x^7}{7!} +$

$\qquad\quad = x + \dfrac{x^3}{6} + \dfrac{3x^5}{40} + \dfrac{5x^7}{112} +$

$\arccos x:\quad f(0) = \dfrac{\pi}{2}$

$\dfrac{dy}{dx} = \dfrac{1}{\sqrt{1 - x^2}}$

$\arccos x = \dfrac{\pi}{2} - x - \dfrac{x^3}{6} - \dfrac{3x^5}{40} - \dfrac{5x^7}{112} -$

Convergence of a series

The basic binomial series showed how to assess the convergence of a series (see "Calculating logarithms" in chapter 24). There you used a specific series to find a certain quantity, rather than using a general series to find a function. The convergence here concerns the range of values for the function, for which the series does converge, or for which it can be used.

Notice the ratios between successive coefficients in the ϵ^x series. They are the harmonic series of whole numbers, which eventually reach infinity (at infinity). However large x is made, provided it is still finite, the series will eventually converge.

In the basic log series [for $\log (1 + x)$], the ratio between successive coefficients is a fraction that has as its numerator, the denominator of the next earlier term and as its denominator, the numerator of the next term. As the series approaches an infinite number of terms, this ratio becomes 1 (i.e., the coefficients cease to converge). So, the only way that this series can converge indefinitely is for x to be less than 1.

In the modified log series, the coefficients again approach unit ratio. Although the condition for continued convergence is still that x must be less than unity, the quantity for which you are finding the log is: $(1 + x)/(1 - x)$, which reaches infinity only when x reaches unity. So, this series can find logs up to infinity.

$$\text{Log } \varepsilon^x = 1 + x + \frac{x^2}{2!} + \frac{x^3}{3!} + \frac{x^4}{4!} + \dotsc + \frac{x^\infty}{\infty!}$$

Coefficient ratios $\Big\}$ $1 \quad \frac{1}{2} \quad \frac{1}{3} \quad \frac{1}{4} \quad \frac{1}{5} \qquad \frac{1}{\infty}$

Converges provided $\boxed{x < \infty}$

$$\text{Log } \varepsilon(1 + x) = x - \frac{x^2}{2} + \frac{x^3}{3} - \frac{x^4}{4} + \dotsc + \frac{x^\infty}{\infty}$$

Coefficient ratios $\Big\}$ $\frac{1}{2} \quad \frac{2}{3} \quad \frac{3}{4} \quad \frac{4}{5} \qquad \frac{\infty - 1}{\infty} \ [=1]$

Converges provided $\boxed{x < 1 \text{ or } (1 + x) < 2}$

$$\text{Log}_\varepsilon \left(\frac{1 + x}{1 - x} \right) = 2 \left[x + \frac{x^3}{3} + \frac{x^5}{5} + \frac{x^7}{7} + \dotsc \frac{x^\infty}{\infty} \right]$$

Coefficient ratios $\Big\}$ $\frac{1}{3} \quad \frac{3}{5} \quad \frac{5}{7} \quad \frac{7}{9} \qquad \frac{\infty - 2}{\infty} \ [=1]$

Converges provided $\boxed{\begin{array}{c} x < 1 \text{ or} \\ \left(\dfrac{1 + x}{1 - x} \right) < \infty \end{array}}$

A useful conversion

Chapter 23 showed that the exponential function with an imaginary index is identical to a complex quantity that consists of a real cosine term and an imaginary sine term. This fact can be used more easily by writing the imaginary exponential function in simpler form, represented by y. Be careful to remember that y is never a simple quantity, but always a complex one of the form ε^{ix}. Similarly, $1/y$ is its conjugate complex quantity, ε^{-ix}.

After identifying cos x in the form: $1/2(y + 1/y)$, and sin x in the form: $(1/2i)(y - 1/y)$, notice that ϵ^{inx} is the same as y^n and ϵ^{-inx} is the same as $1/y^n$, and the real and imaginary parts are functions of nx, not of x^n. With this fact, you can use some useful conversions. By the same way of writing, cos nx is $1/2(y^n + 1/y^n)$, and sin nx is $(1/2i)(y^n - 1/y^n)$.

A Useful Conversion

$$\epsilon^{ix} = \cos x + i\sin x = y$$

$$\epsilon^{-ix} = \cos x - i\sin x = \frac{1}{y}$$

Where y is a simpler way of writing the complex ϵ^{ix}

Then, cos$x$ $= \frac{1}{2}\left(y + \frac{1}{y}\right)$

and $\sin x = \frac{1}{2i}\left(y - \frac{1}{y}\right)$

$$\epsilon^{inx} = \cos nx + i\sin nx = y^n$$

$$\epsilon^{-inx} = \cos nx - i\sin nx = \frac{1}{y^n}$$

Then, cos nx $= \frac{1}{2}\left(y^n + \frac{1}{y^n}\right)$

$\sin nx = \frac{1}{2i}\left(y^n - \frac{1}{y^n}\right)$

Power/multiple conversions

These $(y + 1/y)$ and $(y - 1/y)$ forms can easily make conversions from powers of trigonometric functions to their multiples, and vice versa. It becomes merely a matter of expanding those binomials and pairing off from the ends.

From this method, it becomes obvious why even powers of both sine and cosine functions result in cosine multiple functions. With odd powers, the sine functions produce multiple sine functions, while the cosine functions produce multiple cosine functions. As you will notice, the results are exactly the same as those obtained in Part 3 by a much more lengthy method.

The form: $(y + 1/y)$, etc., makes conversion of a compound power combination easier to follow through. Take $\sin^2 x \cos^4 x$. Each factor, $\sin^2 x$ or $\cos^4 x$, can be converted to multiple functions by either method.

The forms, $-1/4(y^2 - 2 + 1/y^2)$ and $1/16(y^4 + 4y^2 + 6 + 4/y^2 + 1/y^4)$, can be multiplied out in one extra line and then converted back to multiple form. The work becomes very short.

Even Powers

$$\cos^4 x = \frac{1}{16}\left(y + \frac{1}{y}\right)^4 = \frac{1}{16}\left(y^4 + 4y^2 + 6 + \frac{4}{y^2} + \frac{1}{y^4}\right)$$

$$= \frac{1}{8}\left[\cos 4x + 4\cos 2x + 3\right]$$

$$\sin^4 x = \frac{1}{16}\left(y - \frac{1}{y}\right)^4 = \frac{1}{16}\left(y^4 - 4y^2 + 6 - \frac{4}{y^2} + \frac{1}{y^4}\right)$$

$$= \frac{1}{8}\left[\cos 4x - 4\cos 2x + 3\right]$$

Odd Powers

$$\cos^5 x = \frac{1}{32}\left(y + \frac{1}{y}\right)^5 = \frac{1}{32}\left(y^5 + 5y^3 + 10y + \frac{10}{y} + \frac{5}{y^3} + \frac{1}{y^5}\right)$$

$$= \frac{1}{16}\left[\cos 5x + 5\cos 3x + 10\cos x\right]$$

$$\sin^5 x = \frac{1}{32i}\left(y - \frac{1}{y}\right)^5 = \frac{1}{32i}\left(y^5 - 5y^3 + 10y - \frac{10}{y} + \frac{5}{y^3} - \frac{1}{y^5}\right)$$

$$= \frac{1}{16}\left[\sin 5x - 5\sin 3x + 10\sin x\right]$$

$\sin^2 x \cos^4 x$

$$\sin^2 x = -\frac{1}{4}\left(y - \frac{1}{y}\right)^2$$

$$= -\frac{1}{4}\left(y^2 - 2 + \frac{1}{y^2}\right)$$

$$= \frac{1}{2}\left(1 - \cos 2x\right)$$

$$\cos^4 x = \frac{1}{16}\left(y + \frac{1}{y}\right)^4 = \frac{1}{16}\left(y^4 + 4y^2 + 6 + \frac{4}{y^2} + \frac{1}{y^4}\right)$$

$$= \frac{1}{8}\left(\cos 4x + 4\cos 2x + 3\right)$$

$$\sin^2 x\cos^4 x = -\frac{1}{64}\left(y^4 + 4y^2 + 6 + \frac{4}{y^2} + \frac{1}{y^4}\right)\left(y^2 - 2 + \frac{1}{y^2}\right)$$

$$= -\frac{1}{64}\left[y^6 + 2y^4 - y^2 - 4 - \frac{1}{y^2} + \frac{2}{y^4} + \frac{1}{y^6}\right]$$

$$= \frac{1}{32}\left[2 + \cos 2x - 2\cos 4x - \cos 6x\right]$$

Checking the result

You can verify the previous page result by multiplying out equivalent multiple functions for $\sin^2 x$ and $\cos^4 x$. Then, you must find substitutions for $\cos^2 2x$ and the product $\cos 2x \cos 4x$, before you have the answer completely in multiple-function form.

In integral calculus, conversion from power functions to multiple functions makes integration much easier. It splits the power product into simple multiple functions (sum or difference terms) that can be integrated individually.

$$\frac{1}{16}\left(1 - \cos2x\right)\left(3 + 4\cos2x + \cos4x\right)$$

$$= \frac{1}{16}\left[3 + \cos2x + \cos4x - 4\cos^2 2x - \cos2x\cos 4x\right]$$

$$4\cos^2 2x = 2(1 + \cos4x)$$

$$\cos2x\cos4x = \frac{1}{2}\left[\cos6x + \cos2x\right]$$

$$= \frac{1}{16}\left[\begin{array}{l} 3 + \cos2x + \cos4x \\ -2 \qquad\qquad -2\cos4x \\ \quad -\frac{1}{2}\cos2x \qquad -\frac{1}{2}\cos6x \end{array}\right]$$

$$= \frac{1}{32}\left[2 + \cos2x - 2\cos4x - \cos6x\right] \text{Same result} ✔$$

$$\int \sin^2 x\cos^4 x\,dx = \int \frac{1}{32}\left[2 + \cos2x - 2\cos4x - \cos6x\right]dx$$

$$= \frac{x}{16} + \frac{1}{64}\sin2x - \frac{1}{64}\sin4x - \frac{1}{192}\sin6x$$

Integration tools: partial fractions

Another useful tool in integration is partial fractions. An expression such as $(x + 1)/(x^2 - 4)$ would be difficult to integrate, unless it could be "broken down." In this case, the denominator can be a factor, so it is possible to express the complete fraction in partial fractions—each with a simpler numerator (an easier form to integrate).

The easiest method is to assume that $(x + 1)/(x^2 - 4)$ is equal to $a/(x + 2) + b/(x - 2)$, then solve for a and b.

Integrating, the solution takes the form: $1/4 \log (x + 2) + 3/4 \log (x - 2) + a$. That is not its simplest form. Putting the two factors together (adding logs is the same as multiplying the numbers they belong to) and making the constant of integration (c) a factor, the result at least looks much simpler.

$$\boxed{\int \frac{x + 1}{x^2 - 4} \, dx} \qquad \frac{x + 1}{x^2 - 4} \quad \text{must be reducible to} \quad \frac{a}{x + 2} + \frac{b}{x - 2}$$

$$\frac{a}{x + 2} + \frac{b}{x - 2} = \frac{a(x - 2) + b(x + 2)}{x^2 - 4} = \frac{(a + b) \, x + 2 \, (b - a)}{x^2 - 4} \quad \begin{cases} a + b = 1 \\ b - a = \dfrac{1}{2} \end{cases}$$

$$\frac{x + 1}{x^2 - 4} = \frac{1}{4(x + 2)} + \frac{3}{4(x - 2)}$$

$$2 \; b = \frac{3}{2} \quad b = \frac{3}{4}$$

$$2 \; a = \frac{1}{2} \quad a = \frac{1}{4}$$

$$\int \frac{1}{4(x + 2)} \, dx + \int \frac{3}{4(x - 2)} \, dx = \frac{1}{4} \log_\varepsilon (x + 2) + \frac{3}{4} \log_\varepsilon (x - 2) + \log_\varepsilon C$$

$$= \log_\varepsilon C \left[(x + 2) \, (x - 2)^3 \right]^{\frac{1}{4}}$$

More partial fractions

The top example on page 430 is solved just like the previous one—only it is slightly more involved. Each of these examples resulted in denominators with x having a coefficient of 1. What if the denominator takes the form, $ax + b$? The next example tackles this.

Working from the differential side: the derivative of $\log (ax + b)$ is $a/(ax + b)$; to get $1/(ax + b)$ for the derivative, you need to start with $1/a \log (ax + b)$. So, the integral of $1/(ax + b)$ is $1/a \log (ax + b)$, plus the constant of integration, of course, unless it is a definite integral.

Product formula in integration

In differentiation, you found the product formula useful. This formula can be converted for integration, although it's not quite so easy to do. To make the writing easier, use primes to show derivatives, instead of using d/dx every time. Notice that primes should be used only when the independent variable is not ambiguous. However, once you adopt them, you don't have to write the independent variable every time.

$$\int 12 \, \frac{(x-1)(x+2)}{(x-3)(x+3)(x+1)} \, dx \qquad \frac{12(x-1)(x+2)}{(x-3)(x+3)(x+1)} = \frac{12x^2 + 12x - 24}{(x-3)(x+3)(x+1)}$$

$$\frac{a}{x-3} + \frac{b}{x+3} + \frac{c}{x+1} = \frac{a(x^2 + 4x + 3) + b\,(x^2 - 2x - 3) + c\,(x^2 - 9)}{(x-3)(x+3)(x+1)}$$

Coefficients of x^2 : $a + b + c = 12$ ⎤ ⎡$a = 5$
"   " x : $4a - 2b = 12$ ⎥→ ⎢$b = 4$
Numeric: $3a - 3b - 9c = -24$ ⎦ ⎣$c = 3$

$$\int 12 \, \frac{(x-1)(x+2)}{(x-3)(x+3)(x+1)} = \frac{5}{x-3} + \frac{4}{x+3} + \frac{3}{x+1}$$

$$12 \, \frac{(x-1)(x+2)}{(x-3)(x+3)(x+1)} \, dx = 5\log_\varepsilon(x-3) + 4\log_\varepsilon(x+3) + 3\log_\varepsilon(x+1) + \log_\varepsilon C$$

$$= \log_\varepsilon C\,(x-3)^5(x+3)^4(x+1)^3$$

$$\int \frac{1}{ax+b} \, dx \qquad \frac{d}{dx} \log_\varepsilon(ax+b) = \frac{a}{ax+b}$$

$$\text{So,} \quad \frac{d}{dx} \frac{1}{a} \log_\varepsilon(ax+b) = \frac{1}{ax+b}$$

$$\int \frac{1}{ax+b} \, dx = \frac{1}{a} \log_\varepsilon(ax+b)$$

Product formula in integration

Differentiation: $(uv)' = uv' + u'v$

$uv' = (uv)' - u'v$

Integrate: $\int uv'dx = uv - \int u'vdx$

Example: $\int x\cos x \, dx = x\sin x - \int \sin x \, dx$

$= x\sin x + \cos x$

$u = x$
$v' = \cos x$
$v = \sin x$
$u' = 1$

Check ✔ $\frac{d}{dx}(x\sin x + \cos x) = \sin x + x\cos x - \sin x$

$= x\cos x$

First, rearrange the formula so that one factor term is on the left and the other is on the right with the product term. That form is standard for integrals. From there, it's easier to follow an example.

Suppose you must integrate $x \cos x$. Assume that u of the product formula is x, and v' is $\cos x$ on the left side. Now, find v and u' to complete the right side. uv is $x \sin x$ and the remaining expression integrates easily to $\cos x$.

Check by differentiating the answer. First, d/dx of $x \sin x$, using the product formula, is $\sin x + x \cos x$. Then, $d/dx \cos x$ is $-\sin x$. Putting it together, $+\sin x$ cancels $- \sin x$, leaving $x \cos x$, which was what you had to integrate. It checks.

More product formula

Here is an example where the u function isn't so simple. You must integrate ϵx $\sin x \, dx$. This example has one good point: making $u = \epsilon^x$, u' is also ϵ^x.

The formula has a complementary integral on the right, in addition to the product term: $-\epsilon^x \cos x$. Do the same thing and substitute into the first integral. Now the integral you started with is on both ends, except that it is negative on the right. Add it to both sides to double it. Divide both sides by 2 and you have the answer. Again, check with the differentiation product formula to prove that the answer was correct.

$$\int \epsilon^x \sin x \, dx = -\epsilon^x \cos x + \int \epsilon^x \cos x \, dx$$

$$= -\epsilon^x \cos x + \epsilon^x \sin x - \int \epsilon^x \sin x \, dx$$

$$\begin{array}{c} u = \epsilon^x \\ v' = \sin x \\ v = -\cos x \\ u' = \epsilon^x \end{array}$$

$$\int \epsilon^x \cos x \, dx = \epsilon^x \sin x - \int \epsilon^x \sin x \, dx$$

$$2 \int \epsilon^x \sin x \, dx = \epsilon^x \sin x - \epsilon^x \cos x$$

$$\int \epsilon^x \sin x \, dx = \frac{1}{2} \epsilon^x (\sin x - \cos x)$$

$$\begin{array}{c} u = \epsilon^x \\ v' = \cos x \\ v = \sin x \\ u' = \epsilon^x \end{array}$$

Check ✔ $\dfrac{d}{dx} \dfrac{1}{2} \epsilon^x (\sin x - \cos x) = \dfrac{1}{2} \epsilon^x (\sin x - \cos x) + \dfrac{1}{2} \epsilon^x (\cos x + \sin x)$

$$= \epsilon^x \sin x$$

Another one by product formula

Here is another variation of the general method. Take $x^3 \cos x$. Completely integrate the terms by applying the product formula three times, each time reducing the power of x by 1. Substitutions are made at each step and the additional part of that step adds to the result. Finally, terms are collected to simplify it.

Running the differential (in two lines that are added to represent the *udv* and *vdu* separately), the coefficient of sin *x* cancels, leaving $x^3 \cos x$, the number you started with, so the result was correct.

$$\int x^3 \cos x \, dx$$

① $\begin{array}{ll} u = x^3 & v' = \cos x \\ v = \sin x & u' = 3x^2 \end{array}$ ▷ $\int x^3 \cos x \, dx = x^3 \sin x - \int 3x^2 \sin x \, dx$

② $\begin{array}{ll} u = 3x^2 & v' = \sin x \\ v = -\cos x & u' = 6x \end{array}$ ▷ $= x^3 \sin x + 3x^2 \cos x - \int 6x \cos x \, dx$

③ $\begin{array}{ll} u = 6x & v' = \cos x \\ v = \sin x & u' = 6 \end{array}$ ▷ $= x^3 \sin x + 3x^2 \cos x - 6x \sin x + \int 6 \sin x \, dx$

$$= x^3 \sin x + 3x^2 \cos x - 6x \sin x - 6\cos x$$

$$= (x^3 - 6x) \sin x + (3x^2 - 6) \cos x$$

Check ✔ $\dfrac{d}{dx}\left[(x^3 - 6x) \sin x + (3x^2 - 6x) \cos x\right]$

$= \quad (3x^2 - 6) \sin x + (x^3 - 6x) \cos x$

$\dfrac{- (3x^2 - 6) \sin x + \quad 6x \, \cos x}{}$

$= \qquad\qquad\qquad\qquad x^3 \cos x$

Changing the variable

Often changing the variable helps integration. This problem could be tackled by the power/multiple conversion method, but changing the variable is more direct. It has an odd power of cos *x*. Cos *x* is the derivative of sin *x*, so you can write: cos *x dx* = *d* sin *x*. Make sin *x* the variable in place of *x*, which changes the integral (as shown) in the third line. The independent variable is now sin *x*, not *x*. So, $\cos^2 x$ is equal to $1 - \sin^2 x$. Multiplying out, integrate $\sin^2 x - \sin^4 x$, with sin *x* as the variable. This formula is the simplest of all, giving $1/3 \sin^3 x - 1/5 \sin^5 x$. Again, the result checks by differentiation.

Partial fractions will not "work" when the denominator has a surd. To solve it, you need a new function, so that the numerator, with the *dx* and the denominator, each converted to the new variable, simplifies the integral to a form that will integrate readily.

In this case, write $x = \sin \theta$. The denominator is then $\cos \theta$ and dx becomes $\cos \theta \, d\theta$. Substituting all these parts, the cosine terms in numerator and denominator cancel, leaving the very simple integral, $\sin \theta \, d\theta$. The result, $-\cos\theta$ merely must be expressed in terms of the first variable, x. A check shows that the answer is correct.

Making $x = \tan \theta$ works in much the same manner. Follow it through and you'll get the idea of what to look for in deciding on the variable that will help.

$$\int \cos^3 x \sin^2 x \, dx$$

$$\cos x = \frac{d\sin x}{dx} \qquad \cos x \, dx = d\sin x$$

$$
\begin{aligned}
\int \cos^3 x \sin^2 x \, dx \quad &= \quad \int \cos^2 x \sin^2 x \, d\sin x \\
&= \quad \int (1 - \sin^2 x) \sin^2 x \, d\sin x \\
&= \quad \frac{1}{3} \sin^3 x - \frac{1}{5} \sin^5 x
\end{aligned}
$$

Check ✔

$$
\begin{aligned}
\frac{d}{dx}\left[\frac{1}{3} \sin^3 x - \frac{1}{5} \sin^5 x \right] \quad &= \quad (\sin^2 x - \sin^4 x)\cos x \\
&= \quad \sin^2 x \, (1 - \sin^2 x)\cos x \\
&= \quad \sin^2 x \cos^3 x
\end{aligned}
$$

$$\boxed{\int \frac{x}{\sqrt{1 - x^2}} \, dx} \quad \text{If} \quad \boxed{x = \sin \theta}$$

$$\text{then } \sqrt{1 - x^2} = \cos \theta$$

$$\text{und} \qquad dx \quad \cos \theta \, d\theta$$

$$\int \frac{x}{\sqrt{1 - x^2}} \, dx \quad = \int \frac{\sin \theta \cos \theta \, d\theta}{\cos \theta} = \int \sin \theta \, d\theta$$

$$= -\cos \theta$$

$$= -\sqrt{1 - x^2}$$

Check ✔ $\qquad \dfrac{d}{dx} -\sqrt{1 - x^2} = \dfrac{-1}{2\sqrt{1 - x^2}} \cdot -2x$

$$= \frac{x}{\sqrt{1 - x^2}}$$

continued

$$\int \frac{x}{\sqrt{1+x^2}}\, dx \quad \text{If} \quad \boxed{x = \tan\theta}$$

then $\sqrt{1+x^2} = \sec\theta$

and $\qquad dx = \sec^2\theta\, d\theta$

$$\int \frac{x}{\sqrt{1+x^2}}\, dx \;=\; \frac{\tan\theta \cdot \sec^2\theta\, d\theta}{\sec\theta} = \int \tan\theta \sec\theta\, d\theta$$

$$= \int \frac{\sin\theta}{\cos^2\theta}\, d\theta$$

$$\boxed{\sin\theta\, d\theta \;=\; -d\cos\theta} \longrightarrow \;=\int -\frac{d\cos\theta}{\cos^2\theta}$$

$$= \frac{1}{\cos\theta} \;=\; \sec\theta$$

$$= \sqrt{1+x^2}$$

Check ✔ $\quad \dfrac{d}{dx}\sqrt{1+x^2} = \dfrac{1}{2}\cdot \dfrac{1}{\sqrt{1+x^2}} \cdot 2x = \dfrac{x}{\sqrt{1+x^2}}$

Slope on logarithmic scales

When you consider the slope of a curve, the quantities are usually plotted on linear scales—the actual variables. Sometimes the more useful scales are logarithmic. For example, the auditory frequency range is logarithmic, by octaves. Most stimuli, such as hearing, also follow an approximately logarithmic law. Applications of this nature need to be shown on logarithmic scales. Though the quantities marked on the scales might be x and y, the scale used to display them will be logarithmic.

To find the slope of a curve at a point where the function is $y = f(x)$, but x and y are both plotted on log scales, you need to evaluate $d \log y / d \log x$, instead of the more usual dy/dx. Treating this as a function of a function at both ends, the original derivative, dy/dx, must be multiplied by x/y. Notice that nothing is changed if one function is regarded as being x^2 instead of x, provided that x^2 is substituted for x at all points.

For example, $y = 1 - ax^2 + x^4$ represents a curve shown on log scales of x and y. In electronics, x might stand for frequency and y for db. Refer the slope to $\log x^2$ as the variable. Suppose you need the points of maximum slope. Maximum slope is now found by the equating second derivative to zero. The numerator is a quadratic whose roots give the values of x^2, where maximum slope occurs.

The slope (on the log/log scale) is found by substituting these roots into the first derivative. Notice that when $x = 1$, the slope of this equation is 1 on this scale, whatever value a has.

Slope on log scales

$$y = f(x)$$

$$\frac{dy}{dx} = f_1(x) \quad \frac{d\log y}{d\log x} = \frac{d\log y}{dy} \cdot \frac{dy}{dx} \cdot \frac{dx}{d\log x}$$

$$\frac{d\log y}{dy} = \frac{1}{y} \quad \frac{d\log x}{dx} = \frac{1}{x} \quad \frac{dx}{d\log x} = x$$

$$\frac{d\log f(x)}{d\log x} = \frac{x \cdot f_1(x)}{f(x)}$$

$$\frac{d\log f(x^2)}{d\log x^2} = \frac{x^2 f_1(x^2)}{f(x^2)}$$

$$\boxed{y = 1 - ax^2 + x^4}$$

$$\frac{d\log y}{d\log x^2} = \frac{x^2(2^2 x - a)}{x^4 - ax^2 + 1}$$

Maximum slopes $\dfrac{d^2 \log y}{d\log x^2 dx^2} = \dfrac{(x^4 - ax^2 + 1)(4x^2 - a) - (2x^4 - ax^2)(2x^2 - a)}{(x^4 - ax^2 + 1)^2}$

Numerator

$$\begin{array}{l} 4x^6 - 5ax^4 + (4 + a^2)x^2 - a \\ -4x^6 + 4ax^4 \qquad\quad - a^2 x^2 \\ \hline \qquad\quad - ax^4 + 4\;x^2 \qquad - a \end{array} \;\boxed{= 0} \;\Longrightarrow\; x^2 = \frac{2}{a} \pm \sqrt{\frac{4}{a^2} - 1}$$

When $\;x^2 = \dfrac{2}{a} \pm \sqrt{\dfrac{4}{a^2} - 1}\;$ $\;\dfrac{d\log y}{d\log x^2} = \dfrac{x^2(2x^2 - a)}{x^4 - ax^2 + 1} = \dfrac{2x^4 - ax^2}{x^4 - ax^2 + 1}$

$$\boxed{\begin{array}{l} x^4 = \dfrac{8}{a^2} - 1 \pm \dfrac{4}{a^2}\sqrt{4 - a^2} \\[2mm] ax^2 = 2 \pm \sqrt{4 - a^2} \end{array}}$$

$$= \dfrac{\dfrac{16}{a^2} - 4 \pm \left(\dfrac{8}{a^2} - 1\right)\sqrt{4 - a^2}}{\dfrac{8}{a^2} - 2 \pm \left(\dfrac{4}{a^2} - 1\right)\sqrt{4 - a^2}}$$

$$= \dfrac{4\sqrt{4 - a^2} \pm (8 - a^2)}{2\sqrt{4 - a^2} \pm (4 - a^2)} = \dfrac{2\sqrt{4 - a^2} \pm (4 - a^2) + 2\sqrt{4 - a^2} \pm 4}{2\sqrt{4 - a^2} \pm (4 - a^2)}$$

$$\boxed{\begin{array}{c} \text{When } x = 1 \\[1mm] \dfrac{d\log y}{d\log x^2} = \dfrac{2 - a}{2 - a} = 1 \end{array}}$$

$$= 1 + \dfrac{2\sqrt{4 - a^2} \pm 4}{2\sqrt{4 - a^2} \pm (4 - a^2)} = 1 + \dfrac{\pm 2(2 \pm \sqrt{4 - a^2})}{\sqrt{4 - a^2}(2 \pm \sqrt{4 - a^2})}$$

$$= 1 \pm \dfrac{2}{\sqrt{4 - a^2}}$$

A numerical example of slope on log scales

The general method on the previous page will be clearer with a numerical example. Here, $a = 1.8$. The little diagram at the bottom shows what the working means.

If desired, the same values can be used to calculate both the curve itself and its slope at various points, giving a very accurate picture of the curve.

$$y = 1 - 1.8x^2 + x^4$$

$$\frac{d \log y}{d \log x^2} = \frac{x^2(2x^2 - 1.8)}{x^4 - 1.8x^2 + 1}$$

$$= \frac{2x^4 - 1.8x^2}{x^4 - 1.8x^2 + 1}$$

Zero slope: $X^2 = 0$ or $X^2 = 0.9$

Maximum slopes: $\dfrac{d^2 \log y}{d \log x^2 dx^2} = \dfrac{(x^4 - 1.8x^2 + 1)(4x^2 - 1.8) - (2x^4 - 1.8x^2)(2x^2 - 1.8)}{(x^4 - 1.8x^2 + 1)^2}$

Numerator: $4x^6 - 9x^4 + 7.24x^2 - 1.8$

$\underline{\quad\quad - 4x^6 + 7.2x^4 - 3.24x^2 \quad\quad}$

$- 1.8x^4 + 4x^2 - 1.8 = 0$

$$x^2 = \frac{2}{1.8} \pm \sqrt{\frac{4}{3.24} - 1}$$

$$= \frac{2 \pm \sqrt{0.76}}{1.8}$$

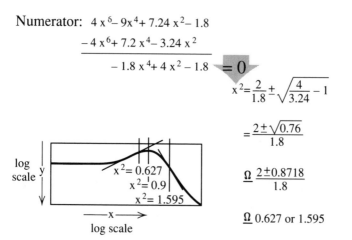

$$\underline{\Omega}\ \frac{2 \pm 0.8718}{1.8}$$

$$\underline{\Omega}\ 0.627 \text{ or } 1.595$$

Labels in diagram:
log scale y
$x^2 = 0.627$
$x^2 = 0.9$
$x^2 = 1.595$
$\longrightarrow x \longrightarrow$
log scale

Making the curve fit parameters

Here, the situation is reversed. Instead of finding the maximum slope for a given equation, find an equation to provide specified maximum slopes. Just work the same equations found in the section, "Slope on logarithmic scales." You can also plot much more detail by having all this extra data.

$$1 \pm \frac{2}{\sqrt{4-a^2}} = 1 \pm 2 \qquad \sqrt{4-a^2} = 1 \qquad a^2 = 3$$

$$a = \sqrt{3}$$

$$y = 1 - \sqrt{3}x^2 + x^4$$

Values of x for maximum slope: $x^2 = \dfrac{2}{\sqrt{3}} \pm \sqrt{\dfrac{4}{3} - 1} = \dfrac{2 \pm 1}{\sqrt{3}}$

$$= \sqrt{3} \text{ or } \frac{1}{\sqrt{3}}$$

Value of x for slope zero: $x^2 = \dfrac{\sqrt{3}}{2}$

Values of curve at these points:

x^2	Approx. x	y	Slope $\left[\dfrac{\log}{\log}\right]$	Approx. y
$\dfrac{1}{\sqrt{3}}$	0.7598	$\dfrac{1}{3}$	-1	0.3333
$\dfrac{\sqrt{3}}{2}$	0.9306	$\dfrac{1}{4}$	0	0.2500
1	1.0000	$2 - \sqrt{3}$	$+1$	0.2679
$\sqrt{3}$	1.3161	1	$+3$	1.0000

Drawing hints

The curve shown in the graph at the top of the next page is drawn from the data in the previous section, calculating just two extra points for $x = 0.5$ and $x = 2$. Notice that, in finding points on a log scale, you can use more convenient measures than the printed scales. For instance, if 5ʺ is a decade of x, 2.5ʺ is a decade of x^2 and 1.25ʺ is a decade of x^4. So, to find a value of root 3 or reciprocal root 3, measure from the center (a reference value of 1) a distance of log 3 on a 2.5ʺ decade, instead of using the 5ʺ measure printed.

Slope is easily aligned by finding points on the x^2 scale in a 4:1 ratio (2:1 ratio on the x scale), then finding points on the y scale that are 4:1 ratio for unity slope (either way, up or down); 16:1 for a slope of 2; 64:1 ratio for a slope of 3 (remembering the scale is logarithmic, so the ratio goes in powers), and so on.

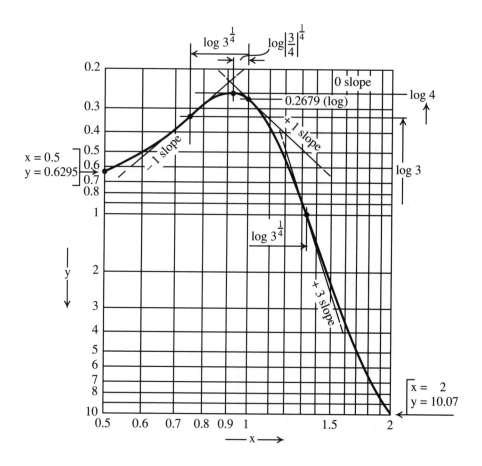

Questions and problems

1. Find the multiple function equivalents of the following products:

$$\cos^3 \times \sin^3 x \qquad \cos^4 \times \sin^3 x$$
$$\cos^5 \times \sin^3 x \qquad \cos^5 \times \sin^5 x$$

From the results, draw a conclusion why such a product, including an odd power of sin x, always has a multiple of sine terms.

2. From two of the results in question 1, also verify the identity: $2 \cos^n x \sin^n x = \sin^n 2x$.

3. Integrate the following expressions, with respect to x:

$$\frac{1 - x}{2x^2 + 5x + 2} \qquad \frac{7x - 1}{3x^2 + 7x + 2}$$

$$\frac{2x^4 + 3x^3 + 8x^2 + 3x + 5}{2x^5 + 2x^4 + 5x^3 + 5x^2 + 2x + 2}$$

4. Integrate the following expressions with respect to x:

$$\epsilon^{ax} \sin bx \qquad x^3 \sin ax$$

5. Integrate the following expressions, with respect to x:

$$\tan x \qquad \sec x \qquad \frac{x}{a^2 + x^2}$$

$$\frac{1}{(x^2 + 2)^{3/2}} \qquad (1 + \cos x)^3 \qquad \log_\epsilon x$$

6. A curve follows an equation of the form $y = 1 + ax^2 + bx^4 + x^6$, where a and b might each have either sign. Find the condition for the curve to have a maximum and minimum for finite values of x.

7. In question 6, find the relationship between constants a and b that will make the value of x^2 for the maximum value of y three times the value of x^2 for the minimum value of y.

8. A curve follows an equation: $y = 1 - ax^2 + x^4$. Find the value of a so that maximum slopes, as measured on log/log scales, are -2 and $+4$.

9. The equation $y = 1 - ax^2 + x^4$ refers the curve to a value of $x - 1$, where the slope of the curve on a log/log scale is unity for all values of a, positive and negative. Find an equation for the same curves, using a new reference x so that zero slope will occur at $x = 1$ and the constant a will have the same significance (for positive values only). a will have no zero-slope point for negative values.

10. What is the minimum value that a number plus its reciprocal can have? What is the maximum value of a sine or cosine function? How can $\cos x = 1/2(y + 1/y)$ for any but one value of x and y?

11. By reference to a complex quantity plane, show that the magnitude of both y and $1/y$ in the expression of question 10 is always unity, but that real and imaginary proportions in each of the conjugate quantities determines the magnitude of the sum, equal to $\cos x$.

12. Explain the significance of cos nx = $1/2(y^n + 1/y^n)$, with reference to a complex plane.

13. Use Maclaurin's series to expand a series for the expression: log cos x.

14. Find a series for: $e^{\sin x}$. Verify that the 1st 3 terms have the same coefficients as e^x, that the coefficient of x^3 is zero, and that the coefficient of x^4 is $-3!$

15. Find a series for arcsinh x and arctanh x.

16. Find the limiting value of x, beyond which each of the following series do not converge:

arcsin x	arccos x	arctan x	tan x
sinh x	cosh x	arcsinh x	tanh x
arccosh x	arctanh x		

17. Integrals worked throughout this chapter have used x as the independent variable. What difference would it make if the independent variable was (a) t, for time, or (b) f, for frequency?

26
CHAPTER

Development of calculator aids

The slide rule

For many years, the *slide rule* was the calculator most used by engineers and others. Its basic form could multiply and divide. It did so by adding or subtracting lengths that were proportional to logarithms of numbers on the scale with which it was marked. It had movable scales and a cursor to aid in reading where scales paralleled. Many more scales were often provided to read sines, cosines, tangents, logarithms, powers, roots, etc.

To multiply 2.3 by 3.7, the 1 on the movable scale was placed alongside 2.3 on the fixed scale. Then, the cursor was moved to 3.7 on the movable scale and the product was read by the cursor on the fixed scale as 8.51.

For division, reverse the process. To divide 8.51 by 3.7, set the cursor to 8.51 on the fixed scale. Bring the mark for 3.7 on the movable scale to the cursor. Then, read the quotient on the fixed scale opposite 1 on the movable scale.

Slide rules were calibrated with very fine markings, to allow you to calculate, perhaps to 3 significant figures. Modern digital calculators provide many times the accuracy.

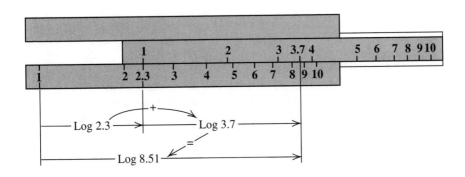

The simple nomogram

A *nomogram* is an alignment chart. Using parallel lines, it works like a slide rule, except that you find the answer by laying a straightedge, such as a ruler, across the chart. Using linear scales, the reference performs addition or subtraction. Using logarithmic scales, like the slide rule, it can perform multiplication or division.

By varying the spacing between the parallel scale lines, the nomogram has a flexibility that the slide rule does not have: different scales accommodate different ranges. Although the scales can be finely marked, reading them with the precision of the slide rule is more difficult. The accuracy of a nomogram is usually not as good as a slide rule.

Development of a Simple Nomogram

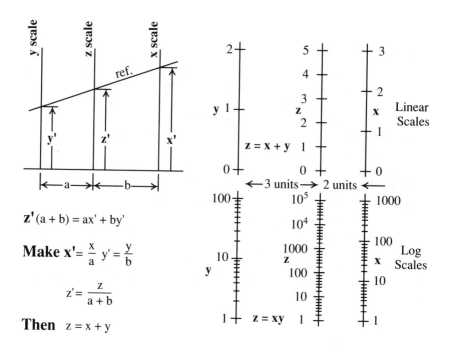

$$z'(a + b) = ax' + by'$$

Make $x' = \dfrac{x}{a}$ $y' = \dfrac{y}{b}$

$$z' = \frac{z}{a + b}$$

Then $z = x + y$

Multi-formula nomograms

At top left is a slightly different-proportioned set of scales for the same operation shown in the previous section, but the sum or product appears on one of the outside lines instead of the middle one (which might not be central). At lower left, the principle extends to a 4-line nomogram where scales can produce two results, z and u—one for $x + y$, the other for $x - y$. The logarithmic scales can represent xy and x/y.

At right is an example of a chart that does this. Notice the length (or range) of the various scales. The spacing of the lines must be calculated to provide correct geometry. Always one scale (in this case, the one for u) must have much less range than the others.

Development of a Two-Formula Nomogram

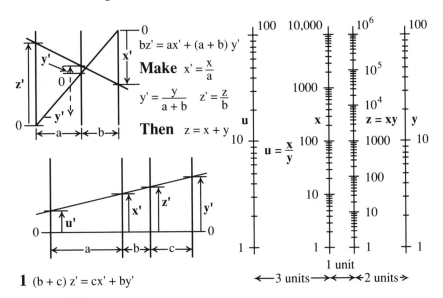

$bz' = ax' + (a + b) y'$

Make $x' = \dfrac{x}{a}$

$y' = \dfrac{y}{a + b}$ $z' = \dfrac{z}{b}$

Then $z = x + y$

$u = \dfrac{x}{y}$

1 $(b + c) z' = cx' + by'$

2 $(b + c) u' = (a + b + c) x' - ay'$

For $z = x + y$ $x' = \dfrac{x}{c}$ $y' = \dfrac{y}{b}$

For $u = x - y$ $x' = \dfrac{x}{a+b+c}$ $y' = \dfrac{y}{a}$

So: $\dfrac{a}{b} = \dfrac{a + b + c}{c}$ $\boxed{a = \left(\dfrac{c + b}{c - b}\right) b}$

The ratio nomogram

So far, nomograms use parallel lines. This new type uses an N configuration. The line joining the parallel lines has a scale that gives the ratio of the quantities on the other two scales. Notice that when the parallel scales have linear scales, the scale on the sloping scale that connects them is nonlinear. If this nomogram is used for other than linear scales (such as logarithmic), the connecting scale presents a changed scale to correspond, which represents the ratio between the functions for the scale's dimensioning.

Development of a Ratio Nomogram

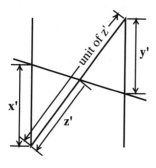
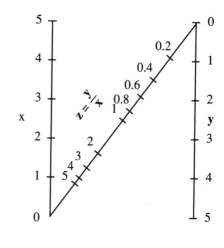

$$\frac{x'}{z'} = \frac{y'}{1 - z'} \quad z'(x' + y') = x'$$

$$z' = \frac{1}{1 + \dfrac{y'}{x'}} \quad \frac{y'}{x'} = \frac{1}{z'} - 1$$

$z = \dfrac{y}{x}$:	0.2	0.4	0.6	0.8	1.0	2.0	3.0	4.0	5.0
$z' = \dfrac{1}{1 + \dfrac{y}{x}}$:	$\dfrac{5}{6}$	$\dfrac{5}{7}$	$\dfrac{5}{8}$	$\dfrac{5}{9}$	$\dfrac{1}{2}$	$\dfrac{1}{3}$	$\dfrac{1}{4}$	$\dfrac{1}{5}$	$\dfrac{1}{6}$

The reciprocal nomogram

By placing the scale lines in a convergent manner, as shown at the top of the facing page, the nomogram construction is adapted for use in reciprocal relationships. Compare the convergent linear scales with the reciprocal scales on the parallel line construction. Of course, this arrangement can be used to make graphic calculations that are more complex than simple reciprocals.

The graphical chart

The graphical chart, shown at the bottom of the facing page, can have two advantages over the simple nomogram. For 4 variables, for which the nomogram form was shown, the graphical chart provides greater range for all variables. As a related secondary benefit, all its areas are equally sensitive. The nomogram tends to compress readability on center scales, and open it on outside scales. This disadvantage disappears with the graphical chart.

Another advantage of the graphical chart is that it is complete in itself. You do not need a straightedge to read it. Like nomograms and slide rules, scales can use any calibration law that is convenient for the purpose: linear, logarithmic, or something else.

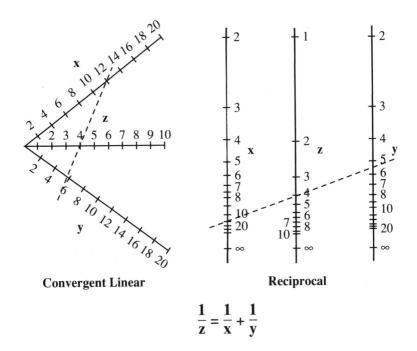

Convergent Linear   **Reciprocal**

$$\frac{1}{z} = \frac{1}{x} + \frac{1}{y}$$

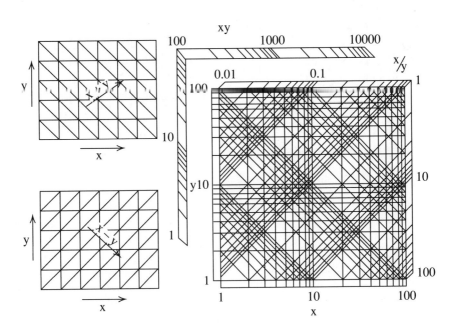

Change of scales in the graphical chart

Graphical charts can improve accuracy where nomograms and simpler graphical charts lose it. However, they can become difficult if not impossible to read; sometimes following one law all the way makes it necessary to cut the scale off too soon or make the chart size infinitely large without logical reason.

Compare the charts shown below. On the left, straight-line scales for horizontal rulings use linear spacing—the values of *y* follow a reciprocal law. Radial-line rulings look linear vertically and reciprocal horizontally. The focus for the radial lines is off-scale on the left.

At right, the same quantities use change of scale to make the whole chart more readable. The top and bottom of the chart use log/log rulings, with appropriate scales. The middle part approaches the advantage of the left arrangement by using curved rulings that are produced by sliding a ruling template sideways. The law is still logarithmic horizontally, but it switches to linear vertically, so it transits from one direction log to the reverse. It removes the focal point, which was just off scale on the left representation, to a theoretically infinite distance.

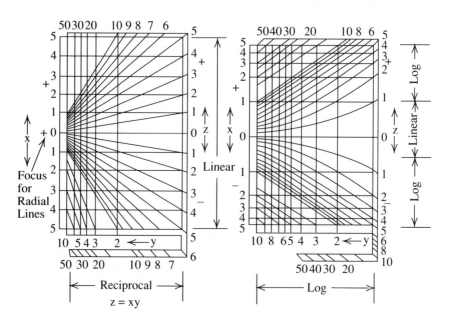

Resolving complex quantities graphically

Graphic ways to resolve complex quantities provide much variety. For positive real values, the simple chart at top left does it. Then reciprocal, going into the other two quadrants, looks like the top right, for magnitude and phase. The real and imaginary components look like the bottom right, making the complete reciprocal lower two quadrants look like the bottom left.

LINEAR GRAPHICAL REPRESENTATION

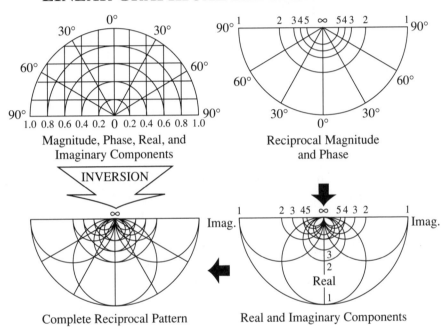

Magnitude, Phase, Real, and
Imaginary Components

Reciprocal Magnitude
and Phase

Complete Reciprocal Pattern

Real and Imaginary Components

Construction for the complex resolution graph

The previous section only showed what the chart looked like; it didn't say how to construct it. Rather obviously, it has a serious discontinuity between the positive real and the negative real half of the circle. Here is a better way before you can get into construction. Magnitudes between 0 and 1 will be in one half circle; between 0 and infinity, they are in the other half.

Instead of having both 0 and infinity at the center, as in the previous section, zero is at the bottom and infinity is at the top. The vertical diameter is the real axis in the previous section, the horizontal diameter the imaginary axis. The vertical diameter remains the real axis, but the imaginary axis the circumference. Then, the component resolution curves are drawn in as shown.

Modified Linear Graphical Representation

Radius of Main Circle is Unity

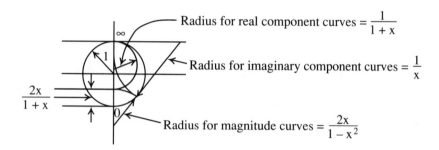

Radius for real component curves $= \dfrac{1}{1+x}$

Radius for imaginary component curves $= \dfrac{1}{x}$

Radius for magnitude curves $= \dfrac{2x}{1-x^2}$

x	$\dfrac{2x}{1+x}$	$\dfrac{2x}{1-x^2}$	$\dfrac{1}{x}$	$\dfrac{1}{1+x}$
0.2	0.333	0.417	5	0.833
0.4	0.571	0.952	2.5	0.714
0.6	0.75	1.875	1.667	0.625
0.8	0.889	4.444	1.25	0.556
1.0	1.000	∞	1.00	0.5
2	1.333	−1.333	0.5	0.333
3	1.5	−0.75	0.333	0.25
4	1.6	−0.533	0.25	0.2
5	1.667	−0.417	0.2	0.167

Modified linear representation

The modified linear representation shows the complete chart, for which construction was shown in the previous section. This form of chart does accommodate all complex quantities—from zero to infinity. The spacing begins to resemble logarithmic (although it is not mathematically), especially around unit magnitude, which is the horizontal diameter.

Other possibilities

Before proceeding to a final idea for a graphical resolution of complex quantities, look at some other possibilities. A slide rule with square-law scales (markings proportional to the square of the value) would do it. The same set of scales would produce magnitude with real and imaginary components, but not the angle.

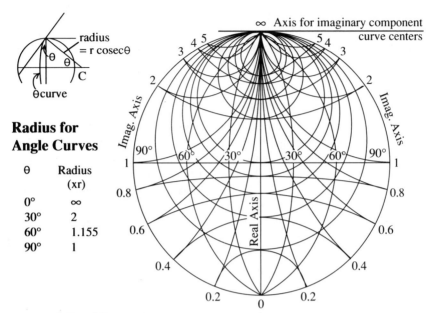

Radius for Angle Curves

θ	Radius (xr)
0°	∞
30°	2
60°	1.155
90°	1

Modified Linear Graphical Representation

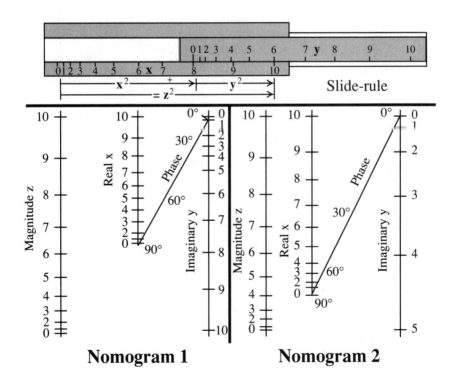

Slide-rule

$$x^2 + y^2 = z^2$$

Nomogram 1

Nomogram 2

The two nomograms, set out on different proportions, do essentially the same functions as the slide rule (but with problems in proportion, which is why more than 1 nomogram is desirable), except that they also provide an angle scale. Notice how the scales change the distribution of values.

Another concept in chart design

Take another look at "Modified linear representation." The scales are open, well-spaced, and almost logarithmic over a considerable area of the chart. They converge to make reading difficult as the values approach 0 or infinity. The remedy is to "open out" the top and bottom of the chart to make it really logarithmic, as shown. Phase is linear, everything else logarithmic. The curves are log sin, made by sliding the template vertically.

Its disadvantage is minor. In theory at least, the chart on page 380 covers values from zero to infinity, if reading near either extreme is impossible. This chart opens up those areas, making them readable, but the chart also never quite reaches zero or infinity. Take your pick!

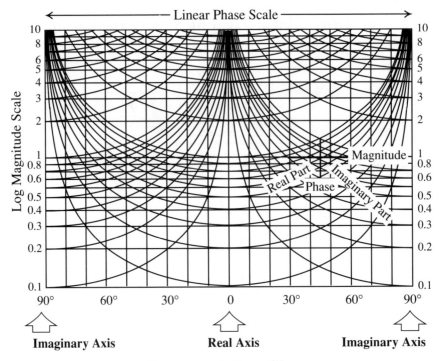

Curves are Log Sine

Duality between types of calculators

Before leaving this subject, observing a duality between the calculators is instructive. The nomogram lines represent variables that are identified by numbers of arrowheads on arrows that mark the direction of increasing value alongside the scales. On the graphical chart to the right, similar arrows show direction of movement for that variable to change.

On the nomogram, the alignment straightedge could pivot about a point on one scale. Such a point on the nomogram becomes a printed line on the chart, on a scale at right angles to the corresponding arrow. The line that represents the straight edge on the nomogram corresponds with a single point where lines intersect on the chart. Angular movements on the chart correspond with a column in space on the nomogram.

The heavily outlined part of the graphical chart corresponds to the area of calculations that the nomogram can make. Thus, the graphical chart covers twice the area in this sense.

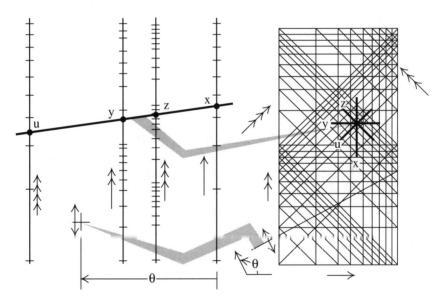

Duality Between Nomogram and Graphical Chart

Waveform synthesis

Any repetitive or periodic waveform can be synthesized from a series of sine waves. Here, over a sawtooth waveform, a series of sine waves are added in succession to show the idea. The first sine wave has the same frequency as the sawtooth. The sine wave bears no resemblance to the sawtooth, except that it crosses the zero line at the same points. Adding a third harmonic (3 times the fundamental) "splits the difference" between the sawtooth and the sine wave. Adding some fifth harmonic again splits the difference and the resulting wave begins to look like the sawtooth. Add enough odd harmonics and you finish up with a sawtooth. But how do you find how much of each harmonic to use?

With Fourier's series, you can write an *f(x)* to describe the wave you want to make. This sawtooth would have an *f(x)* for just one period—a straight line going from +*A* to −*A*, to which you can write a linear equation.

Both the sawtooth and the sine waves that compose the line are symmetrical above and below the zero line over a complete period. However, if you multiply the two together and integrate the product over the whole period, the result will only be zero if that harmonic is not present in the synthesis. At the bottom of the third, the dashed lines show the two waveforms—the sawtooth and third harmonic. The solid line is the two waves multiplied together.

In the first half of the sawtooth, the product crosses the zero line where the fifth harmonic does—the same way. In the second half, it reverses, resulting in a wave that sits more above than below the zero line in both halves, showing that fifth harmonic is part of the sawtooth.

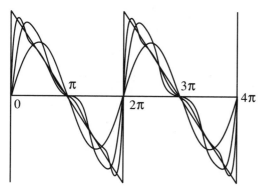

Waveform Synthesis

Any periodic waveform can be synthesized from fundamental and harmonics.

How to calculate their relative magnitudes?

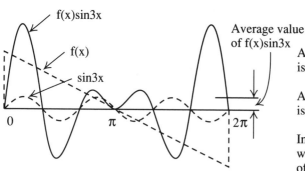

Average value of f(x)sin3x

Average value of f(x) is zero

Average value of sin3x is zero

In whole period of waveform average value of product is zero only if f(x) does NOT contain sin3x

Fourier series

A square wave can be regarded as a switch. It's a simple function that reverses a constant value every half period. Reversing the second harmonic at half the fundamental period results in a reversed wave that is still balanced above and below zero. When this is done to third, the first and second half periods have two half waves up and one down.

This function lets you write the general form for the Fourier series. Each successive harmonic frequency's waveform is multiplied, point by point, by the function for the waveform that you want to synthesize (or analyze). The product curve is integrated over the whole period and divided by π to find the average value.

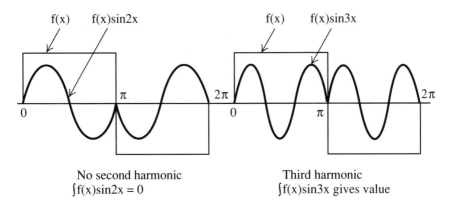

No second harmonic
$\int f(x)\sin 2x = 0$

Third harmonic
$\int f(x)\sin 3x$ gives value

General Form

$$f(x) = a_0 + a_1 \sin x + a_2 \sin 2x + \ldots\ldots\ldots + a_n \sin nx + \ldots\ldots\ldots$$
$$+ b_1 \cos x + b_2 \cos 2x + \ldots\ldots\ldots + b_n \cos nx + \ldots\ldots\ldots$$

a_0 is average value, if other than symmetrical above and below zero line

Fourier Series

$$a_n = \frac{1}{\pi}\int_0^{2\pi} f(x)\sin nx \, dx$$

$$b_n = \frac{1}{\pi}\int_0^{2\pi} f(x)\cos nx \, dx$$

$\left.\vphantom{\begin{array}{c}a\\b\end{array}}\right\}$ Twice average value of product curve

Analyzing the sawtooth

Coming back to the sawtooth, write its function. The second line is the general form for the Fourier series. If the waveform is not balanced above and below the zero line, this constant takes care of the problems. What remains is balanced. The general form now provides for a series of sine terms and a series of cosine terms, each from 1 to infinity. The coefficients use *a* and *b* each with a subscript to match the harmonic number (1 is fundamental). Each is multiplied by the function for the waveshape you are analyzing and integrated over the whole period. The series of coefficients for the sine terms is *2A/n* and the cosine terms yield zeros.

$$f(x)_0^{2\pi} = A(-x)$$

$$f(x) = a_0 + \sum_1^\infty a_n \sin nx + \sum_1^\infty b \cos nx$$

$$\int_0^{2\pi} a_0 \, dx = [a_0 x]_0^{2\pi} = 2\pi a_0; \quad \int_0^{2\pi} f(x) \, dx = \int_0^{2\pi} A(-x) \, dx = \left[A\left(x - \frac{x^2}{2}\right) \right]_0^{2\pi} = 0$$

$$a_0 = 0$$

$$a_n = \frac{1}{}\int_0^{2\pi} A(-x)\sin nx \, dx \qquad = \frac{1}{} \left[\frac{A}{n}(x-)\cos nx - \frac{A}{n}\cos nx \, dx \right]_0^{2\pi}$$

$\int uv' \, dx = uv - \int vu' \, dx$
$u = A(-x)$
$v' = \sin nx$
$v = -\frac{1}{n}\cos nx$
$u' = -A$

$$= \frac{1}{} \left[\frac{A}{n}(x-)\cos nx - \frac{A}{n^2}\sin nx \right]_0^{2\pi}$$

$$= \frac{1}{} \left[\frac{2A}{n} \qquad 0 \right] = \frac{2A}{n}$$

$$b_n = \frac{1}{}\int_0^{2\pi} A(-x)\cos nx \, dx \qquad = \frac{1}{} \left[\frac{A}{n}(-x)\sin nx + \frac{A}{n}\sin nx \, dx \right]_0^{2\pi}$$

$u = A(-x)$
$v' = \cos nx$
$v = -\frac{1}{n}\sin nx$
$u' = -A$

$$= \frac{1}{} \left[\frac{A}{n}(-x)\sin nx - \frac{A}{n^2}\cos nx \right]_0^{2\pi}$$

$$= 0$$

$$\text{Series} = 2A \left[\sin x + \frac{1}{2}\sin 2x + \frac{1}{3}\sin 3x + \frac{1}{4}\sin 4x + \ldots \right]$$

A triangular waveform

A triangular waveform is obviously a cosine series. But first, the function for a triangular wave is in two parts—two straight line sections. Substituting in Fourier, each must be integrated separately. A common factor out front simplifies to $-4A/\pi$. Inside the big brackets is a series of odd harmonic cosine terms, whose coefficients are $1/n^2$.

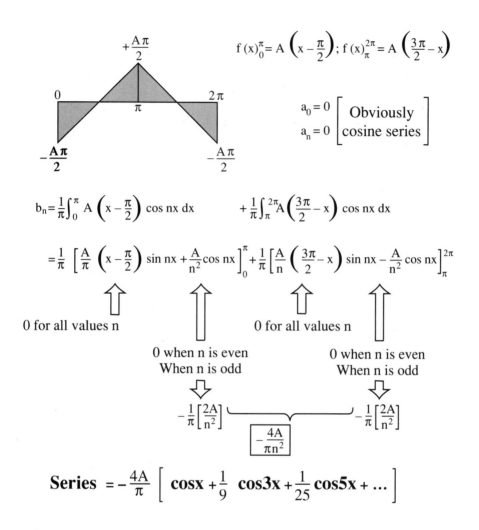

$$f(x)_0^\pi = A \left(x - \frac{\pi}{2} \right); \; f(x)_\pi^{2\pi} = A \left(\frac{3\pi}{2} - x \right)$$

$a_0 = 0$ $\left[\begin{array}{l} \text{Obviously} \\ \text{cosine series} \end{array} \right]$

$a_n = 0$

$$b_n = \frac{1}{\pi} \int_0^\pi A \left(x - \frac{\pi}{2} \right) \cos nx \, dx \qquad + \frac{1}{\pi} \int_\pi^{2\pi} A \left(\frac{3\pi}{2} - x \right) \cos nx \, dx$$

$$= \frac{1}{\pi} \left[\frac{A}{\pi} \left(x - \frac{\pi}{2} \right) \sin nx + \frac{A}{n^2} \cos nx \right]_0^\pi + \frac{1}{\pi} \left[\frac{A}{n} \left(\frac{3\pi}{2} - x \right) \sin nx - \frac{A}{n^2} \cos nx \right]_\pi^{2\pi}$$

0 for all values n   0 for all values n

0 when n is even   0 when n is even
When n is odd   When n is odd

$-\frac{1}{\pi} \left[\frac{2A}{n^2} \right]$ $-\frac{1}{\pi} \left[\frac{2A}{n^2} \right]$

$-\frac{4A}{\pi n^2}$

$$\textbf{Series} \; = -\frac{4A}{\pi} \left[\cos x + \frac{1}{9} \cos 3x + \frac{1}{25} \cos 5x + \ldots \right]$$

A square wave

Look at the square wave (on page 456), which is useful not only as an entity in itself, but as a step toward other waveforms as well. This is obviously a sine series. Like the triangular waveform, *f(x)* has two parts, one for each half wave. Integrating over the whole period, the amplitude is $4A/\pi n$. Putting the $4A/\pi$ outside the brackets, the terms inside are $1/n \sin nx$, with odd values of *n* only.

Relationship between square and triangular

Now that Fourier has given a series for both triangular and square waves, notice a confirming fact. A square wave can be first derivative of a triangular wave. A triangular wave has two constant slopes that coincide with the ups and downs of the square wave.

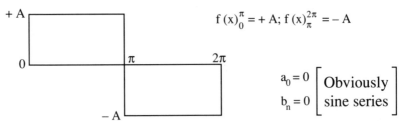

$$f(x)_0^\pi = + A; \, f(x)_\pi^{2\pi} = - A$$

$$a_0 = 0 \quad \begin{bmatrix} \text{Obviously} \\ \text{sine series} \end{bmatrix}$$
$$b_n = 0$$

$$a_n = \frac{1}{\pi}\int_0^\pi A\sin nx\,dx + \frac{1}{\pi}\int_0^{2\pi}(-A)\sin nx\,dx$$

$$= \frac{1}{\pi}\left[-\frac{A}{n}\cos nx\right]_0^\pi - \frac{1}{\pi}\left[-\frac{A}{n}\cos nx\right]_\pi^{2\pi}$$

$$= \frac{1}{\pi}\cdot\frac{2A}{n} + \frac{1}{\pi}\cdot\frac{2A}{n} \quad \text{When n is odd} \qquad 0 \quad \text{When n is even}$$

$$= \frac{4A}{\pi n} \quad \text{When n is odd}$$

$$\text{Series} = \frac{4A}{\pi}\left[\ \sin x + \frac{1}{3}\sin 3x + \frac{1}{5}\sin 5x + ...\right]$$

$$f(x) = -\frac{4A}{\pi}\left[\cos x + \frac{1}{9}\cos 3x + \frac{1}{25}\cos 5x + ...\right]$$

$$f_1(x) = \frac{4A}{\pi}\left[\sin x + \frac{1}{3}\sin 3x + \frac{1}{5}\sin 5x + ...\right]$$

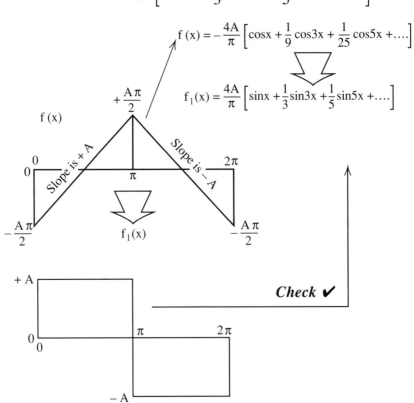

Check ✔

An offset square wave

Suppose the square wave "sits on" the zero line, instead of being symmetrically above and below it. This location produces a constant term. The series turns out just the same, except for the constant term, which offsets the wave.

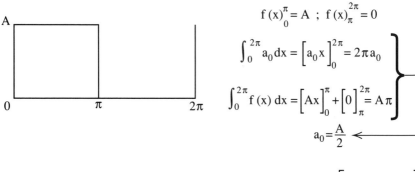

$$f(x)_0^\pi = A \;\; ; \;\; f(x)_\pi^{2\pi} = 0$$

$$\int_0^{2\pi} a_0 \, dx = \left[a_0 x \right]_0^{2\pi} = 2\pi a_0$$

$$\int_0^{2\pi} f(x) \, dx = \left[Ax \right]_0^\pi + \left[0 \right]_\pi^{2\pi} = A\pi$$

$$a_0 = \frac{A}{2}$$

$$a_n = \frac{1}{\pi} \int_0^\pi A \sin nx \, dx + \frac{1}{\pi} \int_0^{2\pi} 0 \sin nx \, dx$$

$$= \frac{1}{\pi} \left[-\frac{A}{n} \cos nx \right]_0^\pi = \frac{1}{\pi} \cdot \frac{2A}{n} \quad \text{When n is odd}$$

$$b_n = 0 \left[\begin{array}{c} \text{Obviously} \\ \text{sine series} \end{array} \right]$$

$$= 0 \quad \text{When n is even}$$

$$\textbf{Series} = \frac{A}{2} + \frac{2A}{\pi} \left[\sin x + \frac{1}{3} \sin 3x + \frac{1}{5} \sin 5x + \dots \right]$$

The square wave as a "switching" function

The square wave, either symmetrical or asymmetrical, can be used in a product function use of the Fourier series. In electrical supplies for electronic equipment, two kinds of supply rectifiers convert alternating current (which is sinusoidal in form) to direct current to power the electronic circuits in the equipment.

A full-wave rectifier turns both half waves of the alternating waveform around, so both "go" the same way. The half-wave rectifier merely accepts one half of the sine wave and blocks the other one.

In terms of Fourier, the full wave is like applying a balanced or symmetrical square wave to the sine wave. The half wave is like applying an asymmetrical square. In each case, the output waveform is the same as if the input sine wave was multiplied by its respective square wave.

You already have the Fourier series for the square wave, so you can use the product expression, term by term. Multiply each term of the square wave series by $\sin x$. Now, each term converts the product sine term (the first one is $\sin^2$) to a difference of cosines. Collect the terms.

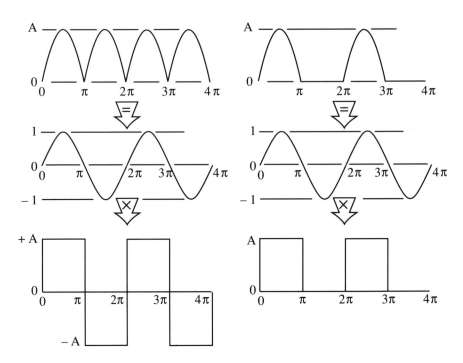

Product Formula Short-Cuts

$f(x) = \sin x$

$\times \dfrac{4A}{\pi}\left[\sin x + \dfrac{1}{3}\sin 3x + \dfrac{1}{5}\sin 5x + \ldots\right]$

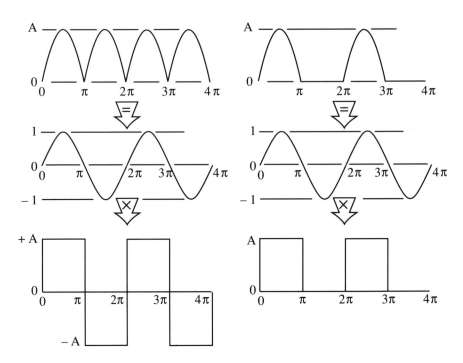

$= \dfrac{4A}{\pi}\left[\sin^2 x + \dfrac{1}{3}\sin x \sin 3x + \dfrac{1}{5}\sin x \sin 5x + \ldots\right]$

$= \dfrac{2A}{\pi}\left[1 - \cos 2x + \dfrac{1}{3}(\cos 2x - \cos 4x) + \dfrac{1}{5}(\cos 4x - \cos 6x) + \ldots\right]$

$= \dfrac{2A}{\pi}\left[1 - \dfrac{2}{3}\cos 2x - \left(\dfrac{1}{3} - \dfrac{1}{5}\right)\cos 4x - \left(\dfrac{1}{5} - \dfrac{1}{7}\right)\cos 6x + \ldots\right]$

$= \dfrac{2A}{\pi} - \dfrac{4A}{\pi}\left[\dfrac{1}{3}\cos 2x + \dfrac{1}{3 \cdot 5}\cos 4x + \dfrac{1}{5 \cdot 7}\cos 6x + \ldots\right]$

f (x) = sinx

$$\times \left\{ \frac{A}{2} + \frac{2A}{\pi} \left[\sin x + \frac{1}{3} \sin 3x + \ldots \right] \right.$$

$$= \frac{A}{2} \sin x + \frac{2A}{\pi} \left[\sin^2 x + \frac{1}{3} \sin x \sin 3x + \frac{1}{5} \sin x \sin 5x + \ldots \right]$$

$$= \frac{A}{\pi} + \frac{A}{2} \sin x - \frac{2A}{\pi} \left[\frac{1}{3} \cos 2x + \frac{1}{3 \cdot 5} \cos 4x + \frac{1}{5 \cdot 7} \cos 6x + \ldots \right]$$

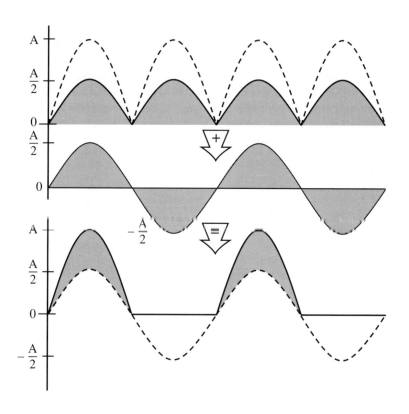

Each series produces a constant, the direct current from the rectifier, plus a new Fourier series, the alternating "ripple" that rides the *dc*.

Notice that the constant from the half wave is half that in the full wave. The half wave also has a sine term that the full wave does not. The amplitude of the cosine series for the half wave is half that of the full wave.

The illustration here verifies the conclusion that was deduced mathematically on the previous page. The dotted wave at the top is the first one. The solid line is the same waveform of half the amplitude. Add this line to an identical sine wave (peak from baseline) amplitude and you have (bottom solid curve) the second curve. It's quite simple to verify the conclusion derived from the calculation.

Series for quadratic curve

The rectified half waves look like part of a quadratic curve. Here, a quadratic curve is used as the basis for Fourier analysis, assuming that, instead of continuing with the quadratic curve, it uses only the part between − and + (repeating this part cyclically). This curve yields a cosine series, but the coefficients are different from the half-wave series.

Here is the synthesis of the quadratic simulation curve. At top left is the first cosine term. At top right are two successively closer approaches. Notice that the first four terms (the constant, fundamental, and the two harmonics) closely approach the quadratic curve, except for the "points," where it levels off to begin the next period. In rectified waveforms, all the terms have the same sign. The quadratic curve has terms with alternate signs. At bottom right, the coefficients are compared.

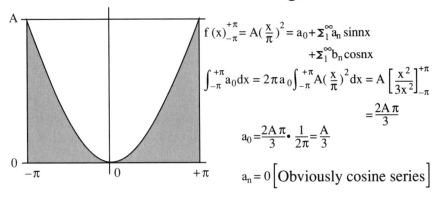

Series for Quadratic Curve

$$f(x)\Big|_{-\pi}^{+\pi} = A\left(\frac{x}{\pi}\right)^2 = a_0 + \sum_1^\infty a_n \sin nx$$
$$+ \sum_1^\infty b_n \cos nx$$

$$\int_{-\pi}^{+\pi} a_0 dx = 2\pi a_0 \int_{-\pi}^{+\pi} A\left(\frac{x}{\pi}\right)^2 dx = A\left[\frac{x^2}{3x^2}\right]_{-\pi}^{+\pi}$$
$$= \frac{2A\pi}{3}$$

$$a_0 = \frac{2A\pi}{3} \cdot \frac{1}{2\pi} = \frac{A}{3}$$

$$a_n = 0 \left[\text{Obviously cosine series}\right]$$

$$b_n = \frac{1}{\pi} \int_{-\pi}^{+\pi} A \left(\frac{x}{\pi}\right)^2 \cos nx \, dx = \frac{1}{\pi} \left[\frac{Ax^2}{n\pi^2} \sin nx - \frac{2Ax}{n^2\pi^2} \sin nx \, dx \right]_{-\pi}^{+\pi}$$

$$= \frac{1}{\pi} \left[\frac{Ax^2}{n\pi^2} \sin nx + \frac{2Ax}{n^2\pi^2} \cos n \, x - \frac{2A}{n^2\pi^2} \cos nx \, dx \right]_{-\pi}^{+\pi}$$

$$= \frac{1}{\pi} \left[\frac{Ax^2}{n\pi^2} \sin nx + \frac{2Ax}{n^2\pi^2} \cos n \, x - \frac{2A}{n^3\pi^2} \sin nx \right]_{-\pi}^{+\pi}$$

$$= \frac{1}{\pi} \left[0 + \frac{4A\pi}{n^2\pi^2}(-1)^n - 0 \right] = \frac{4A}{n^2\pi^2}(-1)^n$$

$$f(x) = \frac{A}{3} + \frac{A}{\pi^2} \left[-4\cos x + \cos 2x - \frac{4}{9}\cos 3x + \frac{1}{4}\cos 4x \ldots \right]$$

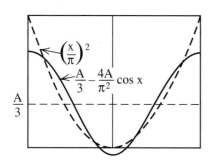

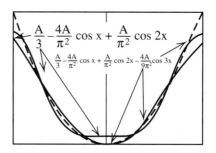

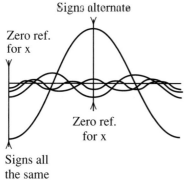

Signs alternate

Zero ref. for x

Zero ref. for x

Signs all the same

Coefficients

	Const.	Fund.	2nd	3rd	4th
1/2 Cosine Series	$\frac{2}{\pi}$	$\frac{4}{3\pi}$	$\frac{4}{15\pi}$	$\frac{4}{35\pi}$	$\frac{4}{63\pi}$
Ratio	$\frac{3}{\pi}$	$\frac{\pi}{3}$	$\frac{4\pi}{15}$	$\frac{9\pi}{35}$	$\frac{16\pi}{63}$ $\blacktriangleright$ $\frac{\pi}{4}$
Quadratic Series	$\frac{2}{3}$	$\frac{4}{\pi^2}$	$\frac{1}{\pi^2}$	$\frac{\cdot 4}{9\pi^2}$	$\frac{1}{4\pi^2}$

From opposite extreme

The finite approach to the infinite

Notice that by adding terms to Fourier, the wave keeps approaching the ultimate waveform, but it would only "get there" if the series continued out to infinity. For many purposes, a finite series can come closer.

Instead of synthesizing the square wave from frequencies, use the transfer characteristic approach from Part 3. By adding extra terms, the "ripples" (overshoot that occurs any time you cut off Fourier at finite frequencies), are avoided. Here, the first three terms are plotted out with the algebra for up to 6 terms.

Here, the coefficients are modified for better comparison and conversions are added to derive the sine series that will produce the linear power series that is derived from the transfer characteristics.

Here, you assemble substitutions (for decimal equivalents of total harmonics for this approach to a square wave) for successive finite numbers of terms—up to 6. Then, Fourier coefficients are compared which, of course, extend to infinity.

Notice that the last term is always much smaller when it is first introduced. On the last line, the 11th harmonic has a coefficient of only 0.0002403, where Fourier has 0.115749. In the same series, the 9th harmonic is more than 3 times its value when it was the last term.

Such series make the slope that theoretically should be vertical in the square wave, nearer to vertical, without producing the "ripples" which cut the Fourier series short of infinity.

Transfer characteristic approach

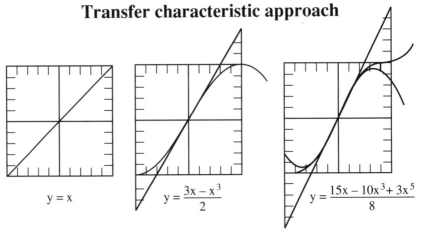

$$y = x$$

$$y = \frac{3x - x^3}{2}$$

$$y = \frac{15x - 10x^3 + 3x^5}{8}$$

Derivation: 2nd term: $y' = k(1 - x^2)$

$$y = k\left(x - \frac{1}{3}x^3\right) \qquad f(1) = \frac{2}{3}k \qquad y = \frac{3x - x^3}{2}$$

3rd term: $y' = k(1 - 2x^2 + x^4)$

$$y = k\left(x - \frac{2}{3}x^3 + \frac{1}{5}x^5\right) \qquad f(1) = \frac{8}{15}k \qquad y = \frac{15x - 10x^3 + 3x^5}{8}$$

4th term: $y' = k(1 - 3x^2 + 3x^4 - x^6)$

$$y = k\left(x - x^3 + \frac{3}{5}x^5 - \frac{1}{7}x^7\right) \quad f(1) = \frac{16}{35}k$$

$$y = \frac{35x - 35x^3 + 21x^5 - 7x^7}{16}$$

5th term: $y' = k(1 - 4x^2 + 6x^4 - 4x^6 + x^8)$

$$y = k\left(x - \frac{4}{3}x^3 + \frac{6}{5}x^5 - \frac{4}{7}x^7 + \frac{1}{9}x^9\right) \quad f(1) = \frac{128}{315}k$$

$$y = \frac{315x - 420x^3 + 378x^5 - 180x^7 + 35x^9}{128}$$

6th term: $y' = k(1 - 5x^2 + 10x^4 - 10x^6 + 5x^8 - x^{10})$

$$y = k\left(x - \frac{5}{3}x^3 + 2x^5 - \frac{10}{7}x^7 + \frac{5}{9}x^9 - \frac{1}{11}x^{11}\right) \quad f(1) = \frac{256}{693}k$$

$$y = \frac{693x - 1155x^3 + 1386x^5 - 990x^7 + 385x^9 - 63x^{11}}{256}$$

Approximations for comparison:

1 term: $y = x$

2 terms: $y = 1.5x - 0.5x^3$

3 terms: $y = 1.875x - 1.25x^3 + 0.375x^5$

4 terms: $y = 2.1875x - 2.1875x^3 + 1.3125x^5 - 0.4375x^7$

5 terms: $y = 2.4609x - 3.2812x^3 + 2.9531x^5 - 1.4062x^7 + 0.2737x^9$

6 terms: $y = 2.7070x - 4.5117x^3 + 5.4141x^5 - 3.8672x^7 + 1.5039x^9 - 0.2461x^{11}$

Conversions: $\sin^3\omega t = \dfrac{3\sin\omega t - \sin 3\omega t}{4}$ $\quad \sin^5\omega t = \dfrac{10\sin\omega t - 5\sin 3\omega t + \sin 5\omega t}{16}$

$$\sin^7\omega t = \frac{35\sin\omega t - 21\sin 3\omega t + 7\sin 5\omega t - \sin 7\omega t}{64}$$

$$\sin^9\omega t = \frac{126\sin\omega t - 84\sin 3\omega t + 36\sin 5\omega t - 9\sin 7\omega t - \sin 9\omega t}{256}$$

$$\sin^{11}\omega t = \frac{462\sin\omega t - 330\sin 3\omega t + 165\sin 5\omega t - 55\sin 7\omega t + 11\sin 9\omega t - \sin 11\omega t}{1024}$$

continued

Substitutions:	sinωt	sin3ωt	sin5ωt	sin7ωt	sin9ωt	sin11ωt	com. den.
2 terms, sinωt	12						8
sin³ωt	– 3	+1					"
total:	9	1					"
	1.125	0.125					decimal
3 terms, sinωt	240						128
sin³ωt	– 120	+40					"
sin⁵ωt	+ 30	– 15	+ 3				"
total:	150	25	3				"
	1.171875	0.195312	0.023437				decimal
4 terms, sinωt	2240						1024
sin³ωt	– 1680	+ 560					"
sin⁵ωt	+ 840	– 420	+ 84				"
sin⁷ωt	– 175	+ 105	– 35	+ 5			"
total:	1225	245	49	5			"
	1.196289	0.239258	0.047852	0.004883			decimal
5 terms, sinωt	80640						32768
sin³ωt	– 80640	+ 26880					"
sin⁵ωt	+ 60480	– 30240	+ 6048				"
sin⁷ωt	– 25200	+ 15120	– 5040	+ 720			"
sin⁹ωt	+ 4410	– 2940	+ 1260	– 315	+ 35		"
total:	39690	8820	2268	405	35		"
	1.211243	0.269165	0.069214	0.012360	0.001068		decimal
6 terms, sinωt	709632						262144
sin³ωt	– 887040	+ 295680					"
sin⁵ωt	+ 887040	– 443520	+ 88704				"
sin⁷ωt	– 554400	+ 332640	– 110880	+ 15840			"
sin⁹ωt	+ 194040	– 129360	+ 55440	– 13860	+ 1540		"
sin¹¹ωt	– 29106	+ 20790	– 10395	+ 3456	– 693	+ 63	"
total:	320166	76230	22869	5436	847	63	"
	1.221336	0.290794	0.087238	0.020737	0.003231	0.0002403	decimal

Fourier, compared: 1.273239   0.424413   0.254648   0.181891   0.141471   0.115749

Questions and problems

1. What will be the law for scale markings on a slide rule to present (a) logarithms, (b) sines of angles (an angle scale is needed here), and (c) tangents of angles—all against the normal logarithmic number scale?

2. A nomogram is to be constructed to present $z = xy$ over the following range: x from 1 to 20; y from 1 to 400; z from 1 to 8000. All scales will be parallel and have the same vertical length. What is the ratio of their spacing?

3. In a presentation of the type developed on page 443, the range of z required is 10^8 while that for u is 10^2. Find the relative horizontal spacing between the u, x, z, and y scales.

4. If the scale construction on page 444 uses logarithmic scales for quantities that are represented on the vertical parallel lines, what relationship would be available on the diagonal scale, and what calculation could the nomogram be used for?

5. In the presentation on page 445, when horizontal and vertical log scales are the same, the diagonal scales are at 45 degrees. What could the chart represent (a) by using rulings at other than 45 degrees, (b) by changing the spacing of the diagonal scales (keeping the horizontal and vertical scales the same)?

6. A rectangular graphical chart is to be used for a quantity whose commonly used range lies from 1 to 1000, but occasional use would extend it from 0 to infinity. What could be done to the scales to allow this function, and how would product or quotient quantity rulings be affected?

7. Derive the formula for scale markings on diagonal phase scales in the two nomograms (shown on page 449). Using trig tables or the trig function on your calculator, calculate the position to be marked (as a fraction of scale length) for angles from 0 to 90 degrees, at 10-degree intervals.

8. Assume that the sloping line on the graphical chart (page 451) identifies with the point on the nomogram below which the chart is not valid. How could this be shown on the nomogram? Determine the formula for this boundary, in terms of variables x and y (i.e., the form it will take).

9. Find the Fourier series for an asymmetrical triangular wave in which $f(x) = 3Ax/4 - A/2$, from $x = 0$ to $x = 4\pi/3$, and $f(x) = 5A\pi/2 - 3Ax/2$, from $x = 4\pi/3$ to $x = 2\pi$. This series will contain both sine and cosine terms. The formula for a and b will change for every value of n, from 1 to 6, before it repeats. Tabulate values of a_n and b_n for this range of values to find the pattern.

10. Find the Fourier series for an asymmetrical square wave with a constant value of $3A/4$—from 0 to $4\pi/3$ and to $-3A/2$, from $4\pi/3$ to 2π.

11. Graphically and algebraically verify that the curve represented by the answer to question 10 is the derivative of that for the answer to question 9.

12. Find a Fourier series for a wave that can be represented by a cubic curve of the form: $f(x) = (x/\pi)^3$, between $x = -\pi$ and $x = +\pi$.

27
CHAPTER

Where to next?

Derivatives and curvature

It now looks as if derivatives, as well as indicating slope, have some connection with curvature. Take the curve for y (shown here) with its derivatives, y' and y''. At the maximum and minimum of y (points a and c), the slope is zero, so y' is zero. At the point of maximum slope (b), y' is at a minimum and y'' is zero. Could the curvature of y be connected with these derivatives?

To be more specific, take a circle (at the right) with its equation (first and second derivative below), and the curves that are superimposed on the circle. You know that the radius of a circle is constant and that it determines the curvature. So far, this effort doesn't look too hopeful.

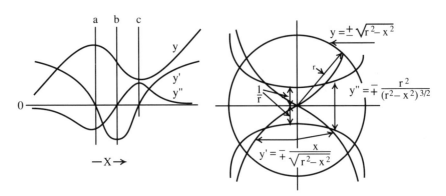

Equation of Circle $y = \pm \sqrt{r^2 - x^2}$ 1st Derivative $y' = \mp \dfrac{x}{\sqrt{r^2 - x^2}}$

2nd Derivative $y'' = \mp \dfrac{r^2}{(r^2 - x^2)^{3/2}}$ When $x = 0$ $y'' = \mp \dfrac{1}{r}$

Curvature and radius

It would be useful to calculate the curvature or radius of a curve at a point from its equation. *Curvature* is the reciprocal of radius. The bigger the radius, the smaller the curvature. The symbol ϱ represents the radius of curvature and $k = 1/\varrho$ for curvature.

In the infinitesimal segment of curve ds, ϱ is $ds/d\theta$ because $ds = \varrho d\theta$. Applying this elemental information to any curve, θ is arctan dy/dx, arctan y'. Then, you can define $d\theta$ in terms of dx by the function of a function method. From the fact that θ is arctan y', d/dy' will be $1/(1 + y'^2)$. Then, dy'/dx is the 2nd derivative of y. Next, ds is defined in terms of dx by Pythagoras, writing it in the form: $(1 + y'^2)^{1/2}$. Substituting, ϱ is given in terms of these expressions for $d\theta$ and ds. k is simply the reciprocal.

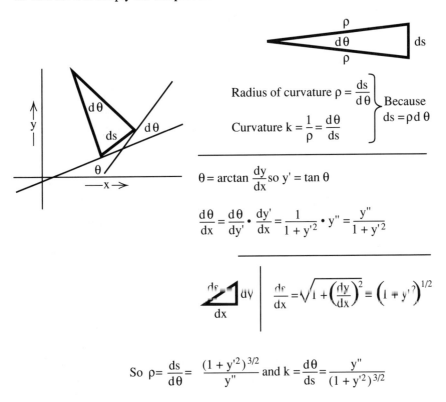

Radius of curvature $\rho = \dfrac{ds}{d\theta}$

Curvature $k = \dfrac{1}{\rho} = \dfrac{d\theta}{ds}$

Because $ds = \rho\, d\theta$

$\theta = \arctan \dfrac{dy}{dx}$ so $y' = \tan \theta$

$$\frac{d\theta}{dx} = \frac{d\theta}{dy'} \cdot \frac{dy'}{dx} = \frac{1}{1+y'^2} \cdot y'' = \frac{y''}{1+y'^2}$$

$$\frac{ds}{dx} = \sqrt{1 + \left(\frac{dy}{dx}\right)^2} = \left(1 + y'^2\right)^{1/2}$$

So $\rho = \dfrac{ds}{d\theta} = \dfrac{(1+y'^2)^{3/2}}{y''}$ and $k = \dfrac{d\theta}{ds} = \dfrac{y''}{(1+y'^2)^{3/2}}$

Radius of a circle derived

The formula finds the radius of a circle, which simplifies down to r, but with a minus sign. The minus sign is included because the formula is based on the derivatives of y. In upper quadrants where y is positive, its derivative is negative—the curve is downwards. In lower quadrants, where y is negative, its derivative is positive—the curve is upwards. So, the sign of curvature, the reciprocal of radius, is always opposite to the sign that the curve itself has at the moment. That is why ϱ is used for the radius of curvature. The symbol r, as the radius of a circle, has no sign.

Radius of Circle

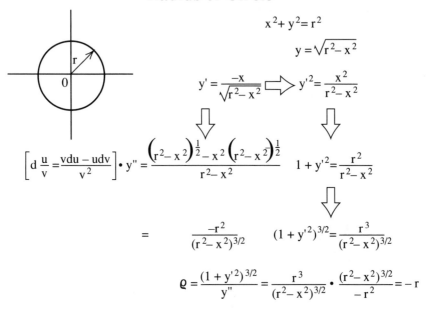

$$x^2 + y^2 = r^2$$

$$y = \sqrt{r^2 - x^2}$$

$$y' = \frac{-x}{\sqrt{r^2 - x^2}} \implies y'^2 = \frac{x^2}{r^2 - x^2}$$

$$\left[d\,\frac{u}{v} = \frac{v\,du - u\,dv}{v^2} \right] \cdot y'' = \frac{\left(r^2 - x^2\right)^{\frac{1}{2}} - x^2\left(r^2 - x^2\right)^{\frac{1}{2}}}{r^2 - x^2} \qquad 1 + y'^2 = \frac{r^2}{r^2 - x^2}$$

$$= \frac{-r^2}{(r^2 - x^2)^{3/2}} \qquad (1 + y'^2)^{3/2} = \frac{r^3}{(r^2 - x^2)^{3/2}}$$

$$\varrho = \frac{(1 + y'^2)^{3/2}}{y''} = \frac{r^3}{(r^2 - x^2)^{3/2}} \cdot \frac{(r^2 - x^2)^{3/2}}{-r^2} = -r$$

Curvature is negative when y is positive and vice versa

Radii of ellipse

An ellipse has a continuously changing radius. It will have extreme values of radius where it passes the ends of its major and minor axes. Here, the formula finds the radius at any point in terms of its half axes, a and b. Note its value at those two points.

Check that each radius is dimensionally correct. Radius at the ends of axis b (the Y axis) is minus a times the ratio a/b. The other extreme is symmetrically opposite, minus b times the ratio b/a. The negative sign has the same significance as for the circle.

Radii of Ellipse

$$\left(\frac{x}{a}\right)^2 + \left(\frac{y}{b}\right)^2 = 1 \qquad y = b\sqrt{1 - \left(\frac{x}{a}\right)^2}$$

$$y' = -\frac{bx}{a^2}\left[1 - \left(\frac{x}{a}\right)^2\right]^{-\frac{1}{2}} = \frac{-bx}{a\left(a^2 - x^2\right)^{-\frac{1}{2}}}$$

$$y'^2 = \frac{b^2 x^2}{a^2\left(a^2 - x^2\right)}$$

$$y'' = \frac{-b\left(a^2 - x^2\right)^{\frac{1}{2}} - bx^2\left(a^2 - x^2\right)^{-\frac{1}{2}}}{a\left(a^2 - x^2\right)}$$

$$\left(1 + y'^2\right) = \frac{a^2 - \left[1 - \dfrac{b^2}{a^2}\right]x^2}{\left(a^2 - x^2\right)}$$

$$= \frac{-bx^2 - b\left(a^2 - x^2\right)}{a\left(a^2 - x^2\right)^{3/2}}$$

$$= \frac{-ab}{\left(a^2 - x^2\right)^{3/2}}$$

$$\varrho = \frac{\left[\dfrac{a^2 - \left[1 - \dfrac{b^2}{a^2}\right]x^2}{\left(a^2 - x^2\right)}\right]^{3/2}}{} \cdot \frac{\left(a^2 - x^2\right)^{3/2}}{-ab} = \frac{\left\{a^2 - \left[1 - \dfrac{b^2}{a^2}\right]x^2\right\}^{3/2}}{-ab}$$

$$\text{When } x = 0 \quad \varrho = -\frac{a^2}{b}$$

$$\text{When } x = a \quad \varrho = -\frac{b^2}{a}$$

Radii of hyperbola

Examine the illustration at the top of the next page. This example of curvature is interesting: the right hyperbola. It is identical with its conjugate ellipse. Notice also that this expression for curvature or radius is positive. In the upper part (as shown) where *y* is positive, curvature is also positive or upward.

Differential equations

You have been using the concept of differential equations and I haven't formally introduced them! The equation: $dy/dx = y$ can be written either way, but solving it needs integration—finding the functional relationship that satisfies the condition stated by the equation. Either way, to be general in our answer, you must add

Radius of Hyperbola

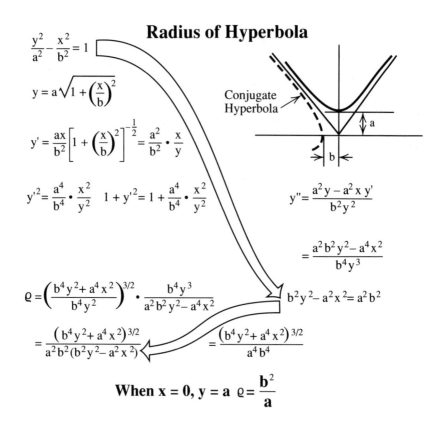

$$\frac{y^2}{a^2} - \frac{x^2}{b^2} = 1$$

$$y = a\sqrt{1 + \left(\frac{x}{b}\right)^2}$$

Conjugate Hyperbola

$$y' = \frac{ax}{b^2}\left[1 + \left(\frac{x}{b}\right)^2\right]^{-\frac{1}{2}} = \frac{a^2}{b^2} \cdot \frac{x}{y}$$

$$y'^2 = \frac{a^4}{b^4} \cdot \frac{x^2}{y^2} \quad 1 + y'^2 = 1 + \frac{a^4}{b^4} \cdot \frac{x^2}{y^2}$$

$$y'' = \frac{a^2 y - a^2 x\, y'}{b^2 y^2}$$

$$= \frac{a^2 b^2 y^2 - a^4 x^2}{b^4 y^3}$$

$$\varrho = \left(\frac{b^4 y^2 + a^4 x^2}{b^4 y^2}\right)^{3/2} \cdot \frac{b^4 y^3}{a^2 b^2 y^2 - a^4 x^2}$$

$$b^2 y^2 - a^2 x^2 = a^2 b^2$$

$$= \frac{\left(b^4 y^2 + a^4 x^2\right)^{3/2}}{a^2 b^2 (b^2 y^2 - a^2 x^2)}$$

$$= \frac{\left(b^4 y^2 + a^4 x^2\right)^{3/2}}{a^4 b^4}$$

When x = 0, y = a $\varrho = \dfrac{b^2}{a}$

Differential Equations

$$\frac{dy}{dx} = y$$

① $\dfrac{dy}{y} = dx$ [Rearranged]

Integrate: $\displaystyle\int \frac{1}{y}\, dy = \int dx$

$$\log_\varepsilon y = x + C$$

② Reciprocal: $\dfrac{dx}{dy} = \dfrac{1}{y}$ $x = \log_\varepsilon y + C$

Forms of Solution:

Here C is the same

$$\log_\varepsilon y = x + \log_\varepsilon C$$
$$x = \log_\varepsilon y - \log_\varepsilon C$$
$$x = \log_\varepsilon \frac{y}{C}$$
$$\varepsilon^x = \frac{1}{C}\, y \quad \text{⟵ Using this form as basis}$$

$$\log_\varepsilon y = x + C$$
$$x = \log_\varepsilon y + C$$
$$x = \log_\varepsilon Cy$$
$$\varepsilon^x = Cy$$
$$y = C\varepsilon^x$$

C is the constant of integration in each case, but not the same constant a constant of integration. Several ways in which the solution can be expressed are shown at the bottom of the facing page. Notice that in all of them, the constant of integration (C) serves the same purpose, although the same value does not always translate.

Second-order equation

The last section featured a first-order equation. Now, take the simplest possible second-order equation, so-called because second is the highest order derivative. If a derivative is raised to a power, it is called the degree of the equation. Thus, the equations on these two pages are first and second order, but both of first degree.

Notice that making: $y = a \sin x$ or $y = a \cos x$ both satisfy this equation. The solution needs to include both. One way is to write: $y = a \cos x + b \sin x$. This equation can be written at least two other ways. A general solution always has 2 constants of integration.

$$\frac{d^2 y}{dx^2} + y = 0 \qquad\qquad \frac{d^2 y}{dx^2} = -y$$

$$\text{If } y = a\sin x \quad \frac{d^2 y}{dx^2} = -a\sin x = -y \left.\vphantom{\begin{array}{c} a \\ a \end{array}}\right\}$$

$$\text{If } y = a\cos x \quad \frac{d^2 y}{dx^2} = -a\cos x = -y$$

General Forms

$$y = a\cos x + b\sin x$$

$$A = \sqrt{a^2 + b^2} \qquad \theta = \tan^{-1}\frac{a}{b} \longleftarrow y = A\sin(x + \theta)$$

$$\alpha = \frac{a - ib}{2} \qquad \beta = \frac{a + ib}{2} \qquad y = \alpha\varepsilon^{ix} + \beta\varepsilon^{-ix}$$

Always 2 Constants of Integration

Homogeneous equations

Take a first-order equation that is in homogeneous form. That means every term in the equation is a combined power function of the dependent and independent variable (in general terms, x and y) so that the sum of their indices is the same for every term. Derivative terms don't count because they are infinitesimal compared to the original variables. In this case, the sum of indices for every term is 2.

This example shows the general method. First, rearrange the equation with dy/dx on one side and everything else on the other. Now, write $y = vx$ and substitute it in the right side for y. This substitution lets you cancel all the x's from the right side. Now, it can be rearranged into a couple of expressions that can be integrated. The form can be changed, but any of the solutions shown satisfies the original equation.

Example $\left(2x^2 + y^2\right) dx + \left(2xy + 3y^2\right) dy = 0$

(1) Rearrange to $\dfrac{dy}{dx} = -\dfrac{2x^2 + y^2}{2xy + 3y_2}$ (2) Write $y = vx$

So $\dfrac{dy}{dx} = x\dfrac{dv}{dx} + v$

(3) Equate (1) and (2) $x\dfrac{dv}{dy} + v = -\dfrac{2x^2 + y^2}{2xy + 3y^2}$

$= -\dfrac{2 + v^2}{2v + 3v^2}$

(4) Rearrange: $\dfrac{\left(2v + 3v^2\right) dv}{2 + 3v^2 + 3v^3} + \dfrac{dx}{x} = 0$

Integrate $\dfrac{1}{3}\log_\varepsilon\left(2 + 3v^2 + 3v^3\right) + \log_\varepsilon x = C_1$

Change form $x^3\left(2 + 3v^2 + 3v^3\right) = C_2$

Substitute $v = \dfrac{y}{x}$ $2x^3 + 3xy^2 + 3y^3 = C_2$

Nonhomogeneous equations

An equation that will not reduce to homogeneous form requires different treatment. The general type can be represented by: $dy/dx + My = N$, where M and N represent any nonhomogeneous functions of x and/or y. This should include about everything you can get in a first-order equation, not already covered in simpler forms.

First, make a substitution $y = uz$, where u and z can both be functions of x. Differentiating and substituting into the original equation to get rid of y (for the moment), you then find a way to break y down to u and z to simplify solution. In the substituted form, if you make the sum of two terms with a factor z equal to zero, you can solve the other two factors of these terms for u. Now, substitute this found value for u into the other two terms of what was a four-term equation and solve $u.dz/dx = N$ for z.

Having obtained expressions for both u and z, multiply them together to get y because you began by making $y = uz$. Thus, you have the solution back into terms of the original variables.

Nonhomogeneous Equations

$$\frac{dz}{dx} + My = N$$

where M and N are nonhomogeneous functions of x and / or y

Method

Write $y = uz$; $\dfrac{dy}{dx} = u\dfrac{dz}{dx} + z\dfrac{du}{dx}$

Substitute in $\dfrac{dy}{dx} + My = N$

$$u\frac{dz}{dx} + z\frac{du}{dx} + Muz = N$$

$$u\frac{dz}{dx} + \left(\frac{du}{dx} + Mu\right)z = N$$

Equate this to 0 ⟵ Leaving $u\dfrac{dz}{dx} = N$

$u =$ Solve both   Substitute $y = uz$ $z =$

Example

$$\frac{dy}{dx} + \frac{y}{x} = y^3 \qquad M = \frac{1}{x} \qquad N = y^3$$

$$\boxed{y = uz}$$

$$u\frac{dz}{dx} + \left(\frac{du}{dx} + \frac{u}{x}\right)z = u^3 z^3$$

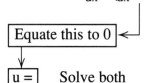

$$\frac{du}{dx} + \frac{u}{x} = 0$$

$$\frac{du}{u} + \frac{dx}{x} = 0$$

$$\text{Log}_\varepsilon u + \log_\varepsilon x = C$$

$$\boxed{ux = C_1}$$

$$u = \frac{C_1}{x}$$

Note: C_1 vanishes

$$\frac{C_1}{x} \cdot \frac{dz}{dx} = \frac{C_1^3 z^3}{x^3}$$

Rearranging ⟹ $\dfrac{dz}{C_1^2 z^3} = \dfrac{dx}{x^2}$

Integrating ⟹ $\dfrac{1}{2C_1^2 z^2} = \dfrac{1}{x} + C_2$

$$z = \frac{y}{u} = \frac{xy}{C_1} \implies \frac{1}{2x^2 y^2} = \frac{1}{x} + C_2$$

$$\boxed{2C_2}\ x^2 y^2 + 2xy^2 - 1 = 0$$

$$Cx^2 y^2 + 2xy^2 - 1 = 0$$

or: $y^2 = \dfrac{1}{cx^2 + 2x}$

Check

$$\frac{dy}{dx} = -\frac{Cx + 1}{(Cx^2 + 2x)^{3/2}}$$

$$\frac{y}{x} = \frac{1}{x\sqrt{Cx^2 + 2x}} = \frac{Cx + 2}{(Cx^2 + 2x)^{3/2}}$$

$$y^3 = \frac{1}{(Cx^2 + 2x)^{3/2}} \qquad \frac{dy}{dx} + \frac{y}{x} = y^3$$

$$\checkmark \quad \frac{-(Cx + 1) + (Cx + 2) = 1}{(Cx^2 + 2x)^{3/2}}$$

Nonhomogeneous first-order example

Here is an example of the general method given in the previous section. Trace it, step by step, to see if it did what was specified. Then, check that the solution satisfied the equation.

Second order again

Now, tackle some second-order equations that are more complicated than the one in "Second-order equations." To make the solution completely general for this form, use constants a and b. The general solution for second-order equations can be written in the form shown, where A and B are constants of integration, and m and n are constants that you must find for the solution. You form a set of values to get a value for m, then a quadratic solution for n. Because n is the square root of a quantity that can be positive or negative, according to values of a and b, n can be real or imaginary.

$$\frac{d^2 y}{dx^2} - a\frac{dy}{dx} + by = 0$$

Solution of form

$$y = A\,\varepsilon^{(m+n)x} + B\varepsilon^{(m-n)x}$$

A and B are integration constants. <u>To find m and n</u>

Coefficients of

$\varepsilon^{(m+n)x}$	$\varepsilon^{(m-n)x}$

$by \longrightarrow Ab$ Bb

$-a\dfrac{dy}{dx} \longrightarrow -Aa(m+n)$ $-Ba(m-n)$

$\dfrac{d^2 y}{dx^2} \longrightarrow A(m^2 + 2mn + n^2)$ $B(m^2 - 2mn + n^2)$

Terms in A

$b - a(m+n) + m^2 + 2mn + n^2 = 0$

Terms in B

$b - a(m-n) + m^2 - 2mn + n^2 = 0$

Subtracted:

$-2an$	$+4mn$	$=0$

$$m = \frac{a}{2}$$

Added:

$$2b - 2am + 2m^2 + 2n^2 = 0$$

Substitute $m = \dfrac{a}{2}$ $m = \dfrac{a^2}{4} \longrightarrow$ $2b - a^2 + \dfrac{a^2}{2} + 2n^2 = 0$

If $\left(\dfrac{a^2}{4} - b\right)$ positive n is real

$$n^2 = \frac{a^2}{4} - b$$

If $\left(\dfrac{a^2}{4} - b\right)$ negative n is imaginary

$$n = \pm\sqrt{\frac{a^2}{4} - b}$$

A special case

This special case lies within the set in the previous section, for which that solution is not valid, You must add something as a factor to one of the general terms. Putting x in the B term satisfies the original equation. However, both terms must use coefficients that produce identical roots, which is implied by the "perfect square" look of the original equation.

If a third-order equation had three identical roots, the second and third would take x and x^2 as factors, and so on, but that is further than this section is going at the moment.

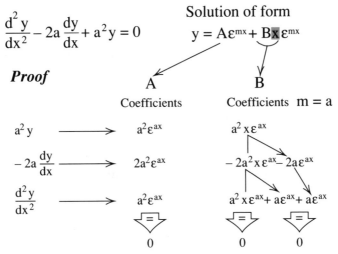

Solution of form

$$\frac{d^2y}{dx^2} - 2a\frac{dy}{dx} + a^2y = 0$$

$$y = Ae^{mx} + Bx\,e^{mx}$$

Proof

A   B

Coefficients   Coefficients $m = a$

a^2y	$a^2\varepsilon^{ax}$	$a^2x\varepsilon^{ax}$
$-2a\dfrac{dy}{dx}$	$2a^2\varepsilon^{ax}$	$-2a^2x\varepsilon^{ax}-2a\varepsilon^{ax}$
$\dfrac{d^2y}{dx^2}$	$a^2\varepsilon^{ax}$	$a^2x\varepsilon^{ax}+a\varepsilon^{ax}+a\varepsilon^{ax}$

0 0 0

With an additional f(x)

Another disqualifying addition to the form is when the 0 on the right is changed to a $f(x)$ (or a constant). Three examples (this is the first) show how to handle that. Add some extra terms to the general-form solution, of the form $y = C + Dx$, then solve, as shown.

$$\frac{d^2y}{dx^2} - a\frac{dy}{dx} + by = f(x)$$ **Case 1**

Example

Solution of:

$$\frac{d^2y}{dx^2} - \frac{dy}{dx} - 2y = 4x$$

$$\frac{d^2y}{dx^2} - \frac{dy}{dx} - 2y = 0 \quad a = 1$$
$$b = -2$$

4x is not a particular value of

$$y = A\varepsilon^{2x} + B\varepsilon^{-x}$$

$$\text{Write } y = C + Dx \qquad \frac{dy}{dx} = D$$

Substitute in original: $\quad -D - 2C - 2Dx = 4x$

$$2C = -D \qquad D = -2 \qquad \text{So, } C = +1$$

Complete Solution $\quad y = A\varepsilon^{2x} + B\varepsilon^{-x} + 1 - 2x$

Two more variations

Here, the nature of the $f(x)$ changes the assumed extra function to match the additional term, enabling the complete solution to be derived.

A similar action occurs when the extra term is a particular value of a term in the solution. That term has a "duplicate," in which a coefficient solves by substitution and uses x as the added factor.

$$\mathbf{\frac{d^2y}{dx^2} - a\,\frac{dy}{dx} + by = f(x)} \qquad \textbf{Case 1}$$

Example

Solution of:

$$\frac{d^2y}{dx^2} - 4\,\frac{dy}{dx} + 3y = 6\varepsilon^{2x}$$

$$\frac{d^2y}{dx^2} - 4\,\frac{dy}{dx} + 3y = 0 \quad a = 4$$
$$b = 3$$

_____

$6\varepsilon^{2x}$ is not a particular value of

$y = A\varepsilon^{3x} + B\varepsilon^{x}$

$$\text{Write } y = C\varepsilon^{2x} \qquad \frac{dy}{dx} = 2C\varepsilon^{2x}$$

$$\frac{d^2y}{dx^2} = 4C\varepsilon^{2x}$$

Substitute in original: $4C - 8C + 3C = 6$

$$-C = 6$$

$$C = -6$$

Complete Solution $\quad y = A\varepsilon^{3x} + B\varepsilon^{x} - 6\varepsilon^{2x}$

$$\frac{d^2 y}{dx^2} - a \frac{dy}{dx} + by = f(x) \qquad \textbf{Case 2}$$

Example

Solution of:

$$\frac{d^2 y}{dx^2} - 2 \frac{dy}{dx} - 3y = 2\varepsilon^{-x}$$

$$\frac{d^2 y}{dx^2} - 2 \frac{dy}{dx} - 3y = 0 \quad a = 2$$

$$b = -3$$

$2 \varepsilon^{-x}$ is a particular value of

$$y = A\varepsilon^{3x} + B\varepsilon^{-x}$$

$$\text{Write } y = Cx\varepsilon^{-x} \qquad \frac{dy}{dx} = C(\varepsilon^{-x} - x\varepsilon^{-x})$$

$$\frac{d^2 y}{dx^2} = C(x\varepsilon^{-x} - 2\varepsilon^{-x})$$

Substitute in original: $C(x - 2) - 2C(1 - x) - 3Cx = 2$

$$Cx + 2Cx - 3Cx = 0$$

$$-2C - 2C \qquad = 2 \qquad C = -\frac{1}{2}$$

Complete Solution $\quad y = A\varepsilon^{3x} + B\varepsilon^{-x} - \frac{1}{2} x\varepsilon^{-x}$

How far can you go?

The study of differential equations digs into ever more complicated forms that are the subject of more advanced mathematics. Here are a few points that will round out the picture as far as you have gone.

The form at the top of the next page, which often occurs in engineering problems, can be converted to the more recognizable form, shown by simple differentiation. Next, is a simple method of solving some first-order equations, where they can be simplified into a form with all the terms in x multiplied by dx.

Higher-order equations

On page 479 is a look at how to solve the simpler higher-order equations, where the terms contain no functions of x, other than derivatives of y. They can contain a term in y, although this example does not. The complete solution could be: $y = A + Be^x + Ce^{i3x} + De^{-i3x}$, which can also be in the form shown, with appropriate substitutions.

Differentiate:

$$a\,\frac{dy}{dx} + by + \int cy\,dx = 0$$

$$a\,\frac{d^2y}{dx^2} + b\,\frac{dy}{dx} + cy = 0$$

$$f(x)\,dx + f(y)\,dy = 0$$

$$f(x)\,dx + \int f(y)\,dy = C$$

Example

$$dy\sqrt{1+x^2} - dx\sqrt{1-y^2} = 0$$

Divide through
by $\sqrt{1+x^2}\sqrt{1-y^2}$

$$\frac{dy}{\sqrt{1-y^2}} - \frac{dx}{\sqrt{1+x^2}} = 0$$

Integrate

$$\int\frac{dy}{\sqrt{1-y^2}} - \int\frac{dx}{\sqrt{1+x^2}} = C$$

Write $\theta = $ arcsiny

Write $\phi = $ arcsinhx

$y = \sin\theta$

$x = \sinh\phi$

$\int d\theta - \int d\phi = C$

$\sqrt{1-y^2} = \cos\theta$

$\sqrt{1+x^2} = \cosh\phi$

$\theta - \phi = C$

$dy = \cos\theta\,d\theta$

$dx = \cosh\phi\,d\phi$

$$\frac{dy}{\sqrt{1-y^2}} = \frac{\cos\theta\,d\theta}{\cos\theta} = d\theta$$

$$\frac{dx}{\sqrt{1+x^2}} = \frac{\cosh\phi\,d\phi}{\cosh\phi}$$

$$= d\phi$$

arcsiny − arcsinhx = C

Alternatively
Write $\phi = $ arctanx

$$\text{arcsiny} - \log_\varepsilon\left(\sqrt{1+x^2} + x\right) = C$$

Higher-Order Equations

Example

$$\frac{d^4y}{dx^4} - \frac{d^3y}{dx^3} + 9\frac{d^2y}{dx^2} - 9\frac{dy}{dx} = 0$$

Write $y = \varepsilon^{rx}$ Substitute: $(r^4 - r^3 + 9r^2 - 9r)\varepsilon^{rx} = 0$

$$r(r-1)(r^2+9) = 0$$

$r = 0$ $y = A\varepsilon^0 = A$

$r = 1$ $y = B\varepsilon^x$

$r = \pm i3$ $y = C\varepsilon^{i3x} + D\varepsilon^{-i3x} = E\cos 3x + F\sin 3x$

$E = C + D$

$F = i(C - D)$

Complete Solution

$$y = A + B\varepsilon^x + E\cos 3x + F\sin 3x$$

Probable values

These two sections tackle a slightly more complicated form of the "find the law" type, introduced in chapter 26. The three variables, x, u and v are related empirically by a law of the form: $x = u^a \cdot v^b$, where a and b are indices (fractional) whose sum is 1. The collected data is shown, taken for three values of x. The next page shows a nomogram construction that focuses on the most probable place to put the line. The position chosen makes $a = 0.35$ and $b = 0.65$.

Now, this problem has been approached by using the graphical chart. The clusters of values for the three values of x indicate the probable slope of the scale for x. Either on the nomogram or on this chart, extra values of x can be drawn in on the appropriate scale.

Probability

In chapter 16, some questions involved small-scale probability. Assuming a reliable type of measurement and proper care in making it, errors still occur. In a national election, votes go in groups, which can be predicted from carefully chosen samples of the population. What is the margin of error? This kind of problem is what the study of probability handles. See the example shown at the bottom of page 481.

Examination of great quantities of empirical data on many different problems shows that error distribution follows a general shape, with only minor discrepancies, unless major parameters are omitted. The general distribution shape

Find the Law of Form
$x = u^a v^b$ $\left[a + b = 1 \right]$

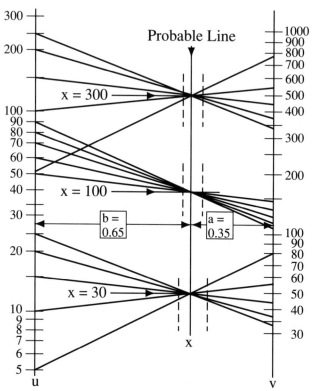

x	u	v
	5	80
	10	54
30	15	45
	20	38
	25	32.5
	50	145
	60	131
100	70	121
	80	112.5
	90	106
	50	750
	100	550
300	150	430
	200	375
	250	335

Finding the Law

$$a = \frac{\tan\theta}{1 + \tan\theta} \qquad b = \frac{\cot\theta}{1 + \cot\theta}$$

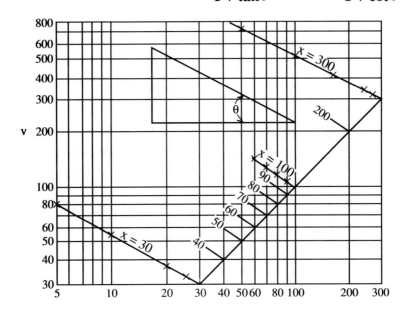

PROBABILITY

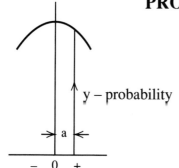

y – probability

a

−　0　+
x
Error

Small values of h
Larger values of h

1. Symmetry means $y = f(x^2)$

    Cannot be $\dfrac{1}{x^2}$ because zero error would be infinitely probable

2. Acceptable form $A\epsilon^{-h^2x^2}$ or $A\epsilon^{-\frac{x^2}{h^2}}$

    Probability error does not exceed a

To find A   is $P_a = \displaystyle\int_{-a}^{+a} A\epsilon^{-\frac{x^2}{h^2}}\,dx$

$$P_\infty = \int_{-\infty}^{\infty} A\epsilon^{-\frac{x^2}{h^2}}\,dx = 1$$

is shown. Empirically, the mathematical laws introduced here, with a constant (*k* or *h*) to indicate the degree of concentration in the vicinity of the median value, provide a satisfactory curve.

The function is logical because the distribution curve is almost always symmetrical, which means errors are equally probable on either side. So, the function must be based on x^2, not on x. Both x^2 and $1/x^2$ are ruled out because they produce illogical curves. Only the formula shown lets a curve be "tailored" to empirical data.

Probability mathematics

This lands us with finding the integral of ϵ^{-x^2}, from negative infinity to positive infinity. First, write *ax* for *x*, which requires *adx* for *dx*. If you multiply both sides by ϵ^{-a^2} and integrate it with respect to *a*, the extra product on the left side corresponds with *P/2A* from the starting point. The integral from zero to infinity is the same, whether the quantity is called *a* or *x*. When the form is identical, it is just the same function with a different letter.

$$P = \int_{-\infty}^{+\infty} A\epsilon^{-x^2}\, dx = 2A\int_0^{\infty} \epsilon^{-x^2} dx$$

1. Write ax for x; adx for dx

$$P = 2A\int_0^{\infty} \epsilon^{-a^2 x^2}\, adx$$

2. Multiply by ϵ^{-a^2}

3. Integrate $\int_0^{\infty} \text{———}\, da$

$$P\epsilon^{-a^2} = 2A\int_8^{\infty} \epsilon^{-a^2(1+x^2)}\, adx$$

$$P\int_0^{\infty} \epsilon^{-a^2} da = 2A\int_{a=0}^{a=\infty}\int_{x=0}^{x=\infty} \epsilon^{-a^2(1+x^2)}\, adadx$$

$$\int_0^{\infty} \epsilon^{-x^2} dx = \int_0^{\infty} \epsilon^{-a^2} da$$
$$= \frac{P}{2A}$$

$$\frac{P^2}{2A} = 2A\int_{a=0}^{a=\infty}\int_{x=0}^{x=\infty} \epsilon^{-a^2(1+x^2)}\, adadx$$

2 a da = da²

$$\frac{P^2}{2A} = A\int_{a=0}^{a=\infty}\int_{x=0}^{x=\infty} \epsilon^{-a^2(1+x^2)}\, da^2 dx$$

$$= A\int_{x=0}^{x=\infty} \frac{1}{1+x^2}\left[\epsilon^{-a^2(1+x^2)}\right]_0^{\infty} dx \qquad = A\int_0^{\infty} \frac{1}{1+x^2}\, dx$$

$$= A\left[\,\arctan x\,\right]_0^{\infty} = \frac{A\pi}{2}$$

So $P = A\sqrt{\pi}$ or $A = \dfrac{P}{\sqrt{\pi}}$

$P = 1$ So $A = \dfrac{1}{\sqrt{\pi}}$

When f (x) is $A\epsilon^{-\frac{x^2}{h^2}}$ $A = \dfrac{1}{h\sqrt{\pi}}$

To integrate the right side with respect to a, change the variable to a^2. This integration takes up the "spare a" and makes the index of e a standard form: a^2 is the variable and $(1 + x^2)$ is a constant (for the time being) because you are integrating with respect to a^2. Putting in values: $a^2 = 0$ and $a^2 = $ infinity, the integrated function has a value of 0 at its upper limit and of 1 at its lower limit. Going through the sign changes, the integral takes the form arctan x, which has a value of $\pi/2$ at its upper limit and 0 at its lower limit.

Rearranging gives the final integral. Working through it all again with the index $(-x^2/h^2)$ leads to the final answer, which represents the constant A, in terms of the sharpness of the concentration factor, h.

Least sum of squares

From a number of readings, assuming that all are taken with equal care, quality of equipment, or whatever controls the expected precision of readings, the probable value is the arithmetic mean of them all. Thus, if 9 out of 10 readings were 20 and 1 was 18, the probability of 20 being the correct reading is 9 times as great as that 18 is correct. The difference of 2 in the odd reading is divided by the total of 10 readings, so the average is 19.8.

PROBABLE VALUE

Arithmetic Mean

Least Sum of Error Squares

Value	10.6	10.5	10.7	
9.0	2.56	2.25	2.89	
10.1	0.25	0.16	0.36	
10.3	0.00	0.04	0.16	
10.5	0.01	—	0.04	Squares of
10.7	0.01	0.04	—	"Errors"
10.8	0.04	0.09	0.01	
10.9	0.09	0.16	0.04	
11.0	0.16	0.25	0.09	
11.2	0.36	0.49	0.25	
11.5	0.81	1.00	0.64	
106.0	4.38	4.48	4.48	Sum of
÷ 10 = 10.6				Squares

Least

Take the ten more random readings listed here. Adding them together and dividing by 10 finds that the average is 10.6. Because the probability of error is a function of the square of the error, the rule for finding the correct reading is that it should make the sum of the squares of the errors least. This is the arithmetic mean. To show it, all the errors from the arithmetic mean have been squared. Squares of errors from the arithmetic mean add up to 4.38. From the other two numbers, each of which differs from the mean by 0.1, the squares of the errors add up to 4.48, justifying the arithmetic mean as the probable correct reading.

Questions and problems

1. Find the curvature of a parabola whose equation is given by: $y = x^{2/4f}$, when $x = 0$, $x = f$, and $x = 2f$. From these results, deduce the relationship between the center of curvature and the curve's focus and axis.

2. Find the curvature of a sine wave. Equation: $y = \sin x$ evaluates it for $x = 0$ and $x = \pi/2$. Show that such a sine wave has unit slope where it crosses the zero line, and unit radius at its maxima and minima.

3. Find the points of maximum curvature on the curve that is given by: $y = 9/16x^3$. Also, find the radius, slope, and position of the center of curvature at these points.

4. Find the expression for radius of curvature when its equation is plotted to logarithmic coordinates. Verify that the radius of curvature is infinite (curvature zero) at points of maximum slope on any such curve.

5. A curve's equation is: $y = 1 + ax^2 - bx^4 + x^6$. Find the relationship between coefficients a and b that is needed for the maximum point on the curve to be equal to its value when x is zero, and the relationship between values of x^2 when this relationship obtains, between the minimum point, maximum slope (to linear scale), and maximum points.

6. An equation for a curve with an even power series up to x^8 has roots where maxima and minima occur, of $x^2 = a$, $x^2 = 1$ and $x^2 = b$. Find the relationship between a and b so that the minimum at $x^2 = 1$ has the same value as when x^2 is zero. Also find the limits of a so that b has a root greater than 1.

7. Solve $dy = y \tan x\, dx$, by the method of page 478, and check your result by substitution.

8. Solve $\dfrac{dy}{dx} + \dfrac{2y}{x} = 2y^2$

9. Solve $x\dfrac{dy}{dx} + y = (1 + x)e^x$

10. Find the solution of the following:

$$\dfrac{d^2y}{dx^2} - 3\dfrac{dy}{dx} + 2y = 0 \qquad\qquad \dfrac{d^2y}{dx^2} + 5\dfrac{dy}{dx} + 4y = 0$$

$$\dfrac{d^2y}{dx^2} + 2\dfrac{dy}{dx} - 3y = 0 \qquad\qquad \dfrac{d^2y}{dx^2} - 10\dfrac{dy}{dx} + 9y = 0$$

11. Find and verify the solution of:

$$\dfrac{d^4y}{dx^4} - 4\dfrac{d^3y}{dx^3} + 4\dfrac{d^2y}{dx^2} - 4\dfrac{dy}{dx} + y = 0$$

12. Find the complete solution of the following:

$$\dfrac{d^2y}{dx^2} - 4y = e^{2x} \qquad\qquad \dfrac{d^2y}{dx^2} - 2\dfrac{dy}{dx} + 5y = 3 \cos x$$

13. An empirical law is of the form: $p = km^a . n^b$, where k is a constant, $a + b = 1$, and p, m, and n are variables. Data from which the law is to be found is tabulated below:

p	m	n	p	m	n	p	m	n
20	2	4.9	100	2	29.00	500	2	174
20	4	4.5	100	4	26.9	500	4	161
20	6	4.3	100	6	25.7	500	6	154
20	8	4.15	100	8	25.0	500	8	149
20	10	4.05	100	10	24.3	500	10	145

Use a graphical method to find probable law.

14. Over a very large count to determine frequency of unrelated events, one of them occurs 25 times every 1000 opportunities, the other occurs 40 times per 1000. What is the probability of the two events coinciding once in every 1000 opportunities? What would be the probable number of coincidences in 10,000 opportunities?

15. One dozen rain gauges are tested where all will probably give the same reading. The readings found are as follows: one at 4.3", one at 4.35", one at 4.4", one at 4.45", two at 4.7", two at 4.75", two at 4.95", one at 5.00", and one at 5.1". What was the probable correct reading?

Index